Design of Quantum Teleportation Schemes

Ri-Gui Zhou • Xiao-Xue Zhang • Lin-Tao Du

Design of Quantum Teleportation Schemes

Springer

Ri-Gui Zhou
College of Information Engineering
Shanghai Maritime University
Shanghai, China

Xiao-Xue Zhang
College of Information Engineering
Shanghai Maritime University
Shanghai, China

Lin-Tao Du
College of Information Engineering
Shanghai Maritime University
Shanghai, China

ISBN 978-3-031-82724-2 ISBN 978-3-031-82725-9 (eBook)
https://doi.org/10.1007/978-3-031-82725-9

This Springer imprint is published by the registered company Springer Nature Switzerland AG
The registered company address is: Gewerbestrasse 11, 6330 Cham, Switzerland

Preface

In the Internet era, big data has become a new form of economic asset like currency and gold. However, while people are enjoying this huge economic asset, how to ensure the safe transmission of secret information is also becoming a major challenge that people have to face. In order to ensure the security of secret information transmission, the classical communication schemes transmit information in ciphertext, and the encryption technology used is based on complex data computation problems. However, with the emergence and development of distributed computing, cloud computing, quantum computing, and other highly parallel technologies, the security of traditional encryption techniques is becoming more and more fragile. Quantum communication is a new type of communication technology based on the principle of quantum mechanics, which realizes the remote transmission of quantum information by means of the non-local nature of entanglement. The unique physical properties such as coherent superposition, entanglement, the no-cloning theorem, and measurement of instantaneous collapse give quantum communication incomparable advantages over classical communication in terms of security and transmission efficiency.

Quantum teleportation is one of the basic technologies of quantum communication, which is based on the unique physical properties of quantum superposition, through unitary transformation and measurement of entangled states, combined with classical communication to complete the remote transmission of information. However, what it transmits is no longer the classical information but the quantum state information. Under the effect of quantum entanglement, with the help of classical information, the quantum state to be transmitted is like an instantaneous transfer, which suddenly disappears in the sender and at the same time suddenly appears in the receiver. Therefore, quantum teleportation, as one of the safe and reliable transmission methods of quantum states between communication parties, will certainly play a vital role in the future quantum communication.

This book provides a variety of schemes related to the various branches of quantum teleportation. There are six chapters in the book. Chapter 1 is an introductory part, which describes the research background, significance of topics, literature review, and research methods. Chapter 2 presents the basic knowledge which provides the techniques of the field of quantum information, such as quantum measurement, quantum gates, and quantum channel noise.

Various quantum teleportation schemes are designed in Chaps. 3, 4, 5, and 6. Chapters 3, 4, and 5 present new schemes in bidirectional quantum teleportation, bidirectional controlled quantum teleportation, and cyclic quantum teleportation, respectively. Besides, considering the influence of noise in quantum channel, three quantum teleportation schemes in noisy environments are proposed in Chap. 6.

These chapters describe the design of quantum teleportation schemes and their corresponding performance analysis. In addition, the simulated experimental results and analysis demonstrate the superiority of the proposed schemes in this book.

Shanghai, China

Ri-Gui Zhou
Xiao-Xue Zhang
Lin-Tao Du

Acknowledgments

This work is supported by the National Natural Science Foundation of China under Grant No. 62172268 and 62302289 and Shanghai Science and Technology Project under Grant No. 23YF1416200.

We thank Chen Qian, RuiQing Xu, YaNan Zhang, and many others for research collaborations and support.

Ri-Gui Zhou would like to thank his wife, Airong Nie, and his son, Hanxuan Zhou, for their understanding and help.

Xiao-Xue Zhang and Lin-Tao Du would like to express their sincere gratitude to their advisor, Prof. Ri-Gui Zhou, for his guidance, nurturing encouragement, and support in every stage of their graduate study. His knowledge, kindness, patience, open-mindedness, and vision have provided us with lifetime benefit.

Competing Interests The authors have no competing interests to declare that are relevant to the content of this manuscript.

Contents

About the Authors

Ri-Gui Zhou received a BS degree from Shandong University, China, in 1997, an MS degree from the Department of Computer Science and Technology of Nanchang Hangkong University, China, in 2003, and a PhD degree from the Department of Computer Science and Technology of Nanjing University of Aeronautics and Astronautics, China, in 2007.

From 2008 to 2010, he was a Postdoctoral Fellow at Tsinghua University, China. From 2010 to 2011, he was a Postdoctoral Fellow at Carleton University, Ottawa, Canada. From 2014 to 2015, he was a Visiting Scholar at North Carolina State University, Raleigh, NC, USA. He is currently a Professor at the College of Information Engineering, Shanghai Maritime University, China. His main research interests include quantum image processing, quantum reversible logic, and the quantum genetic algorithm, among others.

Prof. Zhou is a distinguished member of the China Computer Federation (CCF) and the recipient of the New Century Excellent Talents program, Ministry of Education of China, in 2013.

Xiao-Xue Zhang received a BS degree in Computer Science and Technology from Huaibei Normal University, Anhui, China, in 2019. She is currently pursuing a PhD degree in Information Management and Information System at Shanghai Maritime University, Shanghai, China. Her research interests include quantum communication and quantum computation.

Lin-Tao Du received a BS degree in Information and Computing Science from Shenyang Agricultural University, Liaoning, China, in 2020. She is currently pursuing a PhD degree in Information Management and Information System at Shanghai Maritime University, Shanghai, China. Her research interests include quantum cryptography and secure communication.

Chapter 1
Introduction

Abstract Quantum teleportation represents a significant application form of quantum communication. It is a technology that leverages the distributed quantum entanglement and the conversion of certain physical information to transmit quantum states across arbitrary distances. It constitutes a revolutionary communication mode, characterized by the transmission of not merely classical information but rather the quantum information carried by quantum states. In quantum teleportation, the communicating parties share a pair of entangled particles as the communication channel, enabling the information of an unknown quantum state to be transmitted from the sender to the receiver without physically transferring the particle itself. This technology holds vast potential for application and crucial research value in the fields of quantum information and quantum computation.

1.1 The Concept of Quantum Teleportation

As a new technology in the post-Moore era [1], quantum communication is an important branch of quantum informatics, which is based on the basic principles of quantum mechanics, such as quantum uncertainty principle, quantum no-cloning theorem, and the properties of quantum entanglement; it can achieve absolute secure communication in theory. Because of the high efficiency and absolute security of transmission, quantum communication has become a powerful weapon to ensure the security of network information [2], greatly overturning the security of classical encryption algorithms and changing the development of traditional informatics; it enriches the research methods of cryptography, promotes research in information security-related fields such as national defense and finance, and provides new ideas and strong security guarantee for the research of secret communication and information construction; and it is the leading science to guarantee the future informatization [3].

As a key branch of quantum communication, quantum teleportation (QT) is also the premise of realizing the feasibility of global quantum communication networking. Quantum teleportation is similar in concept to "interstellar travel" in science fiction, to transfer an unknown quantum state to a distant place and realize the spatial

R.-G. Zhou et al., *Design of Quantum Teleportation Schemes*,
https://doi.org/10.1007/978-3-031-82725-9_1

transfer of the quantum state without the need to transfer the physical carrier itself. It has the advantages of high reliability, low communication complexity, and resource saving.

Quantum teleportation is used to transmit quantum information, and the quantum entanglement property is used to realize the hopping transmission of quantum bits, which ensures the reliability and safety of the communication process [4], that is, through the channel formed by quantum entanglement, sent to the receiver, and combined with quantum measurement and unitary transformation to achieve the whole process. In the transfer process, only the quantum state of the particle is transferred, not the particle itself. Therefore, the integrity of information transmission is not affected when the information is transmitted, and it is not necessary to know the definite quantum state before the transmission of the quantum state, and it also does not need to know the geographical location of the receiver that can be directly transmitted to the other quantum state. Since Bennett proposed the first scheme of quantum teleportation in 1993 [5], many subsequent schemes of quantum teleportation have been proposed [6–13]. It can be seen that no matter in theory or practice, researchers have conducted in-depth research and achieved many excellent results.

As an important part of practical quantum technology, quantum teleportation is an absolutely safe and reliable carrier for the transmission of secret information, and it plays an important role in large-scale quantum computation, quantum entanglement, long-distance quantum communication, and quantum computing network. In addition, quantum teleportation is widely used in practical applications, such as cryptography, physical simulation, and quantum sensing, and it could also play an important role in areas such as medical diagnosis, military security, environmental monitoring, and financial transactions; particularly in the financial sector, the application of quantum teleportation has become a hot spot in the research and popularization of financial institutions. Therefore, the study of quantum teleportation is to obtain better information communication methods and security methods and to better apply the advanced ideas of quantum theory, which is of great significance to the development of science and technology.

1.2 Research Status of Quantum Teleportation

Quantum teleportation (QT) usually requires an entangled quantum channel and some classical information to be shared in advance among different remote participants. In 1993, Bennett et al. proposed the first scheme of quantum teleportation [14], in which two classical channels and EPR quantum pairs were used as quantum channels to realize quantum teleportation; the participants, Alice and Bob, share an EPR entanglement pair in advance. Alice passes an unknown quantum state to Bob. Due to quantum non-cloning, Bob needs to transform into the state that Alice initially wants to transmit. Based on this first quantum teleportation scheme, many quantum teleportation schemes with different structures have emerged. In 2000,

Manzoor Ikram et al. considered the teleportation of two-particle and multi-particle entangled states and proposed a teleportation scheme suitable for both high-Q cavity entangled states and field states [15]. In 2004, Gao et al. proposed a simplified scheme for the probability of the unknown quantum state of one particle and two particles and the controllable teleportation, and the new scheme is realized by constructing an efficient quantum logic network consisting of a single-qubit gate, a double-qubit controllable non-gate, a von Neumann measurement, and a classical control operation [16]. In these schemes of ref. [16], only the probability of success can be achieved. In 2005, Zhang et al. put forward a general idea of constructing a quantum teleportation method of multiple qubits, which can be implemented between two remote parties of multiple agents in a control network [17]. In general, the scheme proposed in refs. [15–17] increases the number of transmitted qubits compared with the scheme proposed in ref. [14], and the added controller improves the security; it increases the diversity of scheme design but has the limitation of probability implementation. In addition, literatures [15–17] are traditional one-way transport schemes in which only one sender and one receiver share an entangled state as a quantum channel, by measuring the result, the sender passes the qubit containing the information to the receiver. The most important difference in the design is the quantum channel and the number of transmitted qubits. Therefore, in the design of one-way quantum teleportation, in addition to using EPR pairs as quantum channels, GHZ-like states are used in ref. [18] and W states are used in ref. [19], cluster states are used in ref. [20], and more than one single-particle state is transmitted in ref. [21].

In order to improve the efficiency and security of one-way quantum teleportation, researchers have developed many new schemes by adding bidirectional, controlled, cyclic, symmetric, and asymmetric elements, if one or more controllers are added to improve the security, that is, controlled quantum teleportation (C-QT); with fewer quantum resources and classical resources to transmit more particles, increasing the transmission efficiency, that is, bidirectional quantum teleportation (BQT), cyclic quantum teleportation (CQT), and so on. At present, the classification based on quantum teleportation is becoming more and more extensive. In the following, we will introduce the research status of each branch of quantum teleportation, as shown in Fig. 1.1.

1. Bidirectional Quantum Teleportation

Bidirectional quantum teleportation (BQT), as a subfield of QT, was researched in recent years. Two legitimate users can teleport quantum state to each other. That is to say, both sides of the communication play the role of sender and receiver. In 2014, Duan et al. investigated new applications of the six-particle entanglement channel in bidirectional quantum teleportation; a six-particle entanglement channel shared by Alice, Bob, and Charlie simultaneously transmits an arbitrary single quantum bit state under the control of Charlie [22]. In 2016, Sang et al. proposed a bidirectional quantum teleportation scheme by using a five-qubit cluster state as a quantum channel [23], in which Alice can transmit an arbitrary two-qubit entangled state to Bob, and at the same time, Bob can transmit an arbitrary single-qubit state to Alice. Because the qubits carrying information are not transmitted in quantum channel, it is

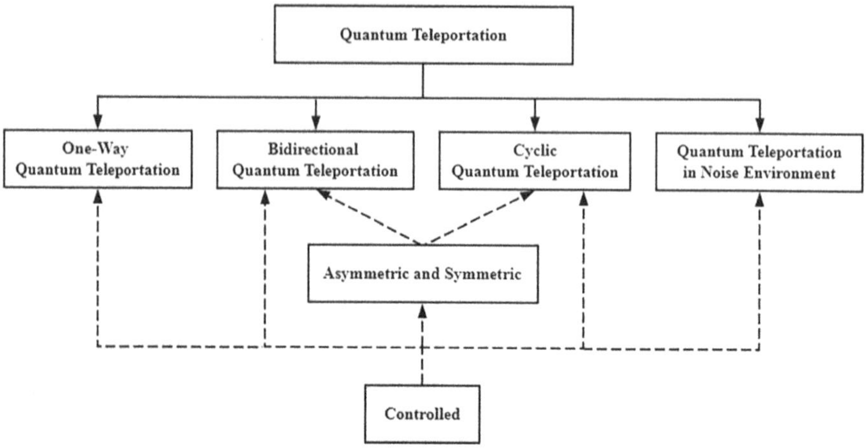

Fig. 1.1 Classification diagram of quantum teleportation

impossible for eavesdropper to intercept the qubits during message transmission process, i.e., intercept of communication. A unique approach for eavesdropper to attack the quantum channel between Alice and Bob is when the controller sends a message to Alice and Bob. In 2017, Gou et al. presented the scheme based on probabilistic resumable bidirectional quantum teleportation [24]. Li et al. in 2019 proposed the bidirectional quantum teleportation scheme, in which two qubits are transmitted to each other or two-thirds of the qubit states are converted to each other by using six-qubit entangled states as quantum channels [25]. In 2022, Wang and Li proposed a bidirectional quantum teleportation scheme in which Alice and Bob transmit to each other a pure entangled EPR pair and a single-particle state, respectively [26].

2. Cyclic Quantum Teleportation

Cyclic quantum teleportation (CQT), a brand-new model of transmission, has been proposed lately. In 2017, Chen et al. put forward the first scheme of cyclic quantum teleportation where three arbitrary unknown single-qubit states were teleported via a six-qubit entangled state among three participants [27]. Recently, Sisodia et al. proposed an optimal scheme for teleportation of an n-qubit quantum state [28], and Sabir et al. presented an efficient scheme for the quantum teleportation of a subclass of tripartite entangled states; both of them have designed the corresponding quantum circuit, and the latter has done the simulation using IBM's real quantum processor [29].

3. Controlled Quantum Teleportation

Controlled quantum teleportation (C-QT) is a branch of quantum teleportation, where the third party or multiparty is introduced into the protocol. In controlled quantum teleportation, controller shared the entanglement resource with senders and also can locally control the quality of the quantum channel, thus affecting the quality of the standard controlled quantum teleportation. Controlled quantum teleportation was firstly presented by Karlsson et al. in 1998 [30]. In 2004, Yang et al. presented a multiparty controlled quantum teleportation scheme to teleport multi-qubit quantum information [31]. Later, various types of controlled quantum teleportation protocols have been introduced [32–35].

Bidirectional controlled quantum teleportation (C-BQT) is the combination of controlled quantum teleportation (C-QT) and bidirectional quantum teleportation (BQT). The first bidirectional controlled quantum teleportation scheme was demonstrated by Zha et al. in 2013 [36] via a five-qubit cluster state as a quantum channel. Up to now, various bidirectional controlled quantum teleportation schemes based on multi-qubit states, such as a five-qubit state [37], a six-qubit state [38, 39], a seven-qubit state [40–42], and an eight-qubit state [43], have been proposed.

Controlled cyclic quantum teleportation is a combination of cyclic quantum teleportation (CQT) and controlled quantum teleportation (C-QT). In 2018, Sang proposed a scheme in regard to cyclic controlled teleportation using a seven-qubit entangled state under the control of supervisor David soon afterward [44]. After that, other controlled cyclic quantum teleportation has been proposed [45–47].

4. Asymmetric Bidirectional Quantum Teleportation

Asymmetric bidirectional quantum teleportation (ABQT) is a process by which mutual exchange of quantum states of different natures is done through the use of quantum entangled channels. Asymmetric bidirectional quantum teleportation scheme was first created by Zhang et al. [48] in 2015 in which Alice transmits her single-qubit entangled state to Bob and Bob wants to teleport his arbitrary two-qubit to Alice by using a maximally seven-qubit entangled state as a quantum channel. In 2017, Binayak et al. proposed an asymmetric bidirectional quantum teleportation scheme by using a nine-qubit cluster state as a quantum channel [49]. In 2021, Verma Vikram proposed an improved asymmetric bidirectional quantum teleportation scheme based on local CNOT gate operation, Bell state measurement, GHZ state measurement, and unitary operation using a channel consisting of two GHZ states [50]. Compared with the previous bidirectional quantum teleportation scheme, this scheme requires less quantum and classical resources and has higher intrinsic efficiency and lower computational complexity. In 2022, Ma et al. proposed the asymmetric bidirectional quantum teleportation scheme with two GHZ states as the channel. Alice can transmit two arbitrary single-qubit states and one three-qubit entangled state to Bob; on the other hand, Bob can transmit a two-qubit entangled state to Alice without additional quantum and classical bit resources [51]. In 2022, Sisodia proposed asymmetric bidirectional quantum teleportation scheme using a seven-qubit cluster state as a quantum channel. Alice and Bob can send each other a three-particle bit entangled state and a two-particle bit entangled state, respectively [52]. Compared with other schemes, this scheme has higher inherent efficiency and lower operational complexity.

Asymmetric bidirectional controlled quantum teleportation (C-ABQT) is a combination of asymmetric bidirectional quantum teleportation (ABQT) and bidirectional controlled quantum teleportation (C-BQT). In 2015, Zhang et al. first proposed a scheme of bidirectional and asymmetric quantum controlled teleportation using a maximally eight-qubit entangled state [53]. In 2016, Li et al. proposed an asymmetric bidirectional controlled quantum teleportation scheme by using a six-qubit cluster state [54], in which the user Alice can transmit an arbitrary two-qubit entangled state to the user Bob and at the same time the user Bob can transmit an arbitrary single-qubit state to the user Alice. The same year, Hong proposed an asymmetric bidirectional controlled quantum teleportation scheme that uses an entangled state of seven qubits as a quantum channel, and Alice sends a single qubit to Bob, and Bob sends two qubit states to Alice under Charlie's control [55]. Yang et al. presented an asymmetric bidirectional controlled quantum teleportation scheme by using a seven-qubit cluster state [56]. Choudhury et al. present an asymmetric bidirectional controlled quantum teleportation scheme by using a nine-qubit entangled state as a quantum channel. With controller Charlie's cooperation, Alice can transmit an arbitrary three-qubit entangled state to Bob, and Bob can transmit an arbitrary two-qubit entangled state to Alice [57]. Compared with the previous asymmetric transmission scheme, this scheme transmits more qubits. In 2021, Zhang et al. designed an asymmetric bidirectional controlled quantum teleportation scheme, in which an 11-qubit entangled state is used as a quantum channel through the control of a third party, Charlie, and Alice can transmit any two-qubit state to Bob, and Bob can also transmit any three-qubit state to Alice [58].

5. Quantum Teleportation in Noisy Environment

Different from the ideal channel hypothesis, quantum teleportation in noisy environment (NE-QT) has been studied in recent years. In 2016, Li et al. first proposed a bidirectional controlled quantum teleportation scheme in noisy environment, which uses nine-qubit entangled states as carriers to transport particles [59]. In 2017, Yang et al. presented a novel scheme of bidirectional quantum teleportation in noisy channel by performing weak measurement [60]. In addition, two schemes [61, 62] are analyzed in noisy environment.

1.3 The Principle of Quantum Teleportation

The basic principle of quantum teleportation [38] is to use the remote correlation of EPR entangled pairs, that is, to realize the smooth transmission of unknown quantum states based on the non-local property of the entangled state. Both sides of communication can share an EPR entanglement pair before sending information. Alice then takes the information she holds (an unknown quantum state), makes a joint Bell state measurement with one of the particles in her EPR pair, and tells Bob the result of the measurement. Bob then follows the result of the measurement; finally, the unknown quantum state is reproduced on Bob's particles, and then the whole process of teleportation of the unknown quantum state is completed. The basic schematic diagram of the quantum teleportation process is shown in Fig. 1.2.

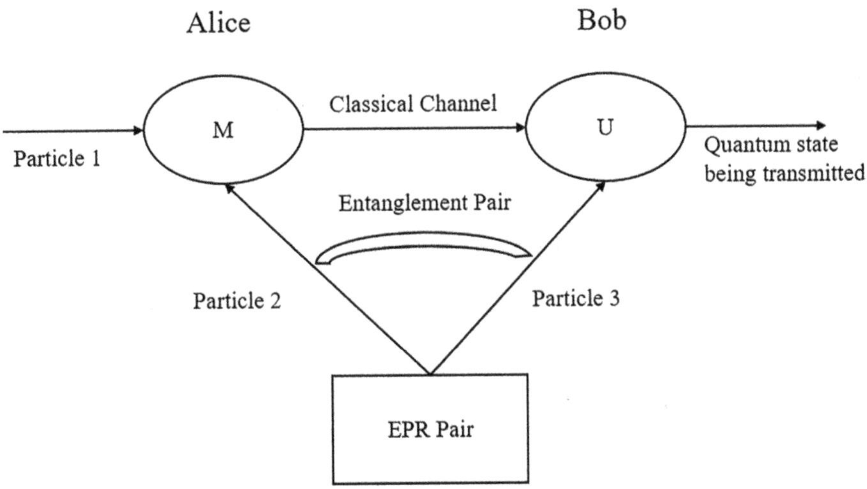

Fig. 1.2 Principle diagram of quantum teleportation

As shown in Fig. 1.2, M in the figure represents the joint Bell state measurement, and U represents a series of unitary operations. The channel from Alice to Bob is a classical channel, and the channel from EPR pair to both sides of communication is a quantum channel. First, quantum entanglement pairs (EPR pairs of particles 2 and 3) are shared between the two parties. The sender then uses the Bell basis to measure the unknown quantum states of particles 1 and 2. Finally, the receiver performs the corresponding unitary operation on particle 3 according to the measured results. The result will be one of the four Bell states, with a quarter of the time. As a result, the system of particles 1 and 2 will collapse on one of the Bell bases when observed, and the entanglement between particles 2 and 3 will disappear. According to the measurement, if the original particle 2 is changed, then the state of particle 3 will also be changed instantaneously, from the original entanglement state to the corresponding deterministic quantum state.

The detailed operation flow is as follows:

Step 1: First, the sender Alice prepares the initial unknown quantum state

$$|\psi_1\rangle = a|0_1\rangle + b|1_1\rangle \tag{1.1}$$

where $|a|^2 + |b|^2 = 1$.

Step 2: Construction of quantum channel—to transmit an unknown quantum state, there must be two other particles in addition to particle 1, called particles 2 and 3, and particles 2 and 3 must be entangled. We can also prepare particles 2 and 3 to the EPR state in advance, as follows:

Table 1.1 Alice's measurement results that correspond to Bob's operation

Alice's measurements of particles 1 and 2	Possible state of particle 3 after measurement	Bob's unitary transformation operation on particle 3 recovery
$\vert\phi^-\rangle_{12}$	$-a\vert 0_3\rangle - b\vert 1_3\rangle$	$-\begin{pmatrix} 1 & 0 \\ 0 & 1 \end{pmatrix}$
$\vert\phi^+\rangle_{12}$	$-a\vert 0_3\rangle + b\vert 1_3\rangle$	$\begin{pmatrix} -1 & 0 \\ 0 & 1 \end{pmatrix}$
$\vert\varphi^-\rangle_{12}$	$a\vert 1_3\rangle + b\vert 0_3\rangle$	$\begin{pmatrix} 0 & 1 \\ 1 & 0 \end{pmatrix}$
$\vert\varphi^+\rangle_{12}$	$a\vert 1_3\rangle - b\vert 0_3\rangle$	$\begin{pmatrix} 0 & 1 \\ -1 & 0 \end{pmatrix}$

$$|\phi_{23}\rangle = \frac{1}{\sqrt{2}}(|0_2\rangle|1_3\rangle - |1_2\rangle|0_3\rangle). \tag{1.2}$$

At this point, since particle 1 is not directly related to particle 2 or particle 3, these particles form a quantum system, and the complex wave function that the system constructs, that is, the quantum state $|\phi_{123}\rangle$, who can write the tensor product of $|\psi_1\rangle$ and $|\phi_{23}\rangle$:

$$|\phi_{123}\rangle = |\psi_1\rangle \otimes |\phi_{23}\rangle = (a|0_1\rangle + b|1_1\rangle) \otimes \frac{1}{\sqrt{2}}(|0_2\rangle|1_3\rangle - |1_2\rangle|0_3\rangle). \tag{1.3}$$

Step 3: Alice has particle 2, and Bob has particle 3. To successfully achieve the whole process of teleportation, Alice needs to take joint Bell-based measurements of particles 1 and 2. Therefore, the quantum system composed of particles 1 and 2 can be described by Bell basis. Therefore, the wave functions of these three particle systems can be described as

$$\begin{aligned} |\phi_{123}\rangle = \frac{1}{2}\Big[&|\phi^-\rangle_{12}(-a|0_3\rangle - b|1_3\rangle) + |\phi^+\rangle_{12}(-a|0_3\rangle + b|1_3\rangle) \\ &+ |\varphi^-\rangle_{12}(a|1_3\rangle + b|0_3\rangle) + |\varphi^+\rangle_{12}(a|1_3\rangle - b|0_3\rangle)\Big]. \end{aligned} \tag{1.4}$$

Step 4: Through the classical channel, Alice transmits the measurement to Bob, and the result of the measurement is a joint measurement of particles 1 and 2. Bob was able to make the correct unitary calculation for particle 3 based on this result, which would allow particle 3 to acquire the "exact replicas" of particle 1, which would restore the original particle state, thus achieving quantum teleportation. The specific observations, the state of the particle after collapse, and its corresponding unitary operation are shown in Table 1.1.

Thus, in the course of communication, only states expressing quantum information are transmitted, not the information particles themselves, so there is no contradiction with quantum no-cloning theorem. From the point of view of communication, the whole process of quantum teleportation is the formation of entangled particle pairs in the information carrier and then the modulation, transmission, and demodulation of the information; the quantum communication channel is composed of entangled particle pairs.

The quantum teleportation process transmits only the state of the original unknown particle (representing the state of particle 1), not the original unknown particle itself. The receiver changes the state of particle 3 (or a particle of a different class from the original particle), and it holds to the same quantum state as particle 1. When the state of particle 3 is the same as the state of particle 1, the state of particle 1 has collapsed. Of course, the quantum teleportation process must rely on the help of classical channels to achieve; therefore, its information transmission rate does not exceed the speed of light. Quantum teleportation has many advantages, for example, its information capacity is beyond imagination and it has great credibility.

1.4 Experimental Progress of Quantum Teleportation

Based on the theoretical research, quantum teleportation has also made many fruitful achievements in experiments.

In 1997, Austrian scientist Zeilinger made the first quantum teleportation experiment, successfully transferring a quantum state from a photon in place A to a photon in place B, and it has greatly promoted the development of quantum teleportation in experiment [63]. In 1998, Boschi has completed the quantum optical experiment of teleportation of unknown pure quantum state and demonstrated the teleportation results of linear and elliptic polarization states [64]. That same year, the Martini team in Italy completed the transport experiment [65]. In 2003, Professor Pan's team successfully realized the experiment of teleportation of free quantum states, which broke the problem of teleportation and laid a solid foundation for the follow-up research in the field of quantum communication in our country [66]. In 2004, the Riedmatten team conducted a long-distance stealth state transfer experiment, based on optical fiber transmission delay and Bell-based measurement operations. In July of the same year, the Chinese Pan Jianwei research group successfully carried out the terminal open quantum teleportation, designed the five-photon entanglement state, and carried out the teleportation through the five-photon entanglement state [67]. In 2005, Pan Jianwei et al. realized the experiment of bidirectional quantum entanglement distribution in Hefei, which was 13 km long and set a world record. In 2006, Pan Jianwei et al. first used six-particle quantum states to conduct entanglement experiments and successfully conducted a two-particle composite system of quantum teleportation experiments. Thus, the quantum secret communication experiment was also carried out smoothly over a total length of 100 km. In 2008, Pan Jianwei et al. first completed the entanglement exchange system with the function of storage and readout and successfully constructed quantum entanglement between two cold atomic systems. In 2013, through a 97 km free-space channel, Ren demonstrated the long-distance quantum teleportation on a low-loss single-link channel and the entanglement distribution on a high-loss double-link channel, and quantum teleportation at the 100 km level is realized [68]. In 2015, academician Pan Jianwei and his colleagues Lu Chaoyang and Liu Naile completed the quantum teleportation of a single particle with multiple degrees of freedom based on spin and orbital angular momentum encoding single photons [69]. In 2016, our country successfully developed and launched the first "Mozi" quantum satellite, which realized quantum teleportation between ground stations [70]. In 2019, Luo et al. proposed a scheme for teleporting arbitrary high-dimensional quantum states and demonstrated the experimental teleporting of the first qutrit state [71]. In 2020, Hu et al. demonstrated the teleportation of three-dimensional quantum states using a single-photon spatial model [72]. Then, Rajiuddin and others demonstrated the scheme of teleporting two-particle entanglement via cluster state by designing suitable quantum circuit on IBM quantum computer [73]. In the same year, Llewellyn et al. demonstrated quantum teleportation and multibody entanglement between chips, laying the foundation for large-scale integrated optical quantum communication and computing [74].

Quantum teleportation has become a conceptual cornerstone in many quantum information processing tasks. In 2006, Sherson et al. demonstrate teleportation between objects of a different nature—light and matter; it can be finessed by teleportation [75]. In 2018, Spyros has shown that Gaussian channels can be faithfully simulated by quantum teleportation [76]. Experimental quantum teleportation has transitioned from proof-of-principle demonstrations to practical applications. To ensure that they can benefit from the potential of quantum technology, multiple countries, enterprises, and research institutions have entered the preliminary stage of commercialization. As an emerging technology, quantum teleportation has been widely applied in numerous fields, including military secure communications, finance, quantum sensing, and medical diagnosis.

National defense, for example, has developed quantum techniques for key areas such as encryption, quantum network, and quantum computer. In 2021, a team of quantum physicists led by Jian-Wei Pan has shown that quantum teleportation allows error correction in quantum computers [77]. Moreover, physicists provide a new conceptual framework, measurement-based quantum computer (MBQC), with the teleportation-based model [78]. In 2022, Tasio et al. adapt quantum teleportation to quantum local area networks (QLANs) in order to reduce the environment-induced entanglement degradation [79]. In medical diagnosis and environmental monitoring, quantum technology transforms early disease diagnosis and detects gas leaks by exploiting quantum sensors. In 2019, Friis showed that teleportation improves the sensitivity of sensing devices, it has potentially been more sensitively than traditional schemes [80]. Financial transactions and machine learning are another example of speeding up the development of quantum teleportation. In 2020, Singh proposed an algorithm to provide a secure mechanism for performing transactions on the blockchain network by adopting quantum digital signatures

(QDS) and quantum teleportation [81]. In 2024, the researchers, from Delft University of Technology, used a two-stage frequency conversion and quantum teleportation process to interconnect different types of quantum hardware components [82].

In the future, with the continuous progress of science and technology, quantum teleportation practice technology will be further improved in terms of transmission distance, transmission speed, and transmission accuracy.

1.5 Outlook for the Future

Quantum teleportation is not only a conceptual protocol but also it serves as a fundamental ingredient to the development of many quantum technologies. Quantum teleportation has been achieved in experiments using many different substrates and technologies, including photonic degrees of freedom [83–85], atomic ensembles [86, 87], nuclear magnetic resonance (NMR) [65], and solid-state systems [88, 89]. However, the four Bell states for photonic qubits are still not completely distinguished using the practical Bell detection. It currently makes the protocol probabilistic. Improving the practical efficiency of the Bell detection is a major issue that needs to be solved urgently.

Fortunately, long-distance quantum teleportation is not affected by the incomplete Bell detection [68]. Satellite-based implementations also can accept this error. A complementary challenge is the techniques of photonic quantum integration in ground-to-satellite or inter-satellite implementations. In the context of long-distance high-fidelity communication, chip-to-chip teleportation applications to integrated quantum photonic. This technology promises new quantum photonic devices and provides a scalable platform for large-scale quantum information processing. On the other hand, short-distance quantum teleportation could be developed using optical modes or circuit quantum electrodynamics (QED) [90]. At the moment, the only drawback of these methods is limited coherence time. This problem can be overcome by integrating circuit QED with atomic ensemble quantum memories. Atomic ensembles have the potential to be good quantum memories [91]. Atomic ensembles are a promising study for scalable multiphoton technology.

In some conceptions of devices, quantum states would have to be transferred between quantum systems of a different nature. Quantum teleportation can provide a transfer of method, for example, a quantum state of light between light and matter [92]. In the future, quantum teleportation, as a basic primitive of a sophisticated connected network, strictly depends on excellent quantum interfaces. Quantum interfaces will be a new way to transfer quantum information between different types of quantum systems, a step that could help interconnect different types of quantum hardware components. Moreover, quantum repeaters would help enable quantum communication across a network via teleportation; it also has the ability to store quantum information. Researchers note that solid-state quantum repeaters have the availability of long coherence times [93]. From this point of view, the research of solid-state systems is an active area.

In 2024, the UK government announced it will invest in five hubs to develop the use of quantum technology in real-word applications, such as medical scanners, secure communication networks, and next-gen positioning systems. For healthcare sector, disease diagnosis needs ultrasensitive quantum sensors, including rapid blood tests and biomedical scanners. Quantum teleportation can improve the sensitivity of quantum sensors and benchmarking devices for quantum computation and simulation [80]. Similarly, teleportation may be a key technology for quantum sensing of positioning systems. It will overcome the key research barriers to quantum sensing, imaging, and timing being used at scale. Thus, quantum teleportation is the groundwork for future cutting-edge science and everyday lives.

1.6 Structure of the Book

The structure of the book is as follows:

1. Chapter 1 is the introduction, which summarizes the concept, principle, and research status of quantum teleportation.
2. Chapter 2 introduces the basic concept of quantum information, including the quantum bits, quantum measurement, quantum fidelity, quantum gates, quantum circuit model, and quantum channel noise.
3. Chapter 3 briefly introduces the research status of bidirectional quantum teleportation, and three bidirectional quantum teleportation schemes with different ground states are proposed, which are quantum bidirectional teleportation $2 \leftrightarrow 2$ or $2 \leftrightarrow 3$ qubit teleportation scheme via a six-qubit entangled state, bidirectional quantum teleportation of a two-qubit state via a four-qubit cluster state, and bidirectional quantum teleportation by using a six-qubit cluster state.
4. Chapter 4 studies the controlled bidirectional quantum teleportation and in turn proposes a controlled bidirectional quantum teleportation of arbitrary single qubit via a non-maximally entangled state, bidirectional quantum controlled teleportation of a 3-qubit state by using GHZ states, and bidirectional controlled quantum teleportation of a 3-qubit state by a new entangled 11-qubit state.
5. Chapter 5 introduces the cyclic quantum teleportation and then proposes cyclic and bidirectional quantum teleportation scheme via pseudo-multi-qubit states and asymmetric cyclic controlled quantum teleportation scheme by using a nine-qubit entangled state.
6. Chapter 6 introduces the theory of quantum teleportation in noisy environment and in turn proposes a novel scheme for bidirectional controlled quantum teleportation of two-qubit states via a seven-qubit entangled state in noisy environment and asymmetric bidirectional controlled teleportation scheme by using a nine-qubit entangled state in noisy environment and enhancing robustness of noisy qutrit teleportation scheme with Markovian memory.

References

1. Shor PW (1994) Algorithms for quantum computation: discrete logarithms and factoring. IEEE Computer Society Press, New Mexico, pp 124–134
2. Takesue H (2016) Quantum communication experiments over optical fiber. Principles and Methods of Quantum Information Technologies, Springer Japan, pp 53–70
3. Yu Y, Man HA, Ateniese G et al (2017) Identity-based remote data integrity checking with perfect data privacy preserving for cloud storage. IEEE Trans Inf Forensics Secur 12(4): 767–778
4. Obando PC, Paula FM, Sarandy MS (2015) Trace-distance correlations for x states and the emergence of the pointer basis in Markovian and non-Markovian regimes. Phys Rev A 92(3): 032307
5. Bennett CH, Brassard G, Crépeau C et al (1993) Teleporting an unknown quantum state via dual classical and Einstein-Podolsky-Rosen channels. Phys Rev Lett 70(13):1895–1899
6. Banaszek K (2000) Optimal quantum teleportation with an arbitrary pure state. Phys Rev A 62(2):024301
7. Roa L, Delgado A, Fuentes-Guridi I (2003) Optimal conclusive teleportation of quantum states. Phys Rev A 68(2):022310
8. Li W-L, Li C-F, Guo G-C (2000) Probabilistic teleportation and entanglement matching. Phys Rev A 61(3):034301
9. Verstraete F, Verschelde H (2003) Optimal teleportation with a mixed state of two qubits. Phys Rev Lett 90(9):097901
10. Man ZX, Xia YJ, An NB (2007) Quantum teleportation of an unknown N-Qubit W-Like state. JETP Lett 85(12):662–666
11. Yang K, Huang L, Yang W et al (2009) Quantum teleportation via GHZ-like state. Int J Theor Phys 48(2):516–521
12. Min L, Nan Z et al (2017) Quantum teleportation of five-qubit state. Int J Theor Phys 56(9): 2710–2715
13. Huang Y, Yang W (2020) Quantum teleportation via qutrit entangled state. Chin J Electron 29(2):228–232
14. Bennett CH, Brassard G et al (1993) Teleporting an unknown quantum state via dual classical and Einstein-Podolsky-Rosen channels. Phys Rev Lett 70:1895–1899
15. Ikram M, Shi-Yao Zhu M, Zubairy S (2000) Quantum teleportation of an entangled state. Phys Rev A 62(2). https://doi.org/10.1103/PhysRevA.62.022307
16. Gao T (2004) Quantum logic networks for probabilistic and controlled teleportation of unknown quantum states*. Commun Theor Phys 42:223–228
17. Zhang ZJ, Man ZX (2005) Many-agent controlled teleportation of multi-qubit quantum information. Phys Lett Sect A Gen At Solid State Phys 341:55–59
18. Yang K, Huang L, Yang W, Song F (2009) Quantum teleportation via GHZ-like state. Int J Theor Phys 48(2):516–521
19. Shi BS, Tomita A (2002) Teleportation of an unknown state by W state. Phys Lett A 296 (Issues):4–5
20. Tian D, Tao Y, Qin M (2008) Teleportation of an arbitrary two-qubit state based on the non-maximally four-qubit cluster state. Sci China, Ser G 51(10):1523–1528
21. Cardoso WB (2008) Teleportation of GHZ-states in QED-Cavities without the explicit Bell-state measurement. Int J Theor Phys 47:977–982
22. Duan YJ, Zha XW (2014) Bidirectional quantum controlled teleportation via a six-qubit entangled state. Int J Theor Phys 53:3780–3786
23. Huang Sang M (2016) Bidirectional quantum teleportation by using five-qubit cluster state. Int J Theor Phys 55(3):1333–1335
24. Gou YT, Shi HL, Wang XH, Liu SY (2017) Probabilistic resumable bidirectional quantum teleportation. Quantum Inf Process 16:1–13

25. Zhou RG, Li X, Qian C et al (2020) Quantum bidirectional teleportation 2 ↔2 or 2 ↔3 qubit teleportation protocol via 6-qubit entangled state. Int J Theor Phys 59:166–172
26. Mengting W, Sheng LH (2022) Bidirectional quantum teleportation using a five-qubit cluster state as a quantum channel. Quantum Inf Process 21(2):44
27. Chen Y, Du J, Liu S, Wang X (2017) Cyclic quantum teleportation. Quantum Inf Process 16(8): 1–9
28. Sisodia M, Shukla A, Thapliyal K, Pathak A (2017) Design and experimental realization of an optimal scheme for teleportation of an n-qubit quantum state. Quantum Inf Process 16(12):1–19
29. Sabir DJM (2018) Efficient schemes for the quantum teleportation of a sub-class of tripartite entangled states. Quantum Inf Process 17(7):1–11
30. Karlsson A, Bourennane M (1998) Quantum teleportation using three-particle entanglement. Phys Rev A, Gen Phys 58(6):4394–4400
31. Yang C-P, Han S (2005) A scheme for the teleportation of multiqubit quantum information via the control of many agents in a network. Phys Lett A, Gen Phys 343(4):267–273
32. Wang Y-H, Song H-S (2008) Preparation of partially entangled W state and deterministic multi-controlled teleportation. Opt Commun 281(3):489–493
33. Song-Song L et al (2008) Controlled teleportation using four-particle cluster state. Commun Theor Phys 50(3):633
34. Yang XWYSG (2009) Controlled teleportation against uncooperation of part of supervisors. Quantum Inf Process 8(4):319–330
35. Tian-Yin W, Qiao-Yan W (2011) Controlled quantum teleportation with Bell states. Chinese Phys B 20(4):040307
36. Zha XW, Zou ZC, Qi JX, Song HY (2013) Bidirectional quantum controlled teleportation via five-qubit cluster state. Int J Theor Phys 52:1740–1744
37. Chen Y (2014) Bidirectional controlled quantum teleportation by using five-qubit entangled state. Int J Theor Phys 53:1454–1458. https://doi.org/10.1007/s10773-013-1943-2
38. Tan X, Zhang X, Song T (2016) Deterministic quantum teleportation of a particular six-qubit state using six-qubit cluster state. Int J Theor Phys 55:155–160
39. Zhou R-G, Xu R, Lan H, Bidirectional quantum teleportation by using six-qubit cluster state (2019) IEEE Access 7:1–1
40. Sang M (2016) Huang: Bidirectional quantum controlled teleportation by using a seven-qubit entangled state. Int J Theor Phys 55:380–383
41. Duan YJ, Zha XW, Sun XM, Xia JF (2014) Bidirectional quantum controlled teleportation via a maximally seven-qubit entangled state. Int J Theor Phys 53:2697–2707
42. Hong W (2016) qin: Asymmetric bidirectional controlled teleportation by using a seven-qubit entangled state. Int J Theor Phys 55:384–387
43. Sadeghi Zadeh MS, Houshmand M, Aghababa H (2017) Bidirectional teleportation of a two-qubit state by using eight-qubit entangled state as a quantum channel. Int J Theor Phys 56:2101–2112
44. Zhi-wen S (2018) Cyclic controlled teleportation by using a seven-qubit entangled state. Int J Theor Phys 57:3835–3838
45. Shao ZL, Long Y (2019) Circular controlled quantum teleportation by a genuine seven-qubit entangled state. Int J Theor Phys 58(13):1957–1967
46. Li YH, Qiao Y, Sang MH et al (2019) Controlled cyclic quantum teleportation of an arbitrary two-qubit entangled state by using a ten-qubit entangled state. Int J Theor Phys 58:1541–1545
47. Sun S, Li L, Zhang H (2020) Quantum cyclic controlled teleportation of unknown states with arbitrary number of qubits by using seven-qubit entangled channel. Int J Theor Phys 59(13): 1017–1030
48. Zhang D, Zha XW, Duan YJ (2015) Bidirectional and asymmetric quantum controlled teleportation. Int J Theor Phys 54(5):1711–1719
49. Choudhury BS, Samanta S (2017) Asymmetric bidirectional 3 ↔ 2 qubit teleportation protocol between Alice and Bob via 9-qubit cluster state. Int J Theor Phys 56(10):3285–3296

50. Vikram V (2021) Bidirectional quantum teleportation by using two GHZ-states as the quantum channel. IEEE Commun Lett 25(3):936–939
51. Mingzi M, Xiaoguang C, Songrong Q (2022) An efficient asymmetric bidirectional quantum teleportation protocol by using coherent bit channel. Int J Theor Phys 61(3):82
52. Rui D, Sheng LH (2022) Asymmetric bidirectional quantum teleportation via seven-qubit cluster state. Int J Theor Phys 61(7):187
53. Zhang D, Wei X, Wei Z, Yan L (2015) Bidirectional and asymmetric quantum controlled teleportation via maximally eight-qubit entangled state. Quantum Inf Process 14:3835–3844
54. Hua Li Y, Ping Nie L, Lan Li X, Huang Sang M (2016) Asymmetric bidirectional controlled teleportation by using six-qubit cluster state. Int J Theor Phys 55(6):3008–3016
55. Hong WQ (2016) Asymmetric bidirectional controlled teleportation by using a seven-qubit entangled state. Int J Theor Phys 55:384–387
56. Yang YQ, Zha XW, Yu Y (2016) Asymmetric bidirectional controlled teleportation via seven-qubit cluster state. Int J Theor Phys 55(10):4197–4204
57. Choudhury BS, Samanta S (2017) Asymmetric bidirectional 3⇔ 2 qubit teleportation protocol between Alice and Bob via 9-qubit cluster state. Int J Theor Phys 56(10):3285–3296
58. Huo G, Zhang T, Zha X et al (2021) Controlled asymmetric bidirectional quantum teleportation of two- and three-qubit states. Quantum Inf Process 20(1):1–11
59. Li YH, Jin X-m (2016) Bidirectional controlled teleportation by using nine-qubit entangled state in noisy environments. Quantum Inf Process 15:929–945
60. Yang G, Lian BW, Nie M, Jin J (2017) Bidirectional multi-qubit quantum teleportation in noisy channel aided with weak measurement. Chinese Phys B 26:040305
61. Sarvaghad-Moghaddam M (2018) Bidirectional controlled quantum teleportation using eight-qubit quantum channel in noisy environments. Quantum Phys 59:3156–3173
62. Hou K, Bao DQ, Zhu CJ, Yang Y-p (2019) Controlled teleportation of an arbitrary two-qubit entanglement in noises environment. Quantum Inf Process 18:1–19
63. Bouwmeester D, Pan JW, Mattle K et al (1997) Experimental quantum teleportation. Nature 390(6660):575–579
64. Boschi D, Branca S, De Martini F et al (1998) Experimental realization of teleporting an unknown pure quantum state via dual classical and Einstein-Podolsky-Rosen channels. Phys Rev Lett 80(6):1121
65. Nielsen MA, Knill E, Laflamme R (1998) Complete quantum teleportation using nuclear magnetic resonance. Nature 396(6706):52–55
66. Pan JW, Gasparoni S, Aspelmeyer M et al (2003) Experimental realization of freely propagating teleported qubits. Nature 421(6924):721–725
67. Zhao Z, Chen YA, Zhang AN et al (2004) Experimental demonstration of five-photon entanglement and open-destination teleportation. Nature 430(6995):54–58
68. Yin J, Ren JG, Lu H et al (2012) Quantum teleportation and entanglement distribution over 100-kilometre free-space channels. Nature 488(7410):185–188
69. Wang XL, Cai XD, Su ZE et al (2015) Quantum teleportation of multiple degrees of freedom of a single photon. Nature 518(7540):516–519
70. Ren JG, Xu P, Yong HL et al (2017) Ground-to-satellite quantum teleportation. Nature 549(7670):70–73
71. Luo Y-H, Zhong H-S, Erhard M et al (2019) Quantum teleportation in high dimensions. Phys Rev Lett 123(7):070505
72. Hu X-M, Zhang C, Liu B-H et al (2020) Experimental high-dimensional quantum teleportation. Phys Rev Lett 125(23):230501
73. Rajiuddin S, Baishya A, Behera BK et al (2020) Experimental realization of quantum teleportation of an arbitrary two-qubit state using a four-qubit cluster state. Quantum Inf Process 19(3): 1–13
74. Llewellyn D, Ding Y, Faruque II et al (2020) Chip-to-chip quantum teleportation and multi-photon entanglement in silicon. Nat Phys 16(2):148–153
75. Sherson JF (2006) Quantum teleportation between light and matter. Nature 443(7111):557–560

76. Tserkis S, Dias J, Ralph TC (2018) Simulation of Gaussian channels via teleportation and error correction of Gaussian states. Phys Rev A 98(5):052335
77. Luo Y, Chen M, Erhard M, Zhong H, Wu D, Tang H, Zhao Q, Wang X, Fujii K, Li L, Liu N, Nemoto K, Munro WJ, Lu C, Zeilinger A, Pan J (2021) Quantum teleportation of physical qubits into logical code spaces. Proc Natl Acad Sci USA 118(36):e2026250118
78. Wei T (2021, March 25) Measurement-based quantum computation. Oxford Research Encyclopedia of Physics. Retrieved November 11, 2024.
79. Gonzalez-Raya T (2022) Open-air microwave entanglement distribution for quantum teleportation. Phys Rev Appl 18(4):Art. no. 044002., APS
80. Friis N, Vitagliano G, Malik M, Huber M (2019) Entanglement certification from theory to experiment. Nat Rev Phys 1(1):72–87
81. Singh S, Rajput NK, Rathi VK et al (2023) Securing blockchain transactions using quantum teleportation and quantum digital signature. Neural Process Lett 55:3827–3842
82. Iuliano M, Slater MC, Stolk AJ et al (2024) Qubit teleportation between a memory-compatible photonic time-bin qubit and a solid-state quantum network node. Npj Quantum Inf 10:107
83. Kim Y-H, Kulik SP, Shih Y (2001) Quantum teleportation of a polarization state with a complete Bell state measurement. Phys Rev Lett 86(7):1370–1373
84. Marcikic I, de Riedmatten H, Tittel W et al (2003) Long-distance teleportation of qubits at telecommunication wavelengths. Nature 421:509–513
85. Metcalf B, Spring J, Humphreys P et al (2014) Quantum teleportation on a photonic chip. Nature Photon 8:770–774
86. Chen YA, Chen S, Yuan ZS et al (2008) Memory-built-in quantum teleportation with photonic and atomic qubits. Nat Phys 4:103–107
87. Bao X, Xu X, Li C, Yuan Z, Lu C, Pan J (2012) Quantum teleportation between remote atomic-ensemble quantum memories. Proc Natl Acad Sci USA 109(50):20347–20351
88. Gao W, Fallahi P, Togan E et al (2013) Quantum teleportation from a propagating photon to a solid-state spin qubit. Nat Commun 4:2744
89. Steffen L, Salathe Y, Oppliger M et al (2013) Deterministic quantum teleportation with feed-forward in a solid state system. Nature 500:319–322
90. Furusawa A et al (1998) Unconditional quantum teleportation. Science 282:706–709
91. Julsgaard B, Sherson J, Cirac J et al (2004) Experimental demonstration of quantum memory for light. Nature 432:482–486
92. Sherson J, Krauter H, Olsson R et al (2006) Quantum teleportation between light and matter. Nature 443:557–560
93. Zhong M, Hedges M, Ahlefeldt R et al (2015) Optically addressable nuclear spins in a solid with a six-hour coherence time. Nature 517:177–180

Chapter 2
The Basic Concept of Quantum Information

Abstract Quantum information refers to the physical information carried by the "state" of quantum systems. It integrates the essence of quantum mechanics and information science, ushering in novel paradigms for computation, encoding, and information transmission. Quantum information represents a mode of information processing that leverages various coherent properties of quantum systems, such as quantum parallelism, quantum entanglement, and the no-cloning theorem, for computation, encoding, and information transmission. As an emerging interdisciplinary field, quantum information is attracting increasing attention from researchers and investors alike, fueled by its unique allure and vast potential for applications. With continuous advancements in technology and expanding applications, it is believed that quantum information will play an increasingly pivotal role in future societies.

2.1 Quantum Bit

In the research of classical information processing, the basic concept used is classical bit, which has only two deterministic states, 0 and 1. In quantum information processing research, the most basic concept used is quantum bit (qubit). But unlike classical bits or the deterministic states of 0 or 1, quantum bits can be in a state of superposition. The classical bits 0 and 1 correspond to quantum bits, represented by $|0\rangle$ and $|1\rangle$. The symbol "$|\rangle$" is a bra–ket notation that represents the state of a quantum bit. A qubit is a unit vector defined in a two-dimensional complex Hilbert space that can be in a state of superposition. In this vector space, there is a set of orthonormal basis states, state $|0\rangle$ and state $|1\rangle$, so the state of a qubit can be represented by a linear superposition of state $|0\rangle$ and state $|1\rangle$, also called a superposition. For example, the quantum state $|\psi\rangle$ in this space can be expressed as follows:

$$|\psi\rangle = \alpha|0\rangle + \beta|1\rangle \tag{2.1}$$

R.-G. Zhou et al., *Design of Quantum Teleportation Schemes*,
https://doi.org/10.1007/978-3-031-82725-9_2

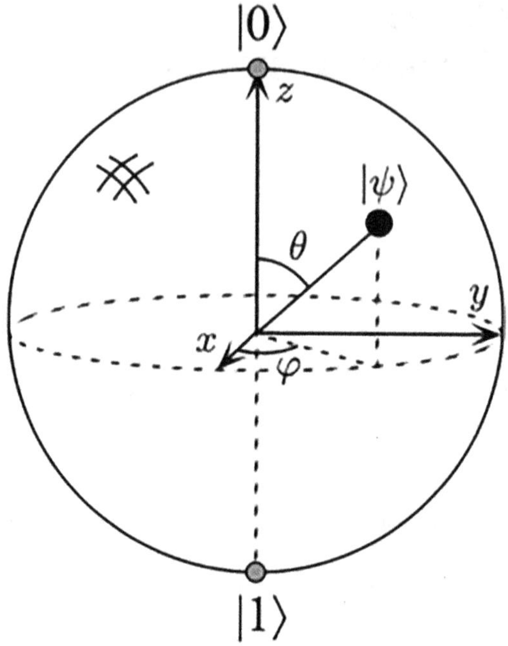

Fig. 2.1 Bloch spherical representation of quantum bits

where α and β are both complex numbers and satisfy the normalization condition $|\alpha|^2 + |\beta|^2 = 1$. For ease of understanding and calculation, states $|0\rangle$ and $|1\rangle$ can be expressed as a column vector in linear algebra:

$$|0\rangle = \begin{pmatrix} 1 \\ 0 \end{pmatrix}, |1\rangle = \begin{pmatrix} 0 \\ 1 \end{pmatrix}. \tag{2.2}$$

So, the quantum state $|\psi\rangle$ can be expressed as

$$|\psi\rangle = \alpha\begin{pmatrix} 1 \\ 0 \end{pmatrix} + \beta\begin{pmatrix} 0 \\ 1 \end{pmatrix} = \begin{pmatrix} \alpha \\ \beta \end{pmatrix}. \tag{2.3}$$

With the Bloch ball, quantum bits can be understood in a more visual way. As shown in Fig. 2.1, the state $|0\rangle$ is at the northern end of the sphere, the state $|1\rangle$ is at the southern end of the sphere, and the quantum state $|\psi\rangle$ can be represented by any point on the sphere.

$$|\psi\rangle = \cos\frac{\theta}{2}|0\rangle + e^{\varphi}\sin\frac{\theta}{2}|1\rangle \tag{2.4}$$

where $0 \leq \theta \leq \pi$, $0 \leq \varphi \leq 2\pi$, and θ and φ are used to define the polarization angle and azimuth angle of any point on the Bloch sphere.

2.2 Quantum Measurement

Quantum measurement, which is the basis of quantum computation and quantum information processing, is the process of obtaining information of quantum system by measuring quantum bits. Quantum measurements measure the quantum system under test by using external measurement devices, and then from the uncertainty of the quantum system to get clear information. After measurement, the state of the unknown quantum bit can be determined. The measurement operation on the line is shown in Fig. 2.1. A set of measurement operators is introduced to describe the effect of the measurement on the quantum system, satisfying the completeness (2.5)

$$\sum_m M_m^+ M_m = I. \tag{2.5}$$

In particular, the measurement operator $\{M_m\}$ acts in the state space of the subsystem to be measured, and the possibility of the measurement result is indicated by subscript m. Before the measurement, the state of the quantum system is $|\psi\rangle$, and after the measurement, the probability that the result is m is

$$P(m) = \langle\psi|M_m^+ M_m|\psi\rangle. \tag{2.6}$$

After the measurement, the state of the quantum system collapses to

$$\frac{M_m \mid \psi\rangle}{\sqrt{\langle\psi|M_m^+ M_m|\psi\rangle}}. \tag{2.7}$$

2.2.1 *Projection Measurement*

The projection measurement can be described by an observable Hermite operator M in the state space of a quantum system, the vector space of the quantum system $|\psi\rangle$ something can be projected onto the eigenspace of P_i by using the Hermite operator P_i. The spectral decomposition of the observed M is

$$M = \sum_m m P_m \tag{2.8}$$

where P_i is a projection of the eigenspace M and has an eigenvalue m, and the measurement corresponds to the eigenvalue m of the observed M; when the state of the measured quantum system is $|\psi\rangle$, the probability of obtaining m can be expressed as

$$p(m) = \langle\psi|P_m|\psi\rangle. \tag{2.9}$$

According to the above (2.9), the measurement result of m can be obtained as follows:

$$\frac{P_m|\psi\rangle}{\sqrt{P(m)}}. \tag{2.10}$$

It is also possible to calculate the average value of M by definition

$$\begin{aligned} E(M) &= \sum_m mP(m) \\ &= \sum_m m\langle\psi|P_m|\psi\rangle = \langle\psi|\left(\sum_m mP_m\right)|\psi\rangle \\ &= \langle\psi|M|\psi\rangle. \end{aligned} \tag{2.11}$$

2.2.2 POVM Measurement

If there is a quantum system with a state $|\psi\rangle$ in which the set of operators $E_i = M_i M_i^*$ for the measurement operation is used, if the completeness condition of the measurement is satisfied, then there are

$$\sum_i E_i = I. \tag{2.12}$$

If E_i is a positive operator, then the measurement is a POVM measurement, where the probability of the result mi is

$$P_{m_i} = \langle\psi|E_i|\psi\rangle. \tag{2.13}$$

2.2.3 Weak Measurement and Reversing Measurement

The result of weak measurement (WM) will not cause the quantum system to collapse completely and become an eigenstate, and the quantum system will remain in an excited state. That is, the state after measurement can be restored by using weak measurement flip. Although weak measurement can only obtain a small part of the information of the quantum system, it can be obtained by continuous weak measurement. The operands of a weak measurement are expressed as

$$M_1 = \sqrt{p}|1\rangle\langle 1| = \begin{pmatrix} 0 & 0 \\ 0 & \sqrt{p} \end{pmatrix}$$

$$M_2 = |0\rangle\langle 0| + \sqrt{1-p}|1\rangle\langle 1| = \begin{pmatrix} 1 & 0 \\ 0 & \sqrt{1-p} \end{pmatrix}. \tag{2.14}$$

Here p denotes the strength of the weak measurement and $0 \leq p \leq 1$. When the operator M_2 of the weak measurement acts on the quantum state, the state being measured does not completely collapse, and then quantum measurement reversal (QMR) technique is used to restore it. The operands of the weak measurement reversal technique are expressed as follows:

$$M_{QMR} = \begin{pmatrix} \sqrt{1-q} & 0 \\ 0 & 1 \end{pmatrix} = \frac{1}{\sqrt{1-q}} X \begin{pmatrix} 1 & 0 \\ 0 & \sqrt{1-q} \end{pmatrix} X. \tag{2.15}$$

Here q denotes the strength of the quantum measurement reversal and $0 \leq q \leq 1$. The $X = |0\rangle\langle 1| + |1\rangle\langle 0|$ represents a bit-flip operation, which means that quantum measurement reversal can be achieved using three local operations: a bit-flip operation, a weak measurement, and a second bit-flip operation.

2.3 Quantum Fidelity

Fidelity is a physical measure of how similar two quantum states are. For two mixed states ρ and σ, a common definition of fidelity is

$$F(\rho, \sigma) \equiv tr\sqrt{\rho^{1/2}\sigma\rho^{1/2}}. \tag{2.16}$$

For a mixed state ρ and a pure state $|\psi\rangle$, the fidelity between the two can be defined as

$$F(|\psi\rangle, \rho) \equiv \sqrt{\langle\psi|\rho|\psi\rangle}. \tag{2.17}$$

2.4 Quantum Gate

In the design of quantum communication scheme, it is often necessary to transform the transmitted quantum states. In this case, it is necessary to use the quantum gate to act on the quantum state, so that the quantum state changes. Based on the linear property, the quantum gate can be expressed in matrix form U. Since the quantum states before and after transformation must satisfy the normalization condition, the matrix U describing the quantum gate should be unitary matrix, and the action of quantum gate on quantum state can also be called unitary operation. A matrix U needs to satisfy a relation $U^*U = I$, where U^* is the conjugate transposition of U. Next, we will introduce the commonly used single-qubit gates, as well as some multi-qubit gates.

2.4.1 *Single-Qubit Gates*

A single-qubit gate can change the state of a single qubit, for example, by converting it from $|0\rangle$ to $|1\rangle$ or by converting it to a superposition state relative to the ground state. Common single-qubit gates include the Pauli matrix (I-gate, X-gate, Y-gate, Z-gate) and Hadamard gate (H-gate).

1. ***I*-Gate**

$$I = |0\rangle\langle 0| + |1\rangle\langle 1| = \begin{bmatrix} 1 & 0 \\ 0 & 1 \end{bmatrix}. \tag{2.18}$$

If a quantum I-gate operation is performed on a single quantum state, after the operation, the quantum state remains unchanged, that is, there is $I|0\rangle = |0\rangle$ and $I|1\rangle = |1\rangle$. The circuit diagram for the I-gate is shown in Fig. 2.2.

2. ***X*-Gate**

$$X = |0\rangle\langle 1| + |1\rangle\langle 0| = \begin{bmatrix} 0 & 1 \\ 1 & 0 \end{bmatrix}. \tag{2.19}$$

If a quantum X-gate operation is performed on a single quantum state, after the operation, the quantum state will be reversed, that is, $X|0\rangle = |1\rangle$ and $X|1\rangle = |0\rangle$. The circuit diagram for the X-gate is shown in Fig. 2.3.

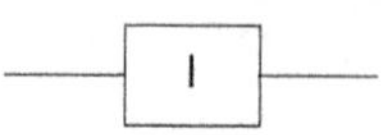

Fig. 2.2 *I*-gate

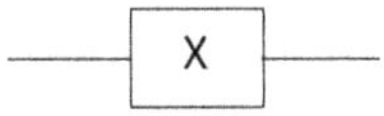

Fig. 2.3 *X*-gate

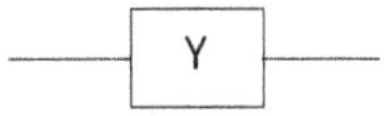

Fig. 2.4 *Y*-gate

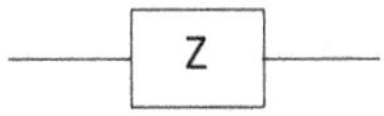

Fig. 2.5 *Z*-gate

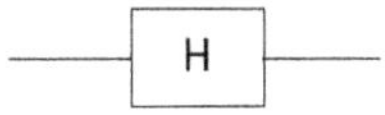

Fig. 2.6 *H*-gate

3. ***Y*-Gate**

$$Y = i(|1\rangle\langle 0| - |0\rangle\langle 1|) = \begin{bmatrix} 0 & -i \\ i & 0 \end{bmatrix}. \tag{2.20}$$

If a quantum *Y*-gate operation is performed on a single quantum state, after the operation, the quantum state will undergo a bit phase reversal, that is, $Y|0\rangle = i|1\rangle$ and $Y|1\rangle = -i|0\rangle$. The circuit diagram for the *Y*-gate is shown in Fig. 2.4.

4. ***Z*-Gate**

$$Z = |0\rangle\langle 0| - |1\rangle\langle 1| = \begin{bmatrix} 1 & 0 \\ 0 & -1 \end{bmatrix}. \tag{2.21}$$

If a quantum *Z*-gate operation is performed on a single quantum state, after the operation, the quantum state will undergo a phase reversal, that is, $Z|0\rangle = |0\rangle$ and $Z|1\rangle = -|1\rangle$. The circuit diagram for the *Z*-gate is shown in Fig. 2.5.

5. **Hadamard Gate (*H*-Gate)**

$$U_H = \begin{bmatrix} \frac{1}{\sqrt{2}} & \frac{1}{\sqrt{2}} \\ \frac{1}{\sqrt{2}} & -\frac{1}{\sqrt{2}} \end{bmatrix}. \tag{2.22}$$

The Hadamard gate, which acts as $H|0\rangle = \frac{1}{\sqrt{2}}(|0\rangle + |1\rangle)$ and $H|1\rangle = \frac{1}{\sqrt{2}}(|0\rangle - |1\rangle)$, transforms two base vectors into two equal-probability superposition states, and the role of the *H*-gate in quantum circuits is to create superposition states; it is commonly used in the front end of quantum circuits. The circuit

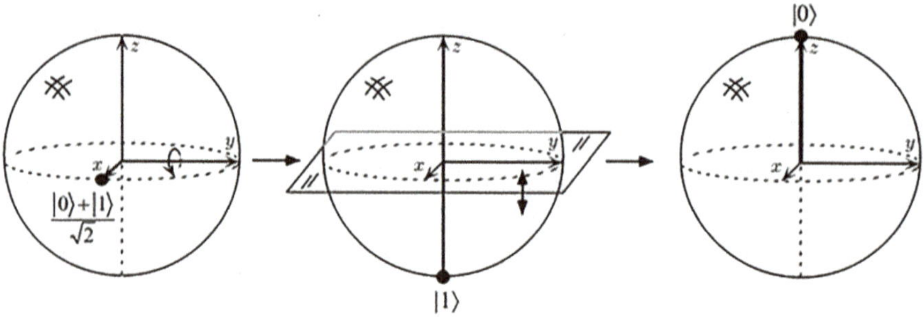

Fig. 2.7 The *H*-gate is displayed on the Bloch sphere

Fig. 2.8 CNOT gate

diagram for the *H*-gate is shown in Figs. 2.6, and 2.7 shows how the *H*-gate acts on the Bloch ball, first rotating 90° around the *Y* axis and then 180° around the *X* axis.

2.4.2 Multi-qubit Gates

Multi-qubit gates can alter interactions between two or more qubits, for example, by entangling them or manipulating them in quantum logic. Common multi-qubit gates include CNOT gate, Toffoli gate, SWAP gate, and so on.

1. **CNOT Gate**

$$U_{CNOT} = \begin{bmatrix} 1 & 0 & 0 & 0 \\ 0 & 1 & 0 & 0 \\ 0 & 0 & 0 & 1 \\ 0 & 0 & 1 & 0 \end{bmatrix}. \tag{2.23}$$

The CNOT gate's circuit diagram is shown in Fig. 2.8, where the solid black circle represents the control qubit and the hollow white circle represents the target qubit. When the control qubit is $|0\rangle$, the CNOT gate acting on it will not change the form of the target qubit, which is $CNOT|00\rangle = |00\rangle$ and $CNOT|01\rangle = |01\rangle$; when the control qubit is $|1\rangle$, the target qubit is changed when the CNOT gate acts, that is, $CNOT|10\rangle = |11\rangle$ and $CNOT|11\rangle = |10\rangle$.

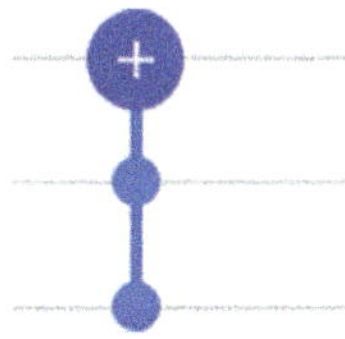

Fig. 2.9 Toffoli gate

Table 2.1 The truth table of the Toffoli gate

Input 0	Input 1	Input 2	Output 0	Output 1	Output 2
0	0	0	0	0	0
0	1	0	0	0	1
0	0	1	0	1	0
0	1	1	0	1	1
1	0	0	1	0	0
1	0	1	1	0	1
1	1	0	1	1	1
1	1	1	1	1	0

2. **Toffoli Gate**

$$U_{Toffoli} = \begin{bmatrix} 1 & 0 & 0 & 0 & 0 & 0 & 0 & 0 \\ 0 & 1 & 0 & 0 & 0 & 0 & 0 & 0 \\ 0 & 0 & 1 & 0 & 0 & 0 & 0 & 0 \\ 0 & 0 & 0 & 1 & 0 & 0 & 0 & 0 \\ 0 & 0 & 0 & 0 & 1 & 0 & 0 & 0 \\ 0 & 0 & 0 & 0 & 0 & 1 & 0 & 0 \\ 0 & 0 & 0 & 0 & 0 & 0 & 1 & 0 \\ 0 & 0 & 0 & 0 & 0 & 0 & 0 & 1 \end{bmatrix}. \quad (2.24)$$

The Toffoli gate is a gate that operates three qubits and is relatively complete for traditional computation. The circuit diagram for the Toffoli gate is shown in Fig. 2.9. If the first two qubits are $|1\rangle$, then a Pauli-X gate operation is performed on the third qubit and vice versa. Since Toffoli gate is a quantum simulation of a traditional logic gate, it can be fully represented as a truth table, as shown in Table 2.1.

Fig. 2.10 Toffoli gate

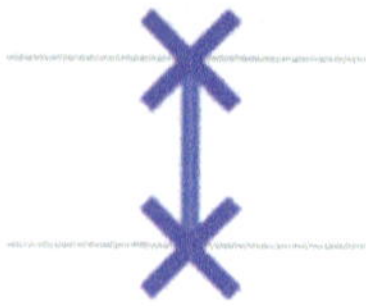

Fig. 2.11 Measuring gate

3. **SWAP Gate**

$$U_{SWAP} = \begin{bmatrix} 1 & 0 & 0 & 0 \\ 0 & 0 & 1 & 0 \\ 0 & 1 & 0 & 0 \\ 0 & 0 & 0 & 1 \end{bmatrix}. \tag{2.25}$$

The SWAP gate is used to manipulate two qubits and can be used to exchange the order of tensor products. The circuit diagram for the SWAP gate is shown in Fig. 2.10. Tensor product is not commutative, that is, $|\psi\rangle_1 \otimes |\psi\rangle_2 \neq |\psi\rangle_2 \otimes |\psi\rangle_1$, but under the action of the SWAP gate, the tensor product can be exchanged, that is, $U_{SWAP}|\psi\rangle_1 \otimes |\psi\rangle_2 = |\psi\rangle_2 \otimes |\psi\rangle_1$.

It should be noted that the quantum gate is reversible, which means that the state of the qubit can be reverted back to the original state by reverse operation. This is an important concept in quantum computing, where all operations must be reversible.

These gates are often used to construct quantum circuits, treating them as single logical units and combining them to build larger quantum algorithms. The choice of quantum gates is critical to the success of quantum computing, so scientists have been studying how to design new gates to better harness the potential of quantum computing in the future.

2.5 Quantum Circuit Model

The quantum circuit is composed of the route representing the evolution of quantum bits and the quantum logic gates acting on the quantum bits. The effect of a quantum circuit is the equivalent of each quantum logic gate acting in turn on a quantum bit. In a real quantum computer, the information of the final state of the quantum system can only be obtained by the measurement operation, so the measurement operation is also a part of the quantum circuit; measuring operations are sometimes called

Fig. 2.12 Single-qubit quantum circuit

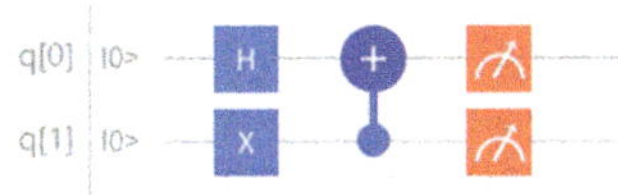

Fig. 2.13 Two-qubit quantum circuit

measuring gates, and the circuit diagram for the measuring gate is shown in Fig. 2.11.

Quantum algorithms are typically represented by the quantum circuit model. Each line in the quantum circuit represents a qubit or classical bit, the symbol on the left side of the line represents the symbol for the qubit, and the operation in the blue box on the line corresponds to a specific unitary transformation, the red icon represents the quantum measurement operation on the corresponding qubit, and the measurement result is saved in the corresponding classical bit.

As shown in Fig. 2.12, this is a simple single-qubit quantum circuit.

Figure 2.13 shows a quantum circuit of two qubits.

2.6 Quantum Channel Noise

In the existing quantum information theory system, there are different noise classifications according to different judgments. The main points are as follows:

1. **Markov Noise.**

If the quantum information in a quantum system is lost only to the environment and the positive feedback effect of the environment on the quantum system is ignored (that is, the quantum information returns to the quantum system from the environment), the interaction between the quantum system and the environment can be described by Markov master equation, which is called Markov noise. In the Markov environment, the entanglement between qubits will gradually decrease with time and may even become zero in a short period of time, with no entanglement between qubits [1, 2].

2. **Correlated Noise and Non-correlated Noise.**

When considering the strong interaction between the system and the environment, that is, the correlation time between the environment and the system is much longer than the evolution time of the system itself, the positive feedback effect of the environment on the system shall be taken into account. In this case, the dynamical evolution of the quantum system cannot be described by Markov master equation, which is called non-Markov environment. In such an environment, quantum entanglement evolves over time into a form of damped oscillations, in which entanglement is restored after a period of disappearance and eventually disappears completely [3]. That is to say, the effects of a non-Markov environment are twofold: on the one hand, because it dissipates, it leads to the decay of entanglement, and on the other hand, because of its associative effects, there is feedback, the quantum entanglement caused by the quantum entanglement. Based on the criterion of energy loss of quantum system, whether the energy of excited state in quantum state is lost to environment can be divided into energy-dissipated noise and non-energy-dissipated noise. It should be noted that both energy-dissipated noise and non-energy-dissipated noise will result in the loss of quantum state information. Noise is divided into correlated noise and non-correlated noise according to the way each particle is subjected to noise. If two particles pass through the same quantum channel, the two particles are not related, that is, the way the noise is not interfering with each other is called non-correlated noise; otherwise, it is called correlated noise.

2.6.1 Correlation Noise

Up to now, the quantum communication through noise channel is mainly concentrated on the non-correlated (no memory) quantum channel. In a non-correlated channel, noise acts independently on each symbol transmitted over the channel. In classical information theory, discrete memoryless channels are well understood as channel capacities, which are invariant in cases involving feedback or shared random correlation. Most of the existing research is also focused on non-correlated channels.

The assumption of non-correlation has some basis in the real communication. According to the instructions of the relevant communication scheme, the time rate of sending signal into the channel is low enough to allow channel reset and prevent signal crosstalk between environments. In this case, it can be assumed that the noise suffered by the particle in the channel is unrelated. Still, it doesn't always make sense. For example, as the speed of signal transmission increases, successive transmissions occur so quickly that the environment can retain "memory" of past events, as shown in Fig. 2.14 [4]. Fiber optics are one such example, and such effects can occur and have been explored experimentally. In many practical applications, it is also unreasonable to assume the existence of non-correlated noise channels, because memory effects need to be taken into account. Therefore, it is necessary to extend the quantum channel theory to the correlated effect and establish a common framework to describe the experiments with correlated and non-correlated noises naturally.

2.6.2 Amplitude-Damping Noise

On the amplitude-damping channel, the interaction of a single qubit with the environment can be expressed as

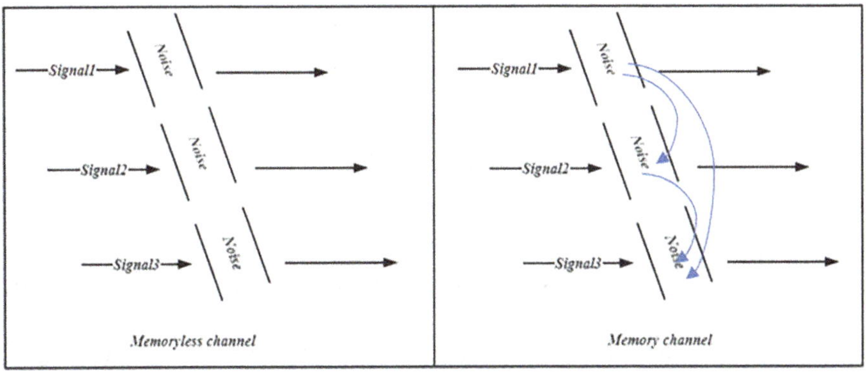

Fig. 2.14 Memory channel and non-memory channel

$$U=\begin{cases} |0\rangle_Q|0\rangle_E\rightarrow|0\rangle_Q|0\rangle_E \\ |1\rangle_Q|0\rangle_E\rightarrow\sqrt{1-p}|1\rangle_Q|0\rangle_E+\sqrt{p}|0\rangle_Q|0\rangle_E \end{cases} \tag{2.26}$$

where Q represents the quantum system, E represents the environmental system, and P represents the probability of amplitude-damping channel.

The dynamic process of amplitude damping is expressed as

$$\xi_{AD}=E_0\rho E_0^\dagger+E_1\rho E_1^\dagger \tag{2.27}$$

where the corresponding channel element is expressed as (2.28) and α is the decoherence coefficient of the amplitude damping, which can be thought of as the probability of losing a photon.

$$E_0=\begin{pmatrix}1 & 0\\ 0 & \sqrt{1-\alpha}\end{pmatrix}, E_1=\begin{pmatrix}0 & \sqrt{\alpha}\\ 0 & 0\end{pmatrix}. \tag{2.28}$$

For an initial state of $\rho=\begin{pmatrix}\rho_{00} & \rho_{01}\\ \rho_{10} & \rho_{11}\end{pmatrix}$, there is after the amplitude-damping channel

$$\xi_{AD}=\sum_{i=0,1}E_i\rho E_i^\dagger=\begin{bmatrix}\rho_{00}+\lambda\rho_{11} & \sqrt{1-\alpha}\rho_{01}\\ \sqrt{1-\alpha}\rho_{10} & \sqrt{1-\alpha}\rho_{11}\end{bmatrix}. \tag{2.29}$$

2.6.3 *Phase-Damping Noise*

Phase-damping channel is a pure quantum mechanical noise process, which describes the loss of quantum information without energy dissipation. According to the unitary freedom of quantum operation, its operand is equivalent to that of phase-flip noise channel.

The dynamic process of phase damping is expressed as

$$\xi_{PD} = E_0 \rho E_0^{\dagger} + E_1 \rho E_1^{\dagger}. \tag{2.30}$$

$E_i (i = 0, 1)$ is the Kraus operator with phase-damping noise, given below (2.31), and α is expressed as the probability that a photon from the system will scatter without loss of energy.

$$E_0 = \begin{pmatrix} 1 & 0 \\ 0 & \sqrt{1-\alpha} \end{pmatrix}, E_1 = \begin{pmatrix} 0 & 0 \\ 0 & \sqrt{\alpha} \end{pmatrix}. \tag{2.31}$$

2.6.4 *Bit-Flip Noise*

The loss of quantum information described by bit-flip noise is not caused by the loss of energy. Similar to the bit flip of classical information, the result of the bit flip is $|0\rangle$ becomes $|1\rangle$ and $|1\rangle$ becomes $|0\rangle$.

The dynamic process of bit flip is expressed as

$$\xi_{BF} = E_0 \rho E_0^{\dagger} + E_1 \rho E_1^{\dagger}. \tag{2.32}$$

$E_i (i = 0, 1)$ is the Kraus operator with bit-flip noise, the corresponding operands are as (2.33), and β is expressed as the probability.

$$E_0 = \sqrt{\beta} \begin{bmatrix} 1 & 0 \\ 0 & 1 \end{bmatrix}, E_1 = \sqrt{1-\beta} \begin{bmatrix} 0 & 1 \\ 1 & 0 \end{bmatrix}. \tag{2.33}$$

2.6.5 *Depolarizing Noise*

Depolarizing noise is a non-energy dissipation noise. This physical process can be expressed as

$$\xi(\rho) = \sum_{i=1}^{4} E_i \rho E_i^{\dagger} \tag{2.34}$$

where ρ is the density matrix of a particle and E_i is the Kraus operator with depolarizing noise, which can be expressed as

$$\begin{aligned} E_0 &= \sqrt{1-\lambda}\begin{pmatrix} 1 & 0 \\ 0 & 1 \end{pmatrix}, E_1 = \sqrt{\frac{\lambda}{3}}\begin{pmatrix} 1 & 0 \\ 0 & 1 \end{pmatrix} \\ E_2 &= \sqrt{\frac{\lambda}{3}}\begin{pmatrix} 1 & 0 \\ 0 & -1 \end{pmatrix}, E_3 = \sqrt{\frac{\lambda}{3}}\begin{pmatrix} 0 & -i \\ i & 0 \end{pmatrix} \end{aligned} \tag{2.35}$$

where λ is the noise intensity.

References

1. Zhang YJ, Man ZX, Xia YJ (2009) Atomic entanglement sudden death in a strongly driven cavity QED system. J Phys B Atomic Mol Phys 42(9):095503
2. Ji YH, Wang ZS, Hu JJ (2011) Influence of interaction between qubits on entanglement sudden death and birth. Int J Theor Phys 50(3):644–653
3. Bellomo B, Franco RL, Compagno G (2007) Non-Markovian effects on the dynamics of entanglement. Phys Rev Lett 99(16):160502
4. Caruso F, Giovannetti V, Lupo C et al (2014) Quantum channels and memory effects. Rev Mod Phys 86(4):1203

Chapter 3
Bidirectional Quantum Teleportation

Abstract Bidirectional quantum teleportation is an advanced technology in the field of quantum communication. It leverages a combination of quantum entanglement and classical communication to achieve the mutual transmission of quantum states between two locations without the need for direct physical connections. This technology not only enables the transmission of a quantum state from one location to another but also allows the received quantum state to be transmitted back to the original location, thereby forming a bidirectional quantum communication link.

3.1 Introduction

Bidirectional quantum teleportation (BQT), as a subfield of QT, was researched in recent years, where two legitimate users can teleport quantum state to each other. That is to say, both sides of the communication play the role of sender and receiver meanwhile.

Recently, BQT has attracted more and more attention of many researchers. In 2013, a bidirectional quantum teleportation scheme was presented by Zha et al. [1]; it is the first time to realize BQT via a five-qubit cluster state. The cluster state is a new multi-particle entangled state proposed by H.J. Briegel and R. Raussendorf [2] in 2001. It is known that the n-particle ($n > 3$) cluster state is both robust against decoherence and maximally connected with the better persistency than the GHZ state. In 2016, Sang [3] proposed a bidirectional quantum teleportation scheme by using a five-qubit cluster state, in which Alice can transmit an arbitrary two-qubit entangled state to Bob and at the same time Bob can teleport an arbitrary single-qubit state to Alice.

In this chapter, three types of BQT schemes are proposed, which uses a six-qubit entangled state, a four-qubit cluster state, and a six-qubit cluster state.

In Sect. 3.2, a new scheme for quantum bidirectional teleportation of two particles in a six-qubit entangled channel is proposed [4], which is based on GHZ measurement and unitary operation. Then, also based on the six-qubit entangled state channel, a scheme for quantum bidirectional teleportation of two to three particles is proposed. The scheme utilizes the Toffoli gate to convert three-qubit states into

R.-G. Zhou et al., *Design of Quantum Teleportation Schemes*,
https://doi.org/10.1007/978-3-031-82725-9_3

two-qubit states and a single qubit, and the rest of the teleportation is the same as the first scheme. Finally, we introduce an auxiliary bit to restore the state of the sender. Both these proposed schemes are more efficient than others in the same situation, which will be compared at the end of the section.

In Sect. 3.3, a novel scheme of bidirectional quantum teleportation is proposed [5], where two qubit states are transferred into single-qubit superposition by Alice and Bob, respectively, using controlled-NOT operation; meanwhile, the four-qubit cluster states are utilized as quantum channel. Our main novelty, which is the significant difference from other schemes of bidirectional (controlled) quantum teleportation, is as the receivers, both legitimate users can reconstruct the target two-qubit states through the corresponding unitary operation and controlled-NOT transmission to realize more efficient teleportation.

In Sect. 3.4, an improved bidirectional quantum teleportation scheme based on a quantum channel of a six-qubit cluster state is proposed [6]. In this scheme, the user Alice can transmit an unknown three-qubit entangled state to the user Bob, and at the same time, the user Bob can transmit an arbitrary single-qubit state to the user Alice by utilizing GHZ state measurement, Bell state measurement, single-qubit von Neumann measurement, and unitary operations. Compared with other schemes in quantum resource consumption, necessary operation complexity including operation difficulty and intensity, classical resource consumption, quantum information bits transmitted, and intrinsic efficiency, our scheme has the remarkable advantages in transmitting more quantum information and possesses higher intrinsic efficiency.

3.2 BQT Schemes via a Six-Qubit Entangled State

How to improve the efficiency in the same teleportation situation is crucial. In this section, two BQT schemes for $2 \leftrightarrow 2$ and $2 \leftrightarrow 3$ qubit teleportation via a six-qubit entangled state are proposed. In the first BQT scheme, Alice and Bob want to send a two-particle entangled state to each other. They do a GHZ measurement with one particle of the channel, respectively. Then, Alice and Bob inform each other of her/his measurement result by the classical channel. Finally, they perform unitary transformations to restore the target state. In the second BQT scheme, Alice is a two-particle entangled state, and Bob is a three-particle entangled state. In this scheme, we utilize a Toffoli gate with the first two particles of Bob's state as the control qubits and the third particle as the target bit. Then, Bob's state becomes a two-qubit entangled state and a single qubit. The rest of operations are the same as the first protocol. But in the end, an auxiliary particle is introduced to restore Bob's state by performing Toffoli gate again with two of Alice's particles.

There are two legitimate users, Alice and Bob. They are both sender and receiver. We suppose Alice wants to send a two-qubit state $|\psi\rangle_A$, and it's also the target state of Bob. Bob wants to send a two-qubit or three-qubit state $|\psi\rangle_B$, and it's also the target state of Alice.

To start with, $|\psi\rangle_A$ and $|\psi\rangle_B$ are defined as

$$\begin{aligned} |\psi\rangle_A &= \alpha\,|00\rangle_{a_1a_2} + \beta\,|11\rangle_{a_1a_2} \\ |\psi\rangle_B &= x\,|00\rangle_{b_1b_2} + y\,|11\rangle_{b_1b_2} \end{aligned} \tag{3.1}$$

where α, β, x, and y are arbitrary coefficients with $|\alpha|^2 + |\beta|^2 = 1$ and $|x|^2 + |y|^2 = 1$.

The quantum channel is a six-qubit entangled state shared by Alice and Bob; qubits

$$|\phi\rangle_{123456} = \frac{1}{2}(|000000\rangle + |001011\rangle + |110100\rangle + |111111\rangle)_{123456} \tag{3.2}$$

1, 5, and 6 belong to Alice, denoted as A_1, A_2, A_3, and qubits 3, 2, and 4 belong to Bob, denoted as B_1, B_2, B_3. The joint state of the senders and the quantum channel is

$$|T\rangle = |\psi\rangle_A \otimes |\psi\rangle_B \otimes |\phi\rangle_{A_1B_2B_1B_3A_2A_3}. \tag{3.3}$$

In the two schemes, we do GHZ measurements on three particles, and the GHZ states are described as

$$\begin{aligned} |\xi_0{}^{\pm}\rangle &= \frac{1}{\sqrt{2}}[|000\rangle \pm |111\rangle] \\ |\xi_1{}^{\pm}\rangle &= \frac{1}{\sqrt{2}}[|100\rangle \pm |011\rangle] \\ |\xi_2{}^{\pm}\rangle &= \frac{1}{\sqrt{2}}[|010\rangle \pm |101\rangle] \\ |\xi_3{}^{\pm}\rangle &= \frac{1}{\sqrt{2}}[|110\rangle \pm |001\rangle]. \end{aligned} \tag{3.4}$$

But during the measurement, we just use $|\xi_0{}^{\pm}\rangle$ and $|\xi_3{}^{\pm}\rangle$ states.

3.2.1 *Transmitting the Two- and Two-Qubit State*

This scheme is designed to three steps as follows, and the process is illustrated in Fig. 3.1.

Alice and Bob do GHZ measurements on qubit pairs (A, 1) and (B, 3), respectively. Then they should tell each other about the measuring result via the classical channel.

In this scheme, the state of joint state $|T\rangle$ is

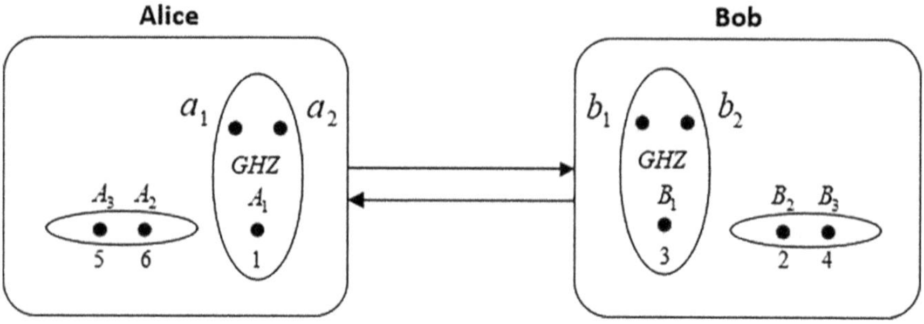

Fig. 3.1 Schematic of the proposed first QBT scheme for transmitting the two- and two-qubit state between Alice and Bob

Table 3.1 GHZ measurement results and the corresponding collapsed states of B_2, B_3, A_2, A_3

Alice's results	Bob's results	The collapsed state of qubits B_2, B_3, A_2, A_3
$\vert \xi_0^+\rangle_{a_1a_2A_1}$	$\vert \xi_0^+\rangle_{b_1b_2B_1}$	$\alpha x\vert 0000\rangle + \beta x\vert 1100\rangle + \alpha y\vert 0011\rangle + \beta y\vert 1111\rangle$
	$\vert \xi_0^-\rangle_{b_1b_2B_1}$	$\alpha x\vert 0000\rangle + \beta x\vert 1100\rangle - \alpha y\vert 0011\rangle - \beta y\vert 1111\rangle$
	$\vert \xi_3^+\rangle_{b_1b_2B_1}$	$\alpha y\vert 0000\rangle + \beta y\vert 1100\rangle + \alpha x\vert 0011\rangle + \beta x\vert 1111\rangle$
	$\vert \xi_3^-\rangle_{b_1b_2B_1}$	$\alpha y\vert 0000\rangle + \beta y\vert 1100\rangle - \alpha x\vert 0011\rangle - \beta x\vert 1111\rangle$
$\vert \xi_0^-\rangle_{a_1a_2A_1}$	$\vert \xi_0^+\rangle_{b_1b_2B_1}$	$\alpha x\vert 0000\rangle - \beta x\vert 1100\rangle + \alpha y\vert 0011\rangle - \beta y\vert 1111\rangle$
	$\vert \xi_0^-\rangle_{b_1b_2B_1}$	$\alpha x\vert 0000\rangle - \beta x\vert 1100\rangle - \alpha y\vert 0011\rangle + \beta y\vert 1111\rangle$
	$\vert \xi_3^+\rangle_{b_1b_2B_1}$	$\alpha y\vert 0000\rangle - \beta y\vert 1100\rangle + \alpha x\vert 0011\rangle - \beta x\vert 1111\rangle$
	$\vert \xi_3^-\rangle_{b_1b_2B_1}$	$\alpha y\vert 0000\rangle - \beta y\vert 1100\rangle - \alpha x\vert 0011\rangle + \beta x\vert 1111\rangle$
$\vert \xi_3^+\rangle_{a_1a_2A_1}$	$\vert \xi_0^+\rangle_{b_1b_2B_1}$	$\beta x\vert 0000\rangle + \alpha x\vert 1100\rangle + \beta y\vert 0011\rangle + \alpha y\vert 1111\rangle$
	$\vert \xi_0^-\rangle_{b_1b_2B_1}$	$\beta x\vert 0000\rangle + \alpha x\vert 1100\rangle - \beta y\vert 0011\rangle - \alpha y\vert 1111\rangle$
	$\vert \xi_3^+\rangle_{b_1b_2B_1}$	$\beta y\vert 0000\rangle + \alpha y\vert 1100\rangle + \beta x\vert 0011\rangle + \alpha x\vert 1111\rangle$
	$\vert \xi_3^-\rangle_{b_1b_2B_1}$	$\beta y\vert 0000\rangle + \alpha y\vert 1100\rangle - \beta x\vert 0011\rangle - \alpha x\vert 1111\rangle$
$\vert \xi_3^-\rangle_{a_1a_2A_1}$	$\vert \xi_0^+\rangle_{b_1b_2B_1}$	$\beta x\vert 0000\rangle - \alpha x\vert 1100\rangle + \beta y\vert 0011\rangle - \alpha y\vert 1111\rangle$
	$\vert \xi_0^-\rangle_{b_1b_2B_1}$	$\beta x\vert 0000\rangle - \alpha x\vert 1100\rangle - \beta y\vert 0011\rangle + \alpha y\vert 1111\rangle$
	$\vert \xi_3^+\rangle_{b_1b_2B_1}$	$\beta y\vert 0000\rangle - \alpha y\vert 1100\rangle + \beta x\vert 0011\rangle - \alpha x\vert 1111\rangle$
	$\vert \xi_3^-\rangle_{b_1b_2B_1}$	$\beta y\vert 0000\rangle - \alpha y\vert 1100\rangle - \beta x\vert 0011\rangle + \alpha x\vert 1111\rangle$

$$\begin{aligned}|T\rangle &= |\psi\rangle_A \otimes |\psi\rangle_B \otimes |\phi\rangle_{A_1B_2B_1B_3A_2A_3} \\ &= (\alpha|00\rangle + \beta|11\rangle)_{a_1a_2} \otimes (x|00\rangle + y|11\rangle)_{b_1b_2} \\ &\otimes \frac{1}{2}(|000000\rangle + |001011\rangle + |110100\rangle + |111111\rangle)\end{aligned} \tag{3.5}$$

After Alice and Bob perform GHZ measurements on three of her/his particles, the rest of the particles A_2, A_3, B_2, and B_3 collapse to the states $|\varphi\rangle$. The collapsed state $|\varphi\rangle$ corresponding to different measurement results is shown in Table 3.1, and the collapse results have 16 possible states with the same possibility. In Table 3.1, $|\xi_0^\pm\rangle$ and $|\xi_3^\pm\rangle$ are GHZ states as mentioned above, and the subscript represents the

Table 3.2 Different results and the corresponding unitary operation of Alice and Bob

Alice's results	Bob's results	The collapsed state of qubits B$_2$, B$_3$, A$_2$, A$_3$	Bob's operations	Alice's operations
$\vert\xi_0^+\rangle_{a_1a_2A_1}$	$\vert\xi_0^+\rangle_{b_1b_2B_1}$	$(\alpha\vert 00\rangle+\beta\vert 11\rangle)(x\vert 00\rangle+y\vert 11\rangle)$	$I_{B_2}\otimes I_{B_3}$	$I_{A_2}\otimes I_{A_3}$
	$\vert\xi_0^-\rangle_{b_1b_2B_1}$	$(\alpha\vert 00\rangle+\beta\vert 11\rangle)(x\vert 00\rangle-y\vert 11\rangle)$	$I_{B_2}\otimes I_{B_3}$	$\sigma_{A_2}^Z\otimes I_{A_3}$
	$\vert\xi_3^+\rangle_{b_1b_2B_1}$	$(\alpha\vert 00\rangle+\beta\vert 11\rangle)(y\vert 00\rangle+x\vert 11\rangle)$	$I_{B_2}\otimes I_{B_3}$	$\sigma_{A_2}^X\otimes\sigma_{A_3}^X$
	$\vert\xi_3^-\rangle_{b_1b_2B_1}$	$(\alpha\vert 00\rangle+\beta\vert 11\rangle)(y\vert 00\rangle-x\vert 11\rangle)$	$I_{B_2}\otimes I_{B_3}$	$-i\sigma_{A_2}^Y\otimes I_{A_3}$
$\vert\xi_0^-\rangle_{a_1a_2A_1}$	$\vert\xi_0^+\rangle_{b_1b_2B_1}$	$(\alpha\vert 00\rangle-\beta\vert 11\rangle)(x\vert 00\rangle+y\vert 11\rangle)$	$\sigma_{B_2}^Z\otimes I_{B_3}$	$I_{A_2}\otimes I_{A_3}$
	$\vert\xi_0^-\rangle_{b_1b_2B_1}$	$(\alpha\vert 00\rangle-\beta\vert 11\rangle)(x\vert 00\rangle-y\vert 11\rangle)$	$\sigma_{B_2}^Z\otimes I_{B_3}$	$\sigma_{A_2}^Z\otimes I_{A_3}$
	$\vert\xi_3^+\rangle_{b_1b_2B_1}$	$(\alpha\vert 00\rangle-\beta\vert 11\rangle)(y\vert 00\rangle+x\vert 11\rangle)$	$\sigma_{B_2}^Z\otimes I_{B_3}$	$\sigma_{A_2}^X\otimes\sigma_{A_3}^X$
	$\vert\xi_3^-\rangle_{b_1b_2B_1}$	$(\alpha\vert 00\rangle-\beta\vert 11\rangle)(y\vert 00\rangle-x\vert 11\rangle)$	$\sigma_{B_2}^Z\otimes I_{B_3}$	$-i\sigma_{A_2}^Y\otimes I_{A_3}$
$\vert\xi_3^+\rangle_{a_1a_2A_1}$	$\vert\xi_0^+\rangle_{b_1b_2B_1}$	$(\beta\vert 00\rangle+\alpha\vert 11\rangle)(x\vert 00\rangle+y\vert 11\rangle)$	$\sigma_{B_2}^X\otimes\sigma_{B_3}^X$	$I_{A_2}\otimes I_{A_3}$
	$\vert\xi_0^-\rangle_{b_1b_2B_1}$	$(\beta\vert 00\rangle+\alpha\vert 11\rangle)(x\vert 00\rangle-y\vert 11\rangle)$	$\sigma_{B_2}^X\otimes\sigma_{B_3}^X$	$\sigma_{A_2}^Z\otimes I_{A_3}$
	$\vert\xi_3^+\rangle_{b_1b_2B_1}$	$(\beta\vert 00\rangle+\alpha\vert 11\rangle)(y\vert 00\rangle+x\vert 11\rangle)$	$\sigma_{B_2}^X\otimes\sigma_{B_3}^X$	$\sigma_{A_2}^X\otimes\sigma_{A_3}^X$
	$\vert\xi_3^-\rangle_{b_1b_2B_1}$	$(\beta\vert 00\rangle+\alpha\vert 11\rangle)(y\vert 00\rangle-x\vert 11\rangle)$	$\sigma_{B_2}^X\otimes\sigma_{B_3}^X$	$-i\sigma_{A_2}^Y\otimes I_{A_3}$
$\vert\xi_3^-\rangle_{a_1a_2A_1}$	$\vert\xi_0^+\rangle_{b_1b_2B_1}$	$(\beta\vert 00\rangle-\alpha\vert 11\rangle)(x\vert 00\rangle+y\vert 11\rangle)$	$-i\sigma_{B_2}^Y\otimes I_{B_3}$	$I_{A_2}\otimes I_{A_3}$
	$\vert\xi_0^-\rangle_{b_1b_2B_1}$	$(\beta\vert 00\rangle-\alpha\vert 11\rangle)(x\vert 00\rangle-y\vert 11\rangle)$	$-i\sigma_{B_2}^Y\otimes I_{B_3}$	$\sigma_{A_2}^Z\otimes I_{A_3}$
	$\vert\xi_3^+\rangle_{b_1b_2B_1}$	$(\beta\vert 00\rangle-\alpha\vert 11\rangle)(y\vert 00\rangle+x\vert 11\rangle)$	$-i\sigma_{B_2}^Y\otimes I_{B_3}$	$\sigma_{A_2}^X\otimes\sigma_{A_3}^X$
	$\vert\xi_3^-\rangle_{b_1b_2B_1}$	$(\beta\vert 00\rangle-\alpha\vert 11\rangle)(y\vert 00\rangle-x\vert 11\rangle)$	$-i\sigma_{B_2}^Y\otimes I_{B_3}$	$-i\sigma_{A_2}^Y\otimes I_{A_3}$

particles being measured. For ease of expression, we have omitted the multiplier factor 1/4.

To transform the result via classical channel, we encode the 16 possible results in 4 classical bits. Then Alice and Bob perform unitary operations on all consequences to restore the target state. Their operations are shown in Table 3.2. In Table 3.2, σ_m^n is an operation mark, m represents the action object, and n represents the operation being performed. I_k represents unit gate transformation, and k represents the action object.

3.2.2 *Transmitting the Two- and Three-Qubit State*

Given the teleportation of two- and three-qubit state, we introduce the Toffoli gate. The Toffoli gate has three input qubits and three output qubits. As is shown in Fig. 3.2, the first two control qubits are not affected by the Toffoli gate, and the third qubit won't change unless the two control qubits are $|1\rangle$ at the same time.

Step 1. In this scheme, Bob has a three-qubit entangled state as

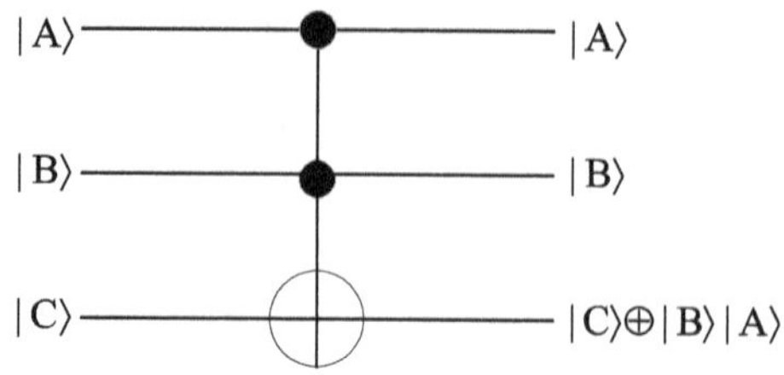

Fig. 3.2 The Toffoli gate's circuit representation

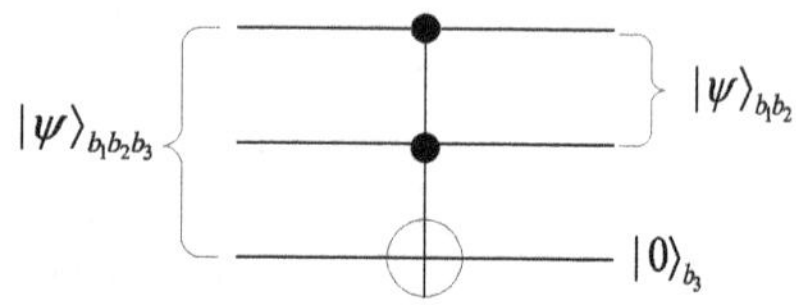

Fig. 3.3 The diagram of transforming Bob's state into a two-qubit state and a single qubit

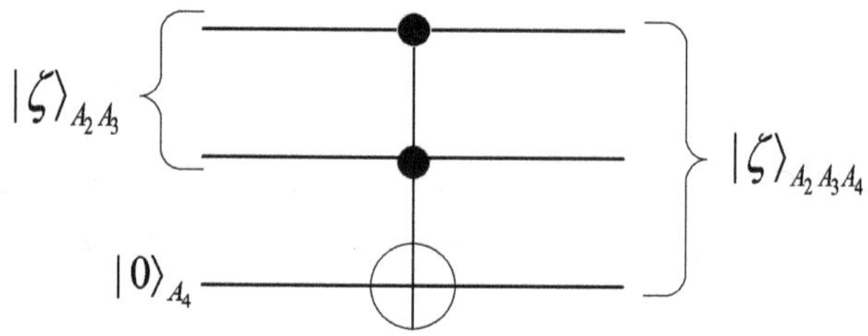

Fig. 3.4 The diagram of restoring Bob's state by a Toffoli gate

$$|\psi\rangle_B = x\,|000\rangle_{b_1b_2b_3} + y\,|111\rangle_{b_1b_2b_3}. \tag{3.6}$$

We perform the Toffoli gate with b_1, b_2 as the control qubits and b_3 as the target bit. The diagram is shown in Fig. 3.3.

Then Bob's state $|\psi\rangle_{b_1b_2b_3}$ becomes a two-qubit entangled state $|\psi\rangle_{b_1b_2}$ and a single qubit $|0\rangle_{b_3}$.

Step 2. The next step is the transmission of $|\psi\rangle_{a_1a_2}$ and $|\psi\rangle_{b_1b_2}$; it's the same process as the first protocol. Firstly, Alice and Bob do GHZ measurements on qubit pairs (a_1, a_2, 1) and (b_1, b_2, 3), respectively. Then, Alice and Bob perform unitary operations to restore her/his target state $|\psi\rangle_{b_1b_2}$ and $|\psi\rangle_{a_1a_2}$.

Step 3. To restore Bob's state, we introduce an auxiliary particle $|0\rangle_{A_4}$. We perform a Toffoli gate with A_2, A_3 as the control qubits and $|0\rangle_{A_4}$ as the target bit. The diagram is shown in Fig. 3.4.

3.2.3 Comparison

It's obvious that efficiency is a very important factor in teleportation. In this subsection, we define the efficiency η as

Table 3.3 Comparison of different protocols with our protocols in efficiency

Scheme	Teleported state of Alice	Teleported state of Bob	Quantum channel	Efficiency (η%)
Ref. [9]	Two-qubit	Two-qubit	Eight-qubit	25
Ref. [3]	Two-qubit	Single-qubit	Five-qubit	30
Ref. [8]	Three-qubit	Two-qubit	Nine-qubit	18.52
Ref. [7]	Two-qubit	Two-qubit	Six-qubit	33.3
First BQT scheme	Two-qubit	Two-qubit	Six-qubit	40
Second BQT scheme	Two-qubit	Three-qubit	Six-qubit	45.5

$$\eta = \frac{c}{p+q+t} \tag{3.7}$$

where c indicates the number of the total particles that Alice and Bob want to teleport, p indicates the number of particles of the quantum channel, q indicates the number of the classical bits that are utilized, and t indicates the number of the auxiliary particles in the scheme.

In the first BQT scheme, the efficiency is $\eta = 4/6 + 4 + 0 = 40\ \%$ ..

To prove this improvement of the efficiency in this scheme, some schemes are introduced for comparison. For example, ref. [7] is a BQT scheme of teleporting two and two qubits via a six-qubit entangled state, which is the same as this scheme. The efficiency of it is $\eta = 4/6 + 6 + 0 = 1/3 \approx 33.3\ \%$.

From this example, we can see the efficiency of it is higher than the scheme in the same situation. More comparisons are shown in Table 3.3.

In the second BQT scheme, the efficiency is $\eta = 5/6 + 4 + 1 = 5/11 \approx 45.5\ \%$ ..

Reference [8] is a BQT scheme of teleporting three and two qubits via a nine-qubit cluster state, and the number of particles that are transmitted is the same as this scheme. The efficiency of it is $\eta = 5/9 + 18 + 0 = 5/27 \approx 18.52\ \%$.

We can see a lot of improvement in efficiency from the scheme of the reference to our scheme. The types of the schemes in Table 3.3 are all BQT.

3.2.4 Summary

In this section, we propose two BQT schemes of two–two and two–three qubits, respectively. By utilizing the GHZ measurement of three particles and unitary operations in our first BQT scheme, we improve the efficiency of teleportation compared with other protocols. In the second BQT scheme, we perform a Toffoli gate on Bob's state to transform a three-qubit state into a two-qubit state and perform a Toffoli gate again to restore Bob's state finally. It's obvious that the efficiency of it also has improved a lot. We wish our schemes can be achieved experimentally in the future. Our schemes are all in a noiseless channel, and we will consider how to achieve our schemes in noisy channels in the next step.

3.3 BQT Scheme of a Two-Qubit State via a Four-Qubit Cluster State

In this section, BQT of a two-qubit state is achieved by utilizing a four-qubit cluster state as a quantum channel; we employ CNOT to transform the target two-qubit state into single-qubit state and make the scheme more efficient. In the last step, Alice and Bob use an auxiliary bit to get the initial two-qubit state individually.

3.3.1 Preparation of Transmissible Target State and Quantum Channel

In this scheme, Alice and Bob, the legitimate communication parties, want to teleport a two-particle state to each other using a four-qubit cluster state. Suppose that Alice and Bob have a two-qubit state, respectively, which are expressed as (3.8):

$$\begin{aligned} \mid \Psi\rangle_{\alpha,\beta} &= a_0 \mid 00\rangle + a_1 \mid 11\rangle \\ \mid \Psi\rangle_{\delta,\gamma} &= b_0 \mid 00\rangle + b_1 \mid 11\rangle \end{aligned} . \tag{3.8}$$

The coefficients a_0, a_1, b_0, and b_1 are unknown, but $|a_0|^2 + |a_1|^2 = 1$ and $|b_0|^2 + |b_1|^2 = 1$.

Alice holds qubits α, β in state $|\Psi\rangle_{\alpha,\beta}$ and Bob owns particles δ, γ in state $|\Psi\rangle_{\delta,\gamma}$. The quantum channel for implementing the bidirectional teleportation, in this scheme, is a four-qubit cluster state, which is generated as follows:

$$\mid \Phi\rangle_{1234} = \frac{1}{2}(|0000\rangle + |0011\rangle + |1100\rangle + |1111\rangle)_{1234}. \tag{3.9}$$

If qubits 1 and 2 belong to Alice and qubits 3 and 4 belong to Bob, respectively, the quantum channel can be expressed as

$$|\Phi_4\rangle_{A_1A_2B_1B_2} = \frac{1}{2}(|0000\rangle + |0011\rangle + |1100\rangle + |1111\rangle)_{A_1A_2B_1B_2}. \tag{3.10}$$

The state of the whole system is given by

$$\mid \Psi\rangle_{\alpha\beta\delta\gamma A_1A_2B_1B_2} = \mid \Psi\rangle_{\alpha,\beta} \otimes \mid \Psi\rangle_{\delta,\gamma} \otimes \mid \Phi\rangle_{A_1A_2B_1B_2}. \tag{3.11}$$

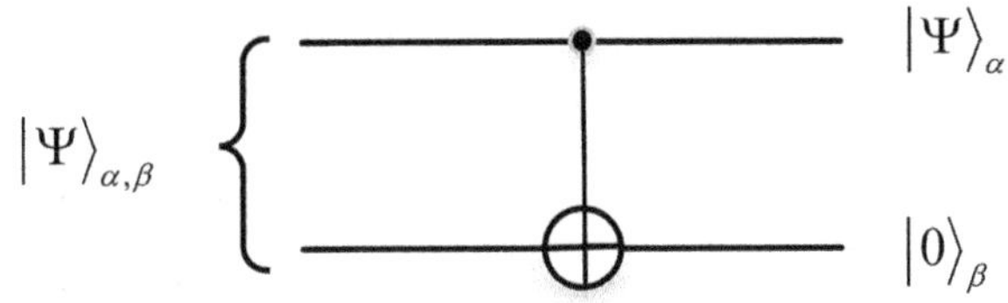

Fig. 3.5 Quantum circuit for obtaining $|\Psi\rangle_{\alpha}$

Alice wants to transmit the quantum state of particles α and β to Bob, and Bob intends to transmit the state of particles δ and γ to Alice.

In order to achieve the QT, Alice and Bob firstly transform their two-qubit states into the form as

$$\begin{aligned} \mathrm{CNOT}\left(a_0|00\rangle_{\alpha,\beta}+a_1|11\rangle_{\alpha,\beta}\right) &= |\,\Psi\rangle_{\alpha}\otimes|\,0\rangle_{\beta} \\ \mathrm{CNOT}\left(b_0|00\rangle_{\delta,\gamma}+b_1|11\rangle_{\delta,\gamma}\right) &= |\,\Psi\rangle_{\delta}\otimes|\,0\rangle_{\gamma} \end{aligned}. \tag{3.12}$$

Here $|\Psi\rangle_{\alpha}=(a_0|\,0\rangle_{\alpha}+a_1|\,1\rangle_{\alpha})$, $|\,\Psi\rangle_{\delta}=(b_0|\,0\rangle_{\delta}+b_1|\,1\rangle_{\delta})$. As is shown in Fig. 3.5, the two-qubit states can be converted into the tensor product of a single bit superposition and 0 bit; the modality can be achieved through controlled-NOT operations on qubits α, β, δ, and γ with qubits α and δ as controlled qubits and β and γ as target qubits. When the input state of circuit is $|\Psi\rangle_{\alpha,\beta}$, the output states of the circuit are $|\Psi\rangle_{\alpha}$ and $|0\rangle_{\beta}$. $|\Psi\rangle_{\delta,\gamma}$ can also be described through the same transformation.

3.3.2 BQT Scheme of Two-Qubit States

Here, Alice and Bob teleport the transformed single-particle superposition state to each other; the state of qubits α, δ, A_1, A_2, B_1, and B_2 will be the form of (3.13).

$$\begin{aligned} |\,\Psi\rangle_{\alpha\delta A_1A_2B_1B_2} &= |\,\Psi\rangle_{\alpha}\otimes|\,\Psi\rangle_{\delta}\otimes|\,\Phi\rangle_{A_1A_2B_1B_2} \\ &= \frac{1}{4}\Big[|\psi^+\rangle_{\alpha A_1}|\psi^+\rangle_{\delta B_2}|\eta^1\rangle_{A_2B_1}+|\psi^+\rangle_{\alpha A_1}|\psi^-\rangle_{\delta B_2}|\eta^2\rangle_{A_2B_1} \\ &\quad+|\psi^+\rangle_{\alpha A_1}|\phi^+\rangle_{\delta B_2}|\eta^3\rangle_{A_2B_1}+|\psi^+\rangle_{\alpha A_1}|\phi^-\rangle_{\delta B_2}|\eta^4\rangle_{A_2B_1} \\ &\quad+|\psi^-\rangle_{\alpha A_1}|\psi^+\rangle_{\delta B_2}|\eta^5\rangle_{A_2B_1}+|\psi^-\rangle_{\alpha A_1}|\psi^-\rangle_{\delta B_2}|\eta^6\rangle_{A_2B_1} \\ &\quad+|\psi^-\rangle_{\alpha A_1}|\phi^+\rangle_{\delta B_2}|\eta^7\rangle_{A_2B_1}+|\psi^-\rangle_{\alpha A_1}|\phi^-\rangle_{\delta B_2}|\eta^8\rangle_{A_2B_1} \\ &\quad+|\phi^+\rangle_{\alpha A_1}|\psi^+\rangle_{\delta B_2}|\eta^9\rangle_{A_2B_1}+|\phi^+\rangle_{\alpha A_1}|\psi^-\rangle_{\delta B_2}|\eta^{10}\rangle_{A_2B_1} \\ &\quad+|\phi^+\rangle_{\alpha A_1}|\phi^+\rangle_{\delta B_2}|\eta^{11}\rangle_{A_2B_1}+|\phi^+\rangle_{\alpha A_1}|\phi^-\rangle_{\delta B_2}|\eta^{12}\rangle_{A_2B_1} \\ &\quad+|\phi^-\rangle_{\alpha A_1}|\psi^+\rangle_{\delta B_2}|\eta^{13}\rangle_{A_2B_1}+|\phi^-\rangle_{\alpha A_1}|\psi^-\rangle_{\delta B_2}|\eta^{14}\rangle_{A_2B_1} \\ &\quad+|\phi^-\rangle_{\alpha A_1}|\phi^+\rangle_{\delta B_2}|\eta^{15}\rangle_{A_2B_1}+|\phi^-\rangle_{\alpha A_1}|\phi^-\rangle_{\delta B_2}|\eta^{16}\rangle_{A_2B_1} \end{aligned} \tag{3.13}$$

where $|\eta^i\rangle_{A_2B_1}(i=1,2,\ldots,16)$ are two-qubit states given by

$$
\begin{aligned}
|\eta^{1}\rangle_{A_2B_1} &= a_0b_0 \mid 00\rangle + a_0b_1 \mid 01\rangle + a_1b_0 \mid 10\rangle + a_1b_1 \mid 11\rangle \\
|\eta^{2}\rangle_{A_2B_1} &= a_0b_0 \mid 00\rangle - a_0b_1 \mid 01\rangle + a_1b_0 \mid 10\rangle - a_1b_1 \mid 11\rangle \\
|\eta^{3}\rangle_{A_2B_1} &= a_0b_1 \mid 00\rangle + a_0b_0 \mid 01\rangle + a_1b_1 \mid 10\rangle + a_1b_0 \mid 11\rangle \\
|\eta^{4}\rangle_{A_2B_1} &= -a_0b_1 \mid 00\rangle + a_0b_0 \mid 01\rangle - a_1b_1 \mid 10\rangle + a_1b_0 \mid 11\rangle \\
|\eta^{5}\rangle_{A_2B_1} &= a_0b_0 \mid 00\rangle + a_0b_1 \mid 01\rangle - a_1b_0 \mid 10\rangle - a_1b_1 \mid 11\rangle \\
|\eta^{6}\rangle_{A_2B_1} &= a_0b_0 \mid 00\rangle - a_0b_1 \mid 01\rangle - a_1b_0 \mid 10\rangle + a_1b_1 \mid 11\rangle \\
|\eta^{7}\rangle_{A_2B_1} &= a_0b_1 \mid 00\rangle + a_0b_0 \mid 01\rangle + a_1b_1 \mid 10\rangle - a_1b_0 \mid 11\rangle \\
|\eta^{8}\rangle_{A_2B_1} &= -a_0b_1 \mid 00\rangle + a_0b_0 \mid 01\rangle - a_1b_1 \mid 10\rangle + a_1b_0 \mid 11\rangle \\
|\eta^{9}\rangle_{A_2B_1} &= a_1b_0 \mid 00\rangle + a_1b_1 \mid 01\rangle + a_0b_0 \mid 10\rangle + a_0b_1 \mid 11\rangle \\
|\eta^{10}\rangle_{A_2B_1} &= a_1b_0 \mid 00\rangle - a_1b_1 \mid 01\rangle + a_0b_0 \mid 10\rangle - a_0b_1 \mid 11\rangle \\
|\eta^{11}\rangle_{A_2B_1} &= a_1b_1 \mid 00\rangle + a_1b_0 \mid 01\rangle + a_0b_1 \mid 10\rangle + a_0b_0 \mid 11\rangle \\
|\eta^{12}\rangle_{A_2B_1} &= -a_1b_1 \mid 00\rangle + a_1b_0 \mid 01\rangle - a_0b_1 \mid 10\rangle + a_0b_0 \mid 11\rangle \\
|\eta^{13}\rangle_{A_2B_1} &= -a_1b_0 \mid 00\rangle - a_1b_1 \mid 01\rangle + a_0b_0 \mid 10\rangle + a_0b_1 \mid 11\rangle \\
|\eta^{14}\rangle_{A_2B_1} &= -a_1b_0 \mid 00\rangle + a_1b_1 \mid 01\rangle + a_0b_0 \mid 10\rangle - a_0b_1 \mid 11\rangle \\
|\eta^{15}\rangle_{A_2B_1} &= -a_1b_1 \mid 00\rangle - a_1b_0 \mid 01\rangle + a_0b_1 \mid 10\rangle + a_0b_0 \mid 11\rangle \\
|\eta^{16}\rangle_{A_2B_1} &= a_1b_1 \mid 00\rangle - a_1b_0 \mid 01\rangle - a_0b_1 \mid 10\rangle + a_0b_0 \mid 11\rangle
\end{aligned}
\tag{3.14}
$$

Alice and Bob perform a Bell state measurement on qubit pairs (α, A_1) and (δ, B_2), respectively, According to Table 3.4, the remaining particles A_2 and B_1 collapse into one of the 16 possible states with the same probability in (3.14).

Then Alice and Bob announce their measurement consequence to each other via classical channel. After that, they can reconstruct the single bit superposition state by means of the corresponding unitary operations, which are displayed in Table 3.5.

Now, Alice and Bob, the two legitimate users, perform CNOT operations to gain the objective two-qubit states once again, where qubits A_2 and B_1 serve as controlled particles and ancillary qubits β' and γ' as target particles. The transformation can be implemented by the following quantum circuit in Fig. 3.6.

Finally, Alice and Bob successfully recover the two-qubit states severally. The technologies we utilize here are feasible within the current research.

3.3.3 Comparison

Though the scheme utilizes two auxiliary qubits when doing CNOT operations, it realizes the bidirectional teleportation of two-qubit states via four-qubit cluster

Table 3.4 The measurement results of users and the corresponding collapsed state

Alice's results	Bob's results	The collapsed state of qubits A_2, B_1
$\vert\psi^{+}\rangle_{\alpha A_1}$	$\vert\psi^{+}\rangle_{\delta B_2}$	$a_0b_0 \mid 00\rangle + a_0b_1 \mid 01\rangle + a_1b_0 \mid 10\rangle + a_1b_1 \mid 11\rangle$
$\vert\psi^{+}\rangle_{\alpha A_1}$	$\vert\psi^{-}\rangle_{\delta B_2}$	$a_0b_0 \mid 00\rangle - a_0b_1 \mid 01\rangle + a_1b_0 \mid 10\rangle - a_1b_1 \mid 11\rangle$
$\vert\psi^{+}\rangle_{\alpha A_1}$	$\vert\phi^{+}\rangle_{\delta B_2}$	$a_0b_1 \mid 00\rangle + a_0b_0 \mid 01\rangle + a_1b_1 \mid 10\rangle + a_1b_0 \mid 11\rangle$
$\vert\psi^{+}\rangle_{\alpha A_1}$	$\vert\phi^{-}\rangle_{\delta B_2}$	$-a_0b_1 \mid 00\rangle + a_0b_0 \mid 01\rangle - a_1b_1 \mid 10\rangle + a_1b_0 \mid 11\rangle$
$\vert\psi^{-}\rangle_{\alpha A_1}$	$\vert\psi^{+}\rangle_{\delta B_2}$	$a_0b_0 \mid 00\rangle + a_0b_1 \mid 01\rangle - a_1b_0 \mid 10\rangle - a_1b_1 \mid 11\rangle$
$\vert\psi^{-}\rangle_{\alpha A_1}$	$\vert\psi^{-}\rangle_{\delta B_2}$	$a_0b_0 \mid 00\rangle - a_0b_1 \mid 01\rangle - a_1b_0 \mid 10\rangle + a_1b_1 \mid 11\rangle$
$\vert\psi^{-}\rangle_{\alpha A_1}$	$\vert\phi^{+}\rangle_{\delta B_2}$	$a_0b_1 \mid 00\rangle + a_0b_0 \mid 01\rangle - a_1b_1 \mid 10\rangle - a_1b_0 \mid 11\rangle$
$\vert\psi^{-}\rangle_{\alpha A_1}$	$\vert\phi^{-}\rangle_{\delta B_2}$	$-a_0b_1 \mid 00\rangle + a_0b_0 \mid 01\rangle - a_1b_1 \mid 10\rangle + a_1b_0 \mid 11\rangle$
$\vert\phi^{+}\rangle_{\alpha A_1}$	$\vert\psi^{+}\rangle_{\delta B_2}$	$a_1b_0 \mid 00\rangle + a_1b_1 \mid 01\rangle + a_0b_0 \mid 10\rangle + a_0b_1 \mid 11\rangle$
$\vert\phi^{+}\rangle_{\alpha A_1}$	$\vert\psi^{-}\rangle_{\delta B_2}$	$a_1b_0 \mid 00\rangle - a_1b_1 \mid 01\rangle + a_0b_0 \mid 10\rangle - a_0b_1 \mid 11\rangle$
$\vert\phi^{+}\rangle_{\alpha A_1}$	$\vert\phi^{+}\rangle_{\delta B_2}$	$a_1b_1 \mid 00\rangle + a_1b_0 \mid 01\rangle + a_0b_1 \mid 10\rangle + a_0b_0 \mid 11\rangle$
$\vert\phi^{+}\rangle_{\alpha A_1}$	$\vert\phi^{-}\rangle_{\delta B_2}$	$-a_1b_1 \mid 00\rangle + a_1b_0 \mid 01\rangle - a_0b_1 \mid 10\rangle + a_0b_0 \mid 11\rangle$
$\vert\phi^{-}\rangle_{\alpha A_1}$	$\vert\psi^{+}\rangle_{\delta B_2}$	$-a_1b_0 \mid 00\rangle - a_1b_1 \mid 01\rangle + a_0b_0 \mid 10\rangle + a_0b_1 \mid 11\rangle$
$\vert\phi^{-}\rangle_{\alpha A_1}$	$\vert\psi^{-}\rangle_{\delta B_2}$	$-a_1b_0 \mid 00\rangle + a_1b_1 \mid 01\rangle + a_0b_0 \mid 10\rangle - a_0b_1 \mid 11\rangle$
$\vert\phi^{-}\rangle_{\alpha A_1}$	$\vert\phi^{+}\rangle_{\delta B_2}$	$-a_1b_1 \mid 00\rangle - a_1b_0 \mid 01\rangle + a_0b_1 \mid 10\rangle + a_0b_0 \mid 11\rangle$
$\vert\phi^{-}\rangle_{\alpha A_1}$	$\vert\phi^{-}\rangle_{\delta B_2}$	$a_1b_1 \mid 00\rangle - a_1b_0 \mid 01\rangle - a_0b_1 \mid 10\rangle + a_0b_0 \mid 11\rangle$

Table 3.5 Relation between the measurement consequences and appropriate unitary operations

Alice's results	Bob's results	The collapsed state of qubits A_2, B_1	Alice's operations	Bob's operations
$\vert\psi^{+}\rangle_{\alpha A_1}$	$\vert\psi^{+}\rangle_{\delta B_2}$	$(a_0\vert 0\rangle + a_1\vert 1\rangle)_{A_2} \otimes (b_0\vert 0\rangle + b_1\vert 1\rangle)_{B_1}$	I_{A_2}	I_{B_1}
$\vert\psi^{+}\rangle_{\alpha A_1}$	$\vert\psi^{-}\rangle_{\delta B_2}$	$(a_0\vert 0\rangle + a_1\vert 1\rangle)_{A_2} \otimes (b_0\vert 0\rangle - b_1\vert 1\rangle)_{B_1}$	I_{A_2}	$\sigma^{Z}_{B_2}$
$\vert\psi^{+}\rangle_{\alpha A_1}$	$\vert\phi^{+}\rangle_{\delta B_2}$	$(a_0\vert 0\rangle + a_1\vert 1\rangle)_{A_2} \otimes (b_1\vert 0\rangle + b_0\vert 1\rangle)_{B_1}$	I_{A_2}	$\sigma^{X}_{B_2}$
$\vert\psi^{+}\rangle_{\alpha A_1}$	$\vert\phi^{-}\rangle_{\delta B_2}$	$(a_0\vert 0\rangle + a_1\vert 1\rangle)_{A_2} \otimes (-b_1\vert 0\rangle + b_0\vert 1\rangle)_{B_1}$	I_{A_2}	$i\sigma^{Y}_{B_2}$
$\vert\psi^{-}\rangle_{\alpha A_1}$	$\vert\psi^{+}\rangle_{\delta B_2}$	$(a_0\vert 0\rangle - a_1\vert 1\rangle)_{A_2} \otimes (b_0\vert 0\rangle + b_1\vert 1\rangle)_{B_1}$	$\sigma^{Z}_{A_2}$	I_{B_1}
$\vert\psi^{-}\rangle_{\alpha A_1}$	$\vert\psi^{-}\rangle_{\delta B_2}$	$(a_0\vert 0\rangle - a_1\vert 1\rangle)_{A_2} \otimes (b_0\vert 0\rangle - b_1\vert 1\rangle)_{B_1}$	$\sigma^{Z}_{A_2}$	$\sigma^{Z}_{B_2}$
$\vert\psi^{-}\rangle_{\alpha A_1}$	$\vert\phi^{+}\rangle_{\delta B_2}$	$(a_0\vert 0\rangle - a_1\vert 1\rangle)_{A_2} \otimes (b_1\vert 0\rangle + b_0\vert 1\rangle)_{B_1}$	$\sigma^{Z}_{A_2}$	$\sigma^{X}_{B_2}$
$\vert\psi^{-}\rangle_{\alpha A_1}$	$\vert\phi^{-}\rangle_{\delta B_2}$	$(a_0\vert 0\rangle + a_1\vert 1\rangle)_{A_2} \otimes (-b_1\vert 0\rangle + b_0\vert 1\rangle)_{B_1}$	I_{A_2}	$i\sigma^{Y}_{B_2}$
$\vert\phi^{+}\rangle_{\alpha A_1}$	$\vert\psi^{+}\rangle_{\delta B_2}$	$(a_1\vert 0\rangle + a_0\vert 1\rangle)_{A_2} \otimes (b_0\vert 0\rangle + b_1\vert 1\rangle)_{B_1}$	$\sigma^{X}_{A_2}$	I_{B_1}
$\vert\phi^{+}\rangle_{\alpha A_1}$	$\vert\psi^{-}\rangle_{\delta B_2}$	$(a_1\vert 0\rangle + a_0\vert 1\rangle)_{A_2} \otimes (b_0\vert 0\rangle - b_1\vert 1\rangle)_{B_1}$	$\sigma^{X}_{A_2}$	$\sigma^{Z}_{B_2}$
$\vert\phi^{+}\rangle_{\alpha A_1}$	$\vert\phi^{+}\rangle_{\delta B_2}$	$(a_1\vert 0\rangle + a_0\vert 1\rangle)_{A_2} \otimes (b_1\vert 0\rangle - b_0\vert 1\rangle)_{B_1}$	$\sigma^{X}_{A_2}$	$-i\sigma^{Y}_{B_2}$
$\vert\phi^{+}\rangle_{\alpha A_1}$	$\vert\phi^{-}\rangle_{\delta B_2}$	$(a_1\vert 0\rangle + a_0\vert 1\rangle)_{A_2} \otimes (-b_1\vert 0\rangle + b_0\vert 1\rangle)_{B_1}$	$\sigma^{X}_{A_2}$	$i\sigma^{Y}_{B_2}$
$\vert\phi^{-}\rangle_{\alpha A_1}$	$\vert\psi^{+}\rangle_{\delta B_2}$	$(-a_1\vert 0\rangle + a_0\vert 1\rangle)_{A_2} \otimes (b_0\vert 0\rangle + b_1\vert 1\rangle)_{B_1}$	$i\sigma^{Y}_{A_1}$	I_{B_1}
$\vert\phi^{-}\rangle_{\alpha A_1}$	$\vert\psi^{-}\rangle_{\delta B_2}$	$(-a_1\vert 0\rangle + a_0\vert 1\rangle)_{A_2} \otimes (b_0\vert 0\rangle - b_1\vert 1\rangle)_{B_1}$	$i\sigma^{Y}_{A_1}$	$\sigma^{Z}_{B_2}$
$\vert\phi^{-}\rangle_{\alpha A_1}$	$\vert\phi^{+}\rangle_{\delta B_2}$	$(-a_1\vert 0\rangle + a_0\vert 1\rangle)_{A_2} \otimes (b_1\vert 0\rangle + b_0\vert 1\rangle)_{B_1}$	$i\sigma^{Y}_{A_1}$	$\sigma^{X}_{B_2}$
$\vert\phi^{-}\rangle_{\alpha A_1}$	$\vert\phi^{-}\rangle_{\delta B_2}$	$(a_1\vert 0\rangle - a_0\vert 1\rangle)_{A_2} \otimes (b_1\vert 0\rangle - b_0\vert 1\rangle)_{B_1}$	$-i\sigma^{Y}_{A_1}$	$-i\sigma^{Y}_{B_2}$

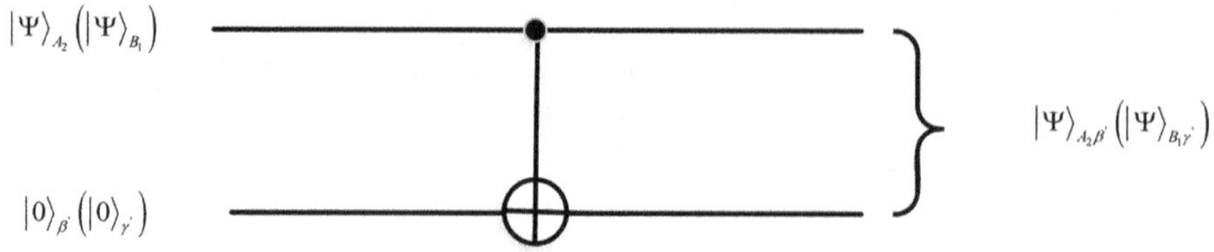

Fig. 3.6 Quantum circuit for recovering original two-qubit states

Table 3.6 Comparison with previous schemes

Scheme	Quantum channel	Teleported states	Efficiency (η%)
Scheme [1]	Five-qubit cluster state	Single-qubit states	40
Scheme [11]	Six-qubit entangled state	Single-qubit states	33.33
Scheme [12]	Seven-qubit entangled state	Single-qubit states	28.5
Scheme [10]	Eight-qubit entangled state	A single-qubit state and a two-qubit state	37.5
Scheme [3]	Five-qubit cluster state	A single-qubit state and a two-qubit state	60
The scheme	Four-qubit cluster state	Two two-qubit states	66.66

states. It is the advantage when compared with schemes [1, 3, 10–12], and the details of the comparison are shown in Table 3.6.

In quantum communication protocols, the formula of efficiency calculation is defined as

$$\eta = \frac{c}{q}. \tag{3.15}$$

Here the qubit efficiency η depends on the ratio of the number of transmitted messages c and the number of the total qubits q.

3.3.4 *Summary*

In this section, a scheme that just makes use of four-qubit cluster states to implement the bidirectional teleportation of two-qubit states. Thereinto, CNOT operation plays an important role, which transforms the two-qubit states into single-qubit state so that the scheme could improve the efficiency compared with other schemes. We hope the new method can be achieved experimentally in the near future.

3.4 BQT by Using a Six-Qubit Cluster State

In this section, an improved scheme for implementing asymmetric bidirectional quantum teleportation using a six-cluster state as a quantum channel is proposed. In BQT scheme, Alice holds an unknown three-qubit entangled state consisting of three qubits A, B, and C, and Bob holds an arbitrary single-qubit state comprised of qubit D. The aim of this scheme is to realize the bidirectional transmission of quantum state between Alice and Bob. To achieve this scheme, first of all, particles in quantum channel need to be distributed to two users, i.e., Alice and Bob, respectively. Next, the two users need to perform proper measurement, that is, Alice performs GHZ state measurement and single-qubit von Neumann measurement on her qubits and Bob performs Bell state measurement on his qubits. Finally, the two users have to perform unitary operations and CNOT operations on their residual states separately to realize the bilateral exchange of the initial quantum states.

3.4.1 *The Selection of Quantum Information Bits Transmitted and Quantum Channel*

(a) **The Selection of Quantum Information Bits Transmitted**

In the ideal environment, BQT can perfectly realize the bilateral transmission of quantum states, which is point to point. Since BQT scheme was first proposed in 2013, many BQT schemes have been proposed. However, in these presented schemes, quantum information bits transferred by two communicating parties are generally two single-qubit states, a single-qubit state and a two-qubit state, and two two-qubit states; the aim of this scheme is to increase the number of quantum information bits transmitted by communicating parties. So, it is an innovational theory to use a three-qubit entangled state and a single-qubit state as quantum information bits transmitted.

But in experimental application, noise is the key issue in quantum teleportation process. Unlike traditional communication, weak quantum signal is transmitted by quantum channel, and it is easily affected by noise. A noisy quantum channel can be a regular communication channel where the input of a pure quantum state can produce a mixed state as output as the data qubits become entangled with the environment. Different noisy environments need different solutions. Our proposed scheme can solve a common noisy environment where quantum channel only causes an error on a single qubit [13].

When quantum information bits transmitted are a three-qubit entangled state $|\varphi\rangle_{ABC} = a_0|000\rangle + a_1|111\rangle$, suppose that quantum channel flips a single qubit, say the first one, which can be presented as $|\varphi\rangle_{ABC}' = a_0|100\rangle + a_1|011\rangle$. In order to correct the false quantum state back into the original quantum state, we should make sure which qubit has been flipped. The method is expressed as follows.

Firstly, comparing the first two qubits, we find they are different, which is not allowed for any valid codeword. Therefore, an error has occurred, and furthermore, it flips either the first or second qubit. Note that measuring the first and second qubit will destroy the superposition in the codeword, so we just measure the difference between them.

Then comparing the first and third qubit, we find that since the first qubit was flipped, it will disagree with the third; if the second qubit had been flipped, the first and third would have agreed. Therefore, the error is narrowed down to the first qubit, and finally we flip it back.

When quantum information bits transmitted are a single-qubit quantum state $|\varphi\rangle_D = b_0|0\rangle + b_1|1\rangle$, suppose that quantum channel flips a single qubit, which can be elaborated as $|\varphi\rangle_D = b_0|1\rangle + b_1|0\rangle$. We can recover the original single-qubit state by using the operation of $\sigma_x = \begin{bmatrix} 0 & 1 \\ 1 & 0 \end{bmatrix}$ on the false quantum state.

This is a quantum error correction scheme for qubit flipping. In order to deal with the phenomenon of qubit flipping effectively, we need to select an initial state, which only includes states $|000\rangle$ and $|111\rangle$. So, an unknown three-qubit entangled state $a_0|000\rangle + a_1|111\rangle$ and a single-qubit state $b_0|0\rangle + b_1|1\rangle$ used as the initial quantum states transmitted can easily handle the phenomenon of qubit flipping in the noisy environment.

(b) **The Selection of Quantum Channel**

Efficiency is the key to judge the feasibility of quantum teleportation; the formula is described as $\tau = q_s/(q_u + b_t)$, where q_s indicates the number of qubits consisting of the quantum information to be transmitted, q_u is the number of qubits used as the quantum channel (except for those chosen for security checking), and b_t represents the number of qubits used as the classical channel. Although b_t has no effect on efficiency, which can also be expressed as $\tau = q_s/q_u$, classical communication that served as the bridge between two communicating parties in quantum teleportation is regarded as an obvious requirement to avoid the occurrence of faster-than-light (FTL) signaling. Therefore, the formula of efficiency characterized as $\tau = q_s/(q_u + b_t)$ is more rational than $\tau = q_s/q_u$.

In this scheme, quantum information bits transmitted in communicating parties are an unknown three-qubit entangled state and a single-qubit state. When the quantum information bits transmitted are finalized, the measurement basis will determine whether quantum information resource in quantum channel is consumed in an economic way. We have summarized it as follows:

1. When the scheme is in the measurement basis consisting of a GHZ state measurement, a Bell state measurement, and a single-qubit state measurement, quantum channel will become a six-qubit entangled state.

2. When the scheme is in the measurement basis comprised of three Bell state measurements and a single-qubit state measurement, quantum channel will become a seven-qubit entangled state.
3. When the scheme is in the measurement basis made up of four Bell state measurements, quantum channel will become an eight-qubit entangled state.

As is clear from the above discussion, selecting a six-qubit entangled state as quantum channel can not only reduce quantum resource consumption (quantum information bits in quantum channel) but also improve intrinsic efficiency. In our scheme, a six-qubit cluster state is selected as quantum channel due to cluster state's unique superiority, namely, maximum connectedness and a high persistency of entanglement.

3.4.2 BQT Scheme

The scheme can be described as follows. The framework of this scheme is illustrated in Fig. 3.7.

Assume that Alice is in the possession of an unknown three-qubit entangled state, which is given by

$$|\varphi\rangle_{ABC} = a_0|000\rangle + a_1|111\rangle \tag{3.16}$$

where a_0 and a_1 are arbitrary complex coefficients with $|a_0|^2 + |a_1|^2 = 1$. At the same time, Bob possesses an arbitrary single-qubit state, which can be expressed as

$$|\varphi\rangle_D = b_0|0\rangle + b_1|1\rangle \tag{3.17}$$

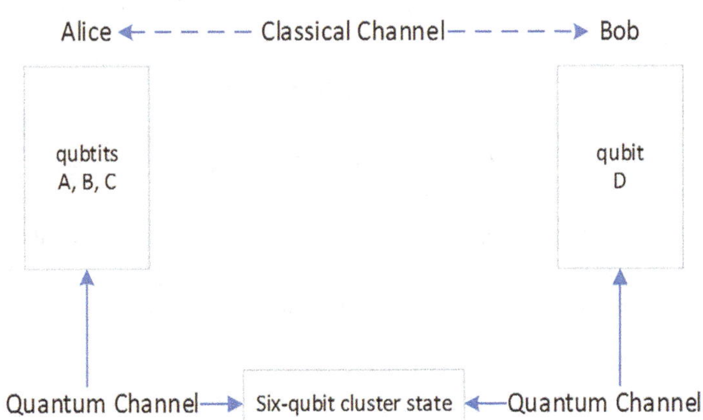

Fig. 3.7. The framework of bidirectional quantum teleportation

where b_0 and b_1 are arbitrary complex coefficients with $|b_0|^2 + |b_1|^2 = 1$. Here, the three qubits A, B, and C belong to Alice, and the single qubit D belongs to Bob.

Now, Alice wants to transmit the three-qubit entangled state to Bob and Bob wants to transmit the single-qubit state to Alice through a single quantum channel consisting of a six-qubit cluster state shared by Alice and Bob. The six-qubit cluster state used as quantum channel has the form

$$|C_6\rangle_{123456} = \frac{1}{2}(|000000\rangle + |000111\rangle + |111000\rangle + |111111\rangle)_{123456} \tag{3.18}$$

where qubits 1 and 6 belong to Alice and qubits 2, 3, 4, and 5 belong to Bob, respectively. Therefore, the initial state of the total system can be defined as

$$\begin{aligned} |\varphi\rangle_{ABCD123456} &= |\varphi\rangle_{ABC} \otimes |\varphi\rangle_D \otimes |C_6\rangle_{123456} \\ &= \frac{1}{2}(a_0|000\rangle + a_1|111\rangle) \otimes (b_0|0\rangle + b_1|1\rangle) \\ &\quad \otimes(|000000\rangle + |000111\rangle + |111000\rangle + |111111\rangle). \end{aligned} \tag{3.19}$$

To implement bidirectional quantum teleportation, the scheme is executed in the following way. Alice firstly performs a GHZ state measurement on her three qubits A, B, and 1 in the measurement basis consisting of $|\xi^{\pm}\rangle_{AB1}$ and $|\zeta^{\pm}\rangle_{AB1}$, which can be characterized as

$$\begin{aligned} |\xi^{\pm}\rangle_{AB1} &= \frac{1}{\sqrt{2}}(|000\rangle \pm |111\rangle)_{AB1} \\ |\zeta^{\pm}\rangle_{AB1} &= \frac{1}{\sqrt{2}}(|001\rangle \pm |110\rangle)_{AB1}. \end{aligned} \tag{3.20}$$

After Alice tells her measurement results to Bob via classical means, Bob performs a Bell state measurement (BSM) on his two qubits D and 5 in the measurement basis made up of $|\phi^{\pm}\rangle_{D5}$ and $|\varphi^{\pm}\rangle_{D5}$, which have the form

$$\begin{aligned} |\phi^{\pm}\rangle_{D5} &= \frac{1}{\sqrt{2}}(|00\rangle \pm |11\rangle)_{D5} \\ |\varphi^{\pm}\rangle_{D5} &= \frac{1}{\sqrt{2}}(|01\rangle \pm |10\rangle)_{D5}. \end{aligned} \tag{3.21}$$

Afterward, Bob informs Alice of his measurement result by classical communication. Based on Bob's measurement result, Alice takes a single-qubit von Neumann measurement on her qubit C in the measurement basis comprised of the following elements:

$$\begin{aligned} |+\rangle_C &= \frac{1}{\sqrt{2}}(|0\rangle + |1\rangle) \\ |-\rangle_C &= \frac{1}{\sqrt{2}}(|0\rangle - |1\rangle). \end{aligned} \tag{3.22}$$

And her measurement result is then conveyed to Bob through a classical channel. Therefore, the initial state of the total system can be transformed into the state

$$\begin{aligned} |\varphi\rangle_{ABCD123456} &= |\varphi\rangle_{ABC} \otimes |\varphi\rangle_D \otimes |C_6\rangle_{123456} \\ &= |\xi^{+}\rangle_{AB1}|\phi^{+}\rangle_{D5}|+\rangle_C|\eta^{1}\rangle_{2346} + |\xi^{+}\rangle_{AB1}|\phi^{+}\rangle_{D5}|-\rangle_C|\eta^{2}\rangle_{2346} \\ &+|\xi^{+}\rangle_{AB1}|\phi^{-}\rangle_{D5}|+\rangle_C|\eta^{3}\rangle_{2346} + |\xi^{+}\rangle_{AB1}|\phi^{-}\rangle_{D5}|-\rangle_C|\eta^{4}\rangle_{2346} \\ &+|\xi^{+}\rangle_{AB1}|\varphi^{+}\rangle_{D5}|+\rangle_C|\eta^{5}\rangle_{2346} + |\xi^{+}\rangle_{AB1}|\varphi^{+}\rangle_{D5}|-\rangle_C|\eta^{6}\rangle_{2346} \\ &+|\xi^{+}\rangle_{AB1}|\varphi^{-}\rangle_{D5}|+\rangle_C|\eta^{7}\rangle_{2346} + |\xi^{+}\rangle_{AB1}|\varphi^{-}\rangle_{D5}|-\rangle_C|\eta^{8}\rangle_{2346} \\ &+|\xi^{-}\rangle_{AB1}|\phi^{+}\rangle_{D5}|+\rangle_C|\eta^{9}\rangle_{2346} + |\xi^{-}\rangle_{AB1}|\phi^{+}\rangle_{D5}|-\rangle_C|\eta^{10}\rangle_{2346} \\ &+|\xi^{-}\rangle_{AB1}|\phi^{-}\rangle_{D5}|+\rangle_C|\eta^{11}\rangle_{2346} + |\xi^{-}\rangle_{AB1}|\phi^{-}\rangle_{D5}|-\rangle_C|\eta^{12}\rangle_{2346} \\ &+|\xi^{-}\rangle_{AB1}|\varphi^{+}\rangle_{D5}|+\rangle_C|\eta^{13}\rangle_{2346} + |\xi^{-}\rangle_{AB1}|\varphi^{+}\rangle_{D5}|-\rangle_C|\eta^{14}\rangle_{2346} \\ &+|\xi^{-}\rangle_{AB1}|\varphi^{-}\rangle_{D5}|+\rangle_C|\eta^{15}\rangle_{2346} + |\xi^{-}\rangle_{AB1}|\varphi^{-}\rangle_{D5}|-\rangle_C|\eta^{16}\rangle_{2346} \\ &+|\zeta^{+}\rangle_{AB1}|\phi^{+}\rangle_{D5}|+\rangle_C|\eta^{17}\rangle_{2346} + |\zeta^{+}\rangle_{AB1}|\phi^{+}\rangle_{D5}|-\rangle_C|\eta^{18}\rangle_{2346} \\ &+|\zeta^{+}\rangle_{AB1}|\phi^{-}\rangle_{D5}|+\rangle_C|\eta^{19}\rangle_{2346} + |\zeta^{+}\rangle_{AB1}|\phi^{-}\rangle_{D5}|-\rangle_C|\eta^{20}\rangle_{2346} \\ &+|\zeta^{+}\rangle_{AB1}|\varphi^{+}\rangle_{D5}|+\rangle_C|\eta^{21}\rangle_{2346} + |\zeta^{+}\rangle_{AB1}|\varphi^{+}\rangle_{D5}|-\rangle_C|\eta^{22}\rangle_{2346} \\ &+|\zeta^{+}\rangle_{AB1}|\varphi^{-}\rangle_{D5}|+\rangle_C|\eta^{23}\rangle_{2346} + |\zeta^{+}\rangle_{AB1}|\varphi^{-}\rangle_{D5}|-\rangle_C|\eta^{24}\rangle_{2346} \\ &+|\zeta^{-}\rangle_{AB1}|\phi^{+}\rangle_{D5}|+\rangle_C|\eta^{25}\rangle_{2346} + |\zeta^{-}\rangle_{AB1}|\phi^{+}\rangle_{D5}|-\rangle_C|\eta^{26}\rangle_{2346} \\ &+|\zeta^{-}\rangle_{AB1}|\phi^{-}\rangle_{D5}|+\rangle_C|\eta^{27}\rangle_{2346} + |\zeta^{-}\rangle_{AB1}|\phi^{-}\rangle_{D5}|-\rangle_C|\eta^{28}\rangle_{2346} \\ &+|\zeta^{-}\rangle_{AB1}|\varphi^{+}\rangle_{D5}|+\rangle_C|\eta^{29}\rangle_{2346} + |\zeta^{-}\rangle_{AB1}|\varphi^{+}\rangle_{D5}|-\rangle_C|\eta^{30}\rangle_{2346} \\ &+|\zeta^{-}\rangle_{AB1}|\varphi^{-}\rangle_{D5}|+\rangle_C|\eta^{31}\rangle_{2346} + |\zeta^{-}\rangle_{AB1}|\varphi^{+}\rangle_{D5}|-\rangle_C|\eta^{32}\rangle_{2346} \end{aligned} \tag{3.23}$$

where $|\eta^{i}\rangle_{2346}(i = 1, 2, 3, ...32)$ are mutually orthometric four-qubit state shown in Table 3.7.

It can be seen from the above formula that after three times measurements, the total quantum system can be collapsed into $|\eta^{i}\rangle_{2346}$. Although $|\eta^{i}\rangle_{2346}$ has 32 possible quantum states, Table 3.7 shows that half of the 32 possible quantum states are the same. So broadly speaking, there are only 16 possible collapsed states in this scheme, which can lessen operation complexity. On the basis of the above measurement results, Alice and Bob perform appropriate unitary transformation on their own qubits, respectively, and the different unitary operations used for different measurement results are summarized in Table 3.8. After performing unitary operations, the collapsed state $|\eta^{i}\rangle_{2346}$ can be evolved into

Table 3.7 The collapsed states composed of qubits 2, 3, 4, and 6 after GHZ state measurement, Bell state measurement, and single-qubit measurement

$\vert\eta^{1}\rangle_{2346} = \vert\eta^{10}\rangle_{2346} = a_0b_0\vert 0000\rangle + a_0b_1\vert 0011\rangle + a_1b_0\vert 1100\rangle + a_1b_1\vert 1111\rangle$
$\vert\eta^{2}\rangle_{2346} = \vert\eta^{9}\rangle_{2346} = a_0b_0\vert 0000\rangle + a_0b_1\vert 0011\rangle - a_1b_0\vert 1100\rangle - a_1b_1\vert 1111\rangle$
$\vert\eta^{3}\rangle_{2346} = \vert\eta^{12}\rangle_{2346} = a_0b_0\vert 0000\rangle - a_0b_1\vert 0011\rangle + a_1b_0\vert 1100\rangle - a_1b_1\vert 1111\rangle$
$\vert\eta^{4}\rangle_{2346} = \vert\eta^{11}\rangle_{2346} = a_0b_0\vert 0000\rangle - a_0b_1\vert 0011\rangle - a_1b_0\vert 1100\rangle + a_1b_1\vert 1111\rangle$
$\vert\eta^{5}\rangle_{2346} = \vert\eta^{14}\rangle_{2346} = a_0b_0\vert 0011\rangle + a_0b_1\vert 0000\rangle + a_1b_0\vert 1111\rangle + a_1b_1\vert 1100\rangle$
$\vert\eta^{6}\rangle_{2346} = \vert\eta^{13}\rangle_{2346} = a_0b_0\vert 0011\rangle + a_0b_1\vert 0000\rangle - a_1b_0\vert 1111\rangle - a_1b_1\vert 1100\rangle$
$\vert\eta^{7}\rangle_{2346} = \vert\eta^{16}\rangle_{2346} = a_0b_0\vert 0011\rangle - a_0b_1\vert 0000\rangle + a_1b_0\vert 1111\rangle - a_1b_1\vert 1100\rangle$
$\vert\eta^{8}\rangle_{2346} = \vert\eta^{15}\rangle_{2346} = a_0b_0\vert 0011\rangle - a_0b_1\vert 0000\rangle - a_1b_0\vert 1111\rangle + a_1b_1\vert 1100\rangle$
$\vert\eta^{17}\rangle_{2346} = \vert\eta^{26}\rangle_{2346} = a_0b_0\vert 1100\rangle + a_0b_1\vert 1111\rangle + a_1b_0\vert 0000\rangle + a_1b_1\vert 0011\rangle$
$\vert\eta^{18}\rangle_{2346} = \vert\eta^{25}\rangle_{2346} = a_0b_0\vert 1100\rangle + a_0b_1\vert 1111\rangle - a_1b_0\vert 0000\rangle - a_1b_1\vert 0011\rangle$
$\vert\eta^{19}\rangle_{2346} = \vert\eta^{28}\rangle_{2346} = a_0b_0\vert 1100\rangle - a_0b_1\vert 1111\rangle + a_1b_0\vert 0000\rangle - a_1b_1\vert 0011\rangle$
$\vert\eta^{20}\rangle_{2346} = \vert\eta^{27}\rangle_{2346} = a_0b_0\vert 1100\rangle - a_0b_1\vert 1111\rangle - a_1b_0\vert 0000\rangle + a_1b_1\vert 0011\rangle$
$\vert\eta^{21}\rangle_{2346} = \vert\eta^{30}\rangle_{2346} = a_0b_0\vert 1111\rangle + a_0b_1\vert 1100\rangle + a_1b_0\vert 0011\rangle + a_1b_1\vert 0000\rangle$
$\vert\eta^{22}\rangle_{2346} = \vert\eta^{29}\rangle_{2346} = a_0b_0\vert 1111\rangle + a_0b_1\vert 1100\rangle - a_1b_0\vert 0011\rangle - a_1b_1\vert 0000\rangle$
$\vert\eta^{23}\rangle_{2346} = \vert\eta^{32}\rangle_{2346} = a_0b_0\vert 1111\rangle - a_0b_1\vert 1100\rangle + a_1b_0\vert 0011\rangle - a_1b_1\vert 0000\rangle$
$\vert\eta^{24}\rangle_{2346} = \vert\eta^{31}\rangle_{2346} = a_0b_0\vert 1111\rangle - a_0b_1\vert 1100\rangle - a_1b_0\vert 0011\rangle + a_1b_1\vert 0000\rangle$

$$a_0b_0|0000\rangle + a_0b_1|0011\rangle + a_1b_0|1100\rangle + a_1b_1|1111\rangle. \tag{3.24}$$

Then Alice and Bob carry out CNOT operation twice. For the first time, Bob performs CNOT operation on qubits 3 and 4, where qubit 3 serves as control qubit and qubit 4 as target qubit. For the second time, Alice performs CNOT operation on qubits 4 and 6, where qubit 6 serves as control qubit and qubit 4 as target qubit. The CNOT operations on qubits 2, 3, 4, and 6 are shown in Fig. 3.8.

Finally, the desired quantum states that are defined as $a_0b_0|0000\rangle + a_0b_1|0001\rangle + a_1b_0|1110\rangle + a_1b_1|1111\rangle$ can be reconstructed successfully. The theoretical quantum circuit of this scheme is shown in Fig. 3.9.

Now, let us take an example to demonstrate the principle of this asymmetric bidirectional quantum teleportation.

Suppose that Alice's measurement outcome is $|\xi^{+}\rangle_{AB1}$ and at the same time Bob's measurement outcome is $|\phi^{+}\rangle_{D5}$; the corresponding collapsed state of qubits C, 2, 3, 4, and 6 is given as

$$|\varphi\rangle_{C2346} = (a_0b_0|00000\rangle + a_0b_1|00011\rangle + a_1b_0|11100\rangle + a_1b_1|11111\rangle)_{C2346}. \tag{3.25}$$

Subsequently, Alice takes a single-qubit von Neumann measurement on her qubit C. If Alice's result is $|+\rangle_C$, then the remaining qubits 2, 3, 4, and 6 are collapsed into the state

Table 3.8 The possible outcomes of measurement and corresponding unitary operations

Alice$_{AB1}$	Bob$_{D5}$	Alice$_{C}$	Measurement results	Unitary transformation
$\vert\xi^{+}\rangle_{AB1}$	$\vert\phi^{+}\rangle_{D5}$	$\vert+\rangle_{C}$	$\vert\eta^{1}\rangle_{2346}$	$I_2 \otimes I_3 \otimes I_4 \otimes I_6$
$\vert\xi^{+}\rangle_{AB1}$	$\vert\phi^{+}\rangle_{D5}$	$\vert-\rangle_{C}$	$\vert\eta^{2}\rangle_{2346}$	$\sigma_2^z \otimes I_3 \otimes I_4 \otimes I_6$
$\vert\xi^{+}\rangle_{AB1}$	$\vert\phi^{-}\rangle_{D5}$	$\vert+\rangle_{C}$	$\vert\eta^{3}\rangle_{2346}$	$\sigma_2^z \otimes \sigma_3^z \otimes \sigma_4^z \otimes I_6$
$\vert\xi^{+}\rangle_{AB1}$	$\vert\phi^{-}\rangle_{D5}$	$\vert-\rangle_{C}$	$\vert\eta^{4}\rangle_{2346}$	$I_2 \otimes \sigma_3^z \otimes \sigma_4^z \otimes I_6$
$\vert\xi^{+}\rangle_{AB1}$	$\vert\varphi^{+}\rangle_{D5}$	$\vert+\rangle_{C}$	$\vert\eta^{5}\rangle_{2346}$	$I_2 \otimes I_4 \otimes \sigma_4^z \otimes \sigma_6^z$
$\vert\xi^{+}\rangle_{AB1}$	$\vert\varphi^{+}\rangle_{D5}$	$\vert-\rangle_{C}$	$\vert\eta^{6}\rangle_{2346}$	$I_2 \otimes \sigma_3^z \otimes \sigma_4^x \otimes \sigma_6^x$
$\vert\xi^{+}\rangle_{AB1}$	$\vert\varphi^{-}\rangle_{D5}$	$\vert+\rangle_{C}$	$\vert\eta^{7}\rangle_{2346}$	$I_2 \otimes I_3 \otimes \sigma_4^x \otimes (-i)\sigma_6^y$
$\vert\xi^{+}\rangle_{AB1}$	$\vert\varphi^{-}\rangle_{D5}$	$\vert-\rangle_{C}$	$\vert\eta^{8}\rangle_{2346}$	$I_2 \otimes \sigma_3^z \otimes \sigma_4^x \otimes (-i)\sigma_6^y$
$\vert\xi^{-}\rangle_{AB1}$	$\vert\phi^{+}\rangle_{D5}$	$\vert+\rangle_{C}$	$\vert\eta^{9}\rangle_{2346}$	$\sigma_2^z \otimes I_3 \otimes I_4 \otimes I_6$
$\vert\xi^{-}\rangle_{AB1}$	$\vert\phi^{+}\rangle_{D5}$	$\vert-\rangle_{C}$	$\vert\eta^{10}\rangle_{2346}$	$I_2 \otimes I_3 \otimes I_4 \otimes I_6$
$\vert\xi^{-}\rangle_{AB1}$	$\vert\phi^{-}\rangle_{D5}$	$\vert+\rangle_{C}$	$\vert\eta^{11}\rangle_{2346}$	$I_2 \otimes \sigma_3^z \otimes \sigma_4^z \otimes I_6$
$\vert\xi^{-}\rangle_{AB1}$	$\vert\phi^{-}\rangle_{D5}$	$\vert-\rangle_{C}$	$\vert\eta^{12}\rangle_{2346}$	$\sigma_2^z \otimes \sigma_3^z \otimes \sigma_4^z \otimes I_6$
$\vert\xi^{-}\rangle_{AB1}$	$\vert\varphi^{+}\rangle_{D5}$	$\vert+\rangle_{C}$	$\vert\eta^{13}\rangle_{2346}$	$I_2 \otimes \sigma_3^z \otimes \sigma_4^x \otimes \sigma_6^x$
$\vert\xi^{-}\rangle_{AB1}$	$\vert\varphi^{+}\rangle_{D5}$	$\vert-\rangle_{C}$	$\vert\eta^{14}\rangle_{2346}$	$I_2 \otimes I_4 \otimes \sigma_4^z \otimes \sigma_6^z$
$\vert\xi^{-}\rangle_{AB1}$	$\vert\varphi^{-}\rangle_{D5}$	$\vert+\rangle_{C}$	$\vert\eta^{15}\rangle_{2346}$	$I_2 \otimes \sigma_3^z \otimes \sigma_4^x \otimes (-i)\sigma_6^y$
$\vert\xi^{-}\rangle_{AB1}$	$\vert\varphi^{-}\rangle_{D5}$	$\vert-\rangle_{C}$	$\vert\eta^{16}\rangle_{2346}$	$I_2 \otimes I_3 \otimes \sigma_4^x \otimes (-i)\sigma_6^y$
$\vert\zeta^{+}\rangle_{AB1}$	$\vert\phi^{+}\rangle_{D5}$	$\vert+\rangle_{C}$	$\vert\eta^{17}\rangle_{2346}$	$\sigma_2^x \otimes \sigma_3^x \otimes I_4 \otimes I_6$
$\vert\zeta^{+}\rangle_{AB1}$	$\vert\phi^{+}\rangle_{D5}$	$\vert-\rangle_{C}$	$\vert\eta^{18}\rangle_{2346}$	$i\sigma_2^y \otimes \sigma_3^x \otimes \sigma_4^z \otimes \sigma_6^z$
$\vert\zeta^{+}\rangle_{AB1}$	$\vert\phi^{-}\rangle_{D5}$	$\vert+\rangle_{C}$	$\vert\eta^{19}\rangle_{2346}$	$\sigma_2^x \otimes \sigma_3^x \otimes I_4 \otimes \sigma_6^z$
$\vert\zeta^{+}\rangle_{AB1}$	$\vert\phi^{-}\rangle_{D5}$	$\vert-\rangle_{C}$	$\vert\eta^{20}\rangle_{2346}$	$i\sigma_2^y \otimes \sigma_3^x \otimes \sigma_4^z \otimes I_6$
$\vert\zeta^{+}\rangle_{AB1}$	$\vert\varphi^{+}\rangle_{D5}$	$\vert+\rangle_{C}$	$\vert\eta^{21}\rangle_{2346}$	$\sigma_2^x \otimes \sigma_3^x \otimes \sigma_4^x \otimes \sigma_6^x$
$\vert\zeta^{+}\rangle_{AB1}$	$\vert\varphi^{+}\rangle_{D5}$	$\vert-\rangle_{C}$	$\vert\eta^{22}\rangle_{2346}$	$i\sigma_2^y \otimes \sigma_3^x \otimes \sigma_4^x \otimes \sigma_6^x$
$\vert\zeta^{+}\rangle_{AB1}$	$\vert\varphi^{-}\rangle_{D5}$	$\vert+\rangle_{C}$	$\vert\eta^{23}\rangle_{2346}$	$\sigma_2^x \otimes \sigma_3^x \otimes i\sigma_4^y \otimes \sigma_6^x$
$\vert\zeta^{+}\rangle_{AB1}$	$\vert\varphi^{-}\rangle_{D5}$	$\vert-\rangle_{C}$	$\vert\eta^{24}\rangle_{2346}$	$\sigma_2^x \otimes i\sigma_3^y \otimes i\sigma_4^y \otimes \sigma_6^x$
$\vert\zeta^{-}\rangle_{ABC}$	$\vert\phi^{+}\rangle_{D5}$	$\vert+\rangle_{C}$	$\vert\eta^{25}\rangle_{2346}$	$i\sigma_2^y \otimes \sigma_3^x \otimes \sigma_4^z \otimes \sigma_6^z$
$\vert\zeta^{-}\rangle_{ABC}$	$\vert\phi^{+}\rangle_{D5}$	$\vert-\rangle_{C}$	$\vert\eta^{26}\rangle_{2346}$	$\sigma_2^x \otimes \sigma_3^x \otimes I_4 \otimes I_6$
$\vert\zeta^{-}\rangle_{ABC}$	$\vert\phi^{-}\rangle_{D5}$	$\vert+\rangle_{C}$	$\vert\eta^{27}\rangle_{2346}$	$i\sigma_2^y \otimes \sigma_3^x \otimes \sigma_4^z \otimes I_6$
$\vert\zeta^{-}\rangle_{ABC}$	$\vert\phi^{-}\rangle_{D5}$	$\vert-\rangle_{C}$	$\vert\eta^{28}\rangle_{2346}$	$\sigma_2^x \otimes \sigma_3^x \otimes I_4 \otimes \sigma_6^z$
$\vert\zeta^{-}\rangle_{ABC}$	$\vert\varphi^{+}\rangle_{D5}$	$\vert+\rangle_{C}$	$\vert\eta^{29}\rangle_{2346}$	$i\sigma_2^y \otimes \sigma_3^x \otimes \sigma_4^x \otimes \sigma_6^x$
$\vert\zeta^{-}\rangle_{ABC}$	$\vert\varphi^{+}\rangle_{D5}$	$\vert-\rangle_{C}$	$\vert\eta^{30}\rangle_{2346}$	$\sigma_2^x \otimes \sigma_3^x \otimes \sigma_4^x \otimes \sigma_6^x$
$\vert\zeta^{-}\rangle_{ABC}$	$\vert\varphi^{-}\rangle_{D5}$	$\vert+\rangle_{C}$	$\vert\eta^{31}\rangle_{2346}$	$\sigma_2^x \otimes i\sigma_3^y \otimes i\sigma_4^y \otimes \sigma_6^x$
$\vert\zeta^{-}\rangle_{ABC}$	$\vert\varphi^{-}\rangle_{D5}$	$\vert-\rangle_{C}$	$\vert\eta^{32}\rangle_{2346}$	$\sigma_2^x \otimes \sigma_3^x \otimes i\sigma_4^y \otimes \sigma_6^x$

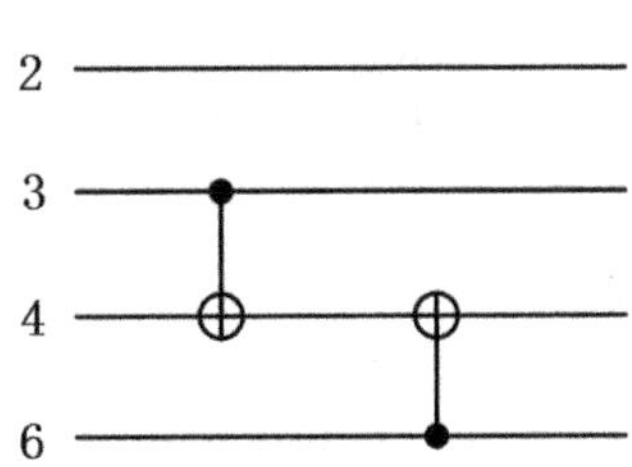

Fig. 3.8 CNOT operations on qubits 2, 3, 4, and 6

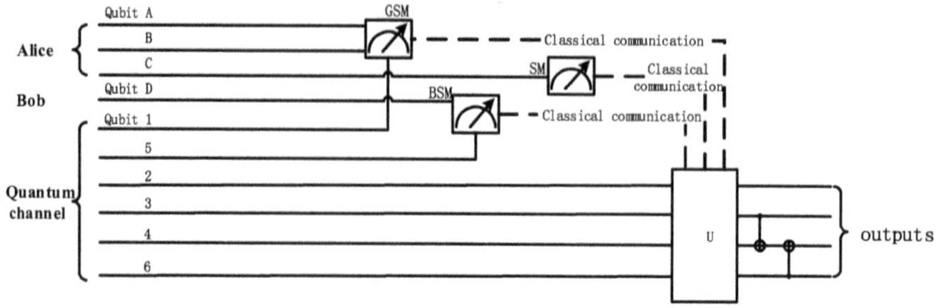

Fig. 3.9 Quantum circuit of bidirectional quantum teleportation by using a six-qubit cluster state in which Alice has an unknown three-qubit entangled state and Bob has an arbitrary single-qubit state

$$|\varphi\rangle_{2346} = (a_0b_0|0000\rangle + a_0b_1|0011\rangle + a_1b_0|1100\rangle + a_1b_1|1111\rangle)_{2346}. \quad (3.26)$$

Then Alice and Bob perform unitary operations, which are I_6 and $I_2 \otimes I_3 \otimes I_4$, respectively.

And after that, Bob needs to carry out a controlled-NOT operation on his qubits with qubit 3 as control qubit and qubit 4 as target qubit; then the state in (3.26) can be evolved into

$$|\varphi'\rangle_{2346} = (a_0b_0|0000\rangle + a_0b_1|0011\rangle + a_1b_0|1110\rangle + a_1b_1|1101\rangle)_{2346}. \quad (3.27)$$

Finally, Alice performs another controlled-NOT operation on her qubit with qubit 6 as control qubit and qubit 4 as target qubit, and then the state becomes the following state:

$$\begin{aligned} |\phi\rangle_{2346} &= (a_0b_0|0000\rangle + a_0b_1|0001\rangle + a_1b_0|1110\rangle + a_1b_1|1111\rangle)_{2346} \\ &= (a_0|000\rangle + a_1|111\rangle)_{234} \otimes (b_0|0\rangle + b_1|1\rangle)_6 \end{aligned} \quad (3.28)$$

Thus, the asymmetric bidirectional quantum teleportation is successfully realized, and the scheme is fully described for all cases in Table 3.8.

3.4.3 *Comparison*

In this scheme, without controller, a six-qubit cluster state is used as quantum channel for the purpose of bipartite exchange of an unknown three-qubit entangled state and an arbitrary single-qubit state. Now, let us make some comparisons among our scheme and other schemes [3, 14]. Comparisons are made from five aspects, namely, the quantum resource consumption, the necessary operation complexity including operation difficulty and intensity, the classical resource consumption, the quantum information bits transmitted, and the intrinsic efficiency, which is defined

Table 3.9 Comparisons between four BQT protocols

Scheme	QRC	NO	CRC	QIBT	
V	EPR	4 SM	4	2	1/3
M	Five-Q CS	2 BSM 1 SM	5	3	3/10
S	Eight-Q ES	8 SM	8	4	1/4
Ours	Six-Q CS	1 BSM 1 GSM 1 SM	5	4	4/11

as $\tau = q_s/(q_u + b_t)$ in Subsection 3.4.1. The comparisons are summarized in Table 3.9.

The VMS, in turn, on behalf of refs. [3, 11, 12]. QRC quantum resource consumption, Q qubit, CS cluster state, ES entangled state, NO necessary operation, CRC classical resource consumption, QIBT quantum information bits transmitted, SM single-qubit measurement, BSM Bell state measurement, GSM GHZ state measurement.

From Table 3.9, the quantum information bits transmitted in V, M, and S schemes are, respectively, two single-qubit states, a two-qubit state and a single-qubit state, and two two-qubit states. With the number of quantum information bits transmitted increasing, quantum resource consumption and classical resource consumption will both become more complex, and efficiencies are correspondingly decreasing in V, M, and S schemes.

This BQT scheme is an innovation and expansion on the basis of the M scheme, so comparing the M scheme with our scheme, we can see several differences, which are shown as follows: (1) the quantum resource consumption in the M scheme is less than our scheme, (2) the operation complexity in our scheme is more difficult than the M scheme, and (3) the remarkable advantages in our scheme are transmitting more quantum information and possessing higher intrinsic efficiency.

3.4.4 Summary

In summary, an ideal of asymmetric bidirectional quantum teleportation by using a six-qubit cluster state as quantum channel is proposed. In our scheme, Alice and Bob are not only senders but also receivers. Alice can transmit an unknown three-qubit entangled state to Bob and Bob can transmit an arbitrary single-qubit state to Alice by using quantum measurements and corresponding unitary operations. What's more, we have not only certified the feasibility of the scheme but also demonstrated the scheme's innovation by comparing our scheme with other schemes in quantum and classical resource consumption, operation complexities, quantum information bits transmitted, and intrinsic efficiencies. Finally, as we all know, quantum secure direct communication (QSDC) considered as an elegant means for secret communication can be used in some urgent circumstances. Based on quantum teleportation, many QSDC schemes have been devised [15, 16]. And the bidirectional quantum teleportation regarded as the expansion of quantum teleportation can also be used to implement bidirectional QSDC; this method can make the implementation of QSDC simple and secure. But at present, bidirectional QSDC can only encode one-bit message on a single qubit and transmit one-bit information every time by using bidirectional quantum teleportation [17], so we hope that more information can be transmitted at a time in bidirectional QSDC by using our proposed bidirectional quantum teleportation in the future.

References

1. Zha XW, Zou ZC, Qi JX, Song HY (2013) Bidirectional quantum controlled teleportation via five-qubit cluster state. Int J Theor Phys 52:1740–1744
2. Briegel HJ, Raussendorf R (2001) Persistent entanglement in arrays of interacting particles. Phys Rev Lett 86:910–913. https://doi.org/10.1103/PhysRevLett.86.910
3. Sang M (2016) Huang: bidirectional quantum teleportation by using five-qubit cluster state. Int J Theor Phys 55:1333–1335
4. Zhou R-G, Li X, Qian C, Ian H (2020) Quantum bidirectional teleportation 2 $\leftrightarrow$ 2 or 2 $\leftrightarrow$ 3 qubit teleportation protocol via 6-qubit entangled state. Int J Theor Phys 59(1):166–172
5. Zhou R-G, Qian C, Ian H (2019) Bidirectional quantum teleportation of two-qubit state via four-qubit cluster state. Int J Theor Phys 58(1):150–156
6. Zhou R, Xu R, Lan H (2019) Bidirectional quantum teleportation by using six-qubit cluster state. IEEE Access 7:744269–744275
7. Hassanpour S, Houshmand M (2016) Bidirectional teleportation of a pure EPR state by using GHZ states. Quantum Inf Process 15:905–912
8. Choudhury BS, Samanta S (2017) Asymmetric bidirectional 3 $\Leftrightarrow$ 2 qubit teleportation protocol between Alice and bob via 9-qubit cluster state. Int J Theor Phys 56:3285–3296
9. Sadeghi Zadeh MS, Houshmand M, Aghababa H (2017) Bidirectional teleportation of a two-qubit state by using eight-qubit entangled state as a quantum channel. Int J Theor Phys 56:2101–2112
10. Zhang D, Zha XW, Li W, Yu Y (2015) Bidirectional and asymmetric quantum controlled teleportation via maximally eight-qubit entangled state. Quantum Inf Process 14:3835–3844
11. Duan YJ, Zha XW (2014) Bidirectional quantum controlled teleportation via a six-qubit entangled state. Int J Theor Phys 53:3780–3786
12. Duan YJ, Zha XW, Sun XM, Xia JF (2014) Bidirectional quantum controlled teleportation via a maximally seven-qubit entangled state. Int J Theor Phys 53:2697–2707
13. Gottesman D (2008) Stabilizer codes and quantum error correction 2008
14. Hassanpour S, Houshniand M (2015) Bidirectional quantum teleportation via entanglement swapping. ICEE 2015 – Proc. 23rd Iran. Conf. Electr. Eng. 10:501–503. https://doi.org/10.1109/IranianCEE.2015.7146267
15. Cao HJ, Song HS (2006) Quantum secure direct communication scheme using a W state and teleportation. Phys Scr 74:572–575
16. Zhang QN, Li CC, Li YH, Nie Y (2013) You: quantum secure direct communication based on four-qubit cluster states. Int J Theor Phys 52:22–27
17. Li YH, Li XL, Sang MH, Nie YY, Wang ZS (2013) Bidirectional controlled quantum teleportation and secure direct communication using five-qubit entangled state. Quantum Inf Process 12:3835–3844

Chapter 4
Controlled Bidirectional Quantum Teleportation

Abstract Controlled bidirectional quantum teleportation is a sophisticated and advanced technology in the field of quantum communication. It refers to the process in which, through a specific control mechanism, quantum states can be transmitted bidirectionally between two or more locations under controlled conditions during quantum communication. This mode of transmission not only relies on quantum entanglement and classical communication but also necessitates the introduction of additional control measures to ensure the accuracy and security of the transmission. Controlled bidirectional quantum teleportation is an essential technology in the quantum communication domain, offering broad application prospects and tremendous development potential. Nevertheless, realizing this technology and its widespread practical application necessitates overcoming a series of technical and practical challenges.

4.1 Introduction

With the development of BQT, the number of qubits in quantum channel is expanded gradually, and the protocols' categories have also been gradually created. One of the branches of BQT is bidirectional controlled quantum teleportation (C-BQT), where a third party or multiparty comes into the protocol. In C-BQT, the supervisor Charlie controlled the entanglement resource with Alice and Bob. Moreover, if Charlie finds insecurity in the process, he can terminate the plan in time, i.e., it has a very important application in the field of secure quantum communication.

In 2013, Li et al. [1] used a five-qubit entangled state as the quantum channel and accordingly proposed the bidirectional controlled quantum teleportation scheme between two participants under the control of a supervisor to transmit one qubit to each other. Sun and Zha [2] proposed a different C-BQT scheme using the six-qubit entangled state as a quantum channel, which has three parties and the supervisor has two control qubits. In the same year, Yan [3] proposed a scheme for bidirectional controlled teleportation of single qubit via a six-qubit cluster state. In 2016, Hong [4] proposed an asymmetric C-BQT (C-ABQT) scheme, which realized more than

R.-G. Zhou et al., *Design of Quantum Teleportation Schemes*,
https://doi.org/10.1007/978-3-031-82725-9_4

single qubit transmitted. It used a seven-qubit entangled state as the quantum channel, and Alice sends one single-qubit state to Bob, while Bob sends a two-qubit state to Alice under the control of the supervisor Charlie.

In this chapter, three C-BQT schemes are proposed.

In Sect. 4.2, a teleportation scheme was developed using a five-qubit non-maximally entangled state as the quantum channel [5]. With Charlie's permission, Alice and Bob could teleport a single-qubit state to each other deterministically. The scheme was introduced in a noiseless environment first. As we all know, when a non-maximally entangled state was used as the quantum channel, teleportation might fail, and the state to be transmitted would be destroyed. To solve this problem, in this scheme, an appropriate unitary transformation was performed on the non-maximally entangled state in advance with the help of an auxiliary qubit. Then, we analyzed the influence of the quantum noise on the fidelity of the desired state using the example of amplitude damping during the qubit distribution. Finally, we utilized the weak measurement and the corresponding reversing measurement to increase fidelity.

In Sect. 4.3, under the control of the supervisor Charlie, Alice transmits an unknown three-qubit entangled state to Bob, and Bob transmits an unknown three-qubit entangled state to Alice [6]. In which a three-qubit entangled state is transformed into a two-qubit entangled state and single-qubit superposition state by using the Toffoli gate and controlled-NOT operation, receivers can by introducing the appropriate unitary transformation and auxiliary particles to reconstruct the initial state.

In Sect. 4.4, an improved C-BQT scheme of the special three-qubit state is proposed [7]. In a little bit more detail, under the control of the third supervisor Charlie, Alice wants to send one special three-qubit entangled state to Bob, and at the meantime, Bob also wants to transmit another special three-qubit entangled state to Alice. In other words, both Alice and Bob can be the sender and receiver simultaneously. To achieve this aim, a specific 11-qubit entangled state is shared among Alice, Bob, and Charlie in advance acting as the quantum channel. Then, Alice and Bob first implement the GHZ state measurement and Bell state measurement, respectively, and following Charlie's single-qubit measurement. Finally, upon the foregoing measurement results, Alice and Bob can, respectively, implement the specific unitary operators on their local particles to recover the initial state transmitted by the other.

4.2 C-BQT Scheme of Arbitrary Single Qubit via a Non-maximally Entangled State

Many schemes have a common ground, in that they all use a maximally entangled state as the quantum channel, by which the desired state can be transmitted with 100% success probability. From this, it can be known that the maximally entangled state is the essential ingredient for the deterministic quantum teleportation. That is, if we want to perform an exact quantum teleportation, the simplest and most direct method is to find a maximally entangled state. However, in practice, maybe a maximally entangled state cannot be obtained or maintained because of the effect of the decoherence caused by the surrounding noisy environment.

Moreover, in [8], the researchers proved that it is impossible to convert a non-maximally entangled state to a maximal one. Thus, it becomes considerably more significant to study the quantum teleportation via a non-maximally entangled state as the quantum channel. This type of teleportation is usually called the probabilistic teleportation because, in this method, a teleportation failure is possible. In 2000, Li et al. [9] proposed a quantum teleportation scheme via a partially entangled quantum channel instead of a maximal one. In this scheme, the maximal probability of successful teleportation can be obtained when the sender performs a measurement satisfying "entanglement matching" to the adopted quantum channel. In the same year, Lu et al. [10] reported the probabilistic quantum teleportation protocols for the two-qubit and three-qubit entangled states via entanglement swapping and pointed out that the success probability of quantum teleportation only depends on the least of the quantum state's coefficients in the quantum channel. In 2002, Agrawal et al. [11] proposed a probabilistic quantum teleportation protocol for an unknown two-qubit state via a non-maximally entangled state. In 2004, Dai et al. [12, 13] presented two probabilistic quantum teleportation schemes for an arbitrary two-qubit state. One adopts two partially entangled three-particle W states as the quantum channel. The other uses a partially entangled three-particle W state and the GHZ state as the quantum channel. In 2008, Han et al. [14] introduced a probabilistic teleportation of an arbitrary two-qubit state via a one-dimensional four-qubit cluster state. In 2016, Choudhury et al. [15] presented a scheme for probabilistically teleporting arbitrary two-qubit states. Different from other schemes, the scheme uses the POVM measurement, which has some operational advantages. In 2017, Gou et al. [16] proposed a probabilistic resumable bidirectional quantum teleportation. In this scheme, the states to be teleported can be rebuilt when the teleportation fails. In 2018, Wei et al. [17] proposed a novel probabilistic teleportation scheme for an unknown single-qubit state with multi-parameter measurements.

Here, we propose a controlled bidirectional teleportation scheme for an arbitrary unknown single-qubit state. In the scheme, we consider a non-maximally five-qubit cluster state as the quantum channel. However, the probability of successful teleportation could not reach 1 when the entangled state used as the quantum channel is not maximally entangled. To solve the problem, we performed an appropriate unitary transformation on the non-maximally entangled state in advance with the help of an auxiliary qubit. The advantage of the scheme was that it could improve a probabilistic teleportation to a deterministic teleportation. Furthermore, we assumed that during the entanglement state distribution, the coherence of the quantum state was broken by the quantum noise by the example of the amplitude damping (AD). Then, we calculated the fidelity of the desired state and tried to use the weak measurement and the reversing measurement to improve the fidelity. The result showed that the weak measurement was a useful method for enhancing the fidelity on the quantum channel affected by the AD noise.

4.2.1 Scheme for Deterministic and Controlled Bidirectional Quantum Teleportation

In this subsection, we elaborate the teleportation scheme in detail. In the present scheme, with Charlie's permission, Alice can transmit an unknown single-qubit state $|\theta\rangle_a$ to Bob; at the same time, Bob can transmit an unknown single-qubit state $|\theta\rangle_b$ to Alice. The two quantum states are given as follows:

$$\begin{aligned} |\theta\rangle_a &= a_0|0\rangle + a_1|1\rangle \\ |\theta\rangle_b &= b_0|0\rangle + b_1|1\rangle \end{aligned} \tag{4.1}$$

where a_0, a_1, b_0, and b_1 are the amplitude of the respective basis state and satisfy the normalized condition $|a_0|^2 + |a_1|^2 = 1$ and $|b_0|^2 + |b_1|^2 = 1$. The quantum channel adopted in the scheme is a non-maximally five-qubit cluster state [18], which may be expressed as follows:

$$|\psi\rangle_{12345} = (x_0|00000\rangle + x_1|00111\rangle + x_2|11010\rangle + x_3|11101\rangle)_{12345} \tag{4.2}$$

where x_0, x_1, x_2, and x_3 are the real coefficients and $|x_0| > |x_1| > |x_2| > |x_3|$. We assumed that the five-qubit state was prepared on Charlie's side. After the preparation, Charlie did not have to send qubits to Alice and Bob immediately.

As the state was not maximally entangled, the teleportation could fail if it was used as the quantum channel directly. It has been demonstrated that the probability of successful teleportation depended on the least of the quantum state's coefficients in the entangled state. Moreover, when the teleportation failed, the quantum state to be teleported would be destroyed. In order to solve the problem, the entangled state had to be modulated by Charlie with an appropriate transformation before it was distributed.

4.2.1.1 Transformations for Entangled State

The non-maximally entangled state $|\psi\rangle_{12345}$ given in (4.2) was changed into an appropriate state, which helped Alice and Bob perform the quantum teleportation deterministically. The entire transformation process consisted of the following three steps:

Step 1: Charlie first performed a *controlled-NOT (CNOT)* gate operation on qubits 2 and 4, in which qubit 2 was the control qubit and qubit 4 was the target qubit. The CNOT gate could be described as follows:

$$\text{CNOT} = |0\rangle_2\langle 0| \otimes I_4 + |1\rangle_2\langle 1| \otimes \sigma_4^x, \sigma^x = \begin{bmatrix} 0 & 1 \\ 1 & 0 \end{bmatrix}. \tag{4.3}$$

If qubit 2 was set to 0, qubit 4 would be left alone, and if qubit 2 was set to 1, qubit 4 would be reversed. After the operation, the original state $|\psi\rangle_{12345}$ changed to the following:

$$|\psi\rangle_{12345} = (x_0|00000\rangle + x_1|00111\rangle + x_2|11000\rangle + x_3|11111\rangle)_{12345}. \tag{4.4}$$

Step 2: Charlie introduced an auxiliary qubit r with the initial state $|0\rangle_r$. Then, Charlie performed a collective unitary transformation on qubits 2, 4, and r under the basis $\{|000\rangle_{24r}, |010\rangle_{24r}, |100\rangle_{24r}, |110\rangle_{24r}, |001\rangle_{24r}, |011\rangle_{24r}, |101\rangle_{24r}, |111\rangle_{24r}\}$. The unitary transformation could be expressed as follows:

$$U_{24r} = \begin{bmatrix} U_1 & U_2 \\ U_2 & -U_1 \end{bmatrix} \tag{4.5}$$

where

$$U_1 = \begin{bmatrix} \frac{x_3}{x_0} & 0 & 0 & 0 \\ 0 & \frac{x_3}{x_1} & 0 & 0 \\ 0 & 0 & \frac{x_3}{x_2} & 0 \\ 0 & 0 & 0 & 1 \end{bmatrix}, U_2 = \begin{bmatrix} \sqrt{1-\frac{x_3^2}{x_0^2}} & 0 & 0 & 0 \\ 0 & \sqrt{1-\frac{x_3^2}{x_1^2}} & 0 & 0 \\ 0 & 0 & \sqrt{1-\frac{x_3^2}{x_2^2}} & 0 \\ 0 & 0 & 0 & 1 \end{bmatrix}. \tag{4.6}$$

After the unitary transformation, the quantum state of qubits 2, 4, and r was converted into the following form:

$$\begin{aligned} |\omega\rangle_{24r} &= U_{24r}[x_0 \ x_1 \ x_2 \ x_3 \ 0 \ 0 \ 0 \ 0]^T \\ &= x_3\left(|00\rangle_{24} + |01\rangle_{24} + |10\rangle_{24} + |11\rangle_{24}\right)|0\rangle_r + \\ &\left(\sqrt{x_0^2 - x_3^2}|00\rangle_{24} + \sqrt{x_1^2 - x_3^2}|01\rangle_{24} + \sqrt{x_2^2 - x_3^2}|10\rangle_{24} + x_3|11\rangle_{24}\right)|1\rangle_r \end{aligned} \tag{4.7}$$

Thus, the whole quantum state of qubits 1, 2, 3, 4, 5, and r would be transformed into a new state given as follows:

$$\begin{aligned} |\psi\rangle_{12345r} &= x_3(|00000\rangle + |00111\rangle + |11000\rangle + |11111\rangle)_{12345}|0\rangle_r \\ &+ \begin{pmatrix} \sqrt{x_0^2 - x_3^2}|00000\rangle + \sqrt{x_1^2 - x_3^2}|00111\rangle \\ + \sqrt{x_2^2 - x_3^2}|11000\rangle + x_3|11111\rangle \end{pmatrix}_{12345} |1\rangle_r \end{aligned} \tag{4.8}$$

Step 3: Charlie carried out a single-qubit measurement on the auxiliary qubit r with the basis $\{|0\rangle, |1\rangle\}$. If the measurement result was $|0\rangle_r$, the whole state would collapse into the following state:

$$\begin{aligned} |\psi\rangle_{12345} &= x_3(|00000\rangle + |00111\rangle + |11000\rangle + |11111\rangle)_{12345} \\ &= (|00000\rangle + |00111\rangle + |11000\rangle + |11111\rangle)_{12345} \end{aligned}. \tag{4.9}$$

Here, for simplicity, the normalization coefficient x_3 was ignored. (4.9) shows that state $|\psi\rangle_{12345}$ was appropriate as the quantum channel, by which Alice and Bob could transmit the desired states to each other deterministically with some measurements and unitary transformations. Next, we will describe the remaining process of the teleportation via state $|\psi\rangle_{12345}$ as the quantum channel, which was assumed in a noiseless environment.

4.2.1.2 Teleportation in Noiseless Environment

In order to realize the teleportation, three participants shared the qubits of the entangled state $|\psi\rangle_{12345}$ first. We assumed that Alice owned qubits 1 and 3, qubits 2 and 4 belonged to Bob, and Charlie possessed qubit 5. Here, we supposed that the quantum channel was in an ideal environment. The quantum state of the whole system composed of qubits a, b, 1, 2, 3, 4, and 5 could be expressed as the following form:

$$\begin{aligned} |sys\rangle &= |\theta\rangle_a \otimes |\theta\rangle_b \otimes |\psi\rangle_{12345} \\ &= (a_0|0\rangle + a_1|1\rangle)_a \otimes (b_0|0\rangle + b_1|1\rangle)_b \\ &\quad \otimes(|00000\rangle + |00111\rangle + |11000\rangle + |11111\rangle)_{12345} \\ &= (a_0b_0|00\rangle + a_0b_1|01\rangle + a_1b_0|10\rangle + a_1b_1|11\rangle)_{ab} \\ &\quad \otimes(|00000\rangle + |00111\rangle + |11000\rangle + |11111\rangle)_{12345} \\ &= a_0b_0|00\rangle_{a1}|00\rangle_{b4}|00\rangle_{32}|0\rangle_5 + a_0b_0|00\rangle_{a1}|01\rangle_{b4}|01\rangle_{23}|1\rangle_5 \\ &\quad +a_0b_0|01\rangle_{a1}|00\rangle_{b4}|10\rangle_{23}|0\rangle_5 + a_0b_0|01\rangle_{a1}|01\rangle_{b4}|11\rangle_{23}|1\rangle_5 \\ &\quad +a_0b_1|00\rangle_{a1}|10\rangle_{b4}|00\rangle_{23}|0\rangle_5 + a_0b_1|00\rangle_{a1}|11\rangle_{b4}|01\rangle_{23}|1\rangle_5 \\ &\quad +a_0b_1|01\rangle_{a1}|10\rangle_{b4}|10\rangle_{23}|0\rangle_5 + a_0b_1|01\rangle_{a1}|11\rangle_{b4}|11\rangle_{23}|1\rangle_5 \\ &\quad +a_1b_0|10\rangle_{a1}|00\rangle_{b4}|00\rangle_{23}|0\rangle_5 + a_1b_0|10\rangle_{a1}|01\rangle_{b4}|01\rangle_{23}|1\rangle_5 \\ &\quad +a_1b_0|11\rangle_{a1}|00\rangle_{b4}|10\rangle_{23}|0\rangle_5 + a_1b_0|11\rangle_{a1}|01\rangle_{b4}|11\rangle_{23}|1\rangle_5 \\ &\quad +a_1b_1|10\rangle_{a1}|10\rangle_{b4}|00\rangle_{23}|0\rangle_5 + a_1b_1|10\rangle_{a1}|11\rangle_{b4}|01\rangle_{23}|1\rangle_5 \\ &\quad +a_1b_1|11\rangle_{a1}|10\rangle_{b4}|10\rangle_{23}|0\rangle_5 + a_1b_1|11\rangle_{a1}|11\rangle_{b4}|11\rangle_{23}|1\rangle_5 \end{aligned}. \tag{4.10}$$

Next, Alice, Bob, and Charlie had to perform some quantum measurements and unitary transformations on their own qubits, including the following four steps:

Step 1: Alice and Bob performed the Bell state measurements on their own qubits (a, 1) and (b, 4) on the basis of $\{|\phi^+\rangle, |\phi^-\rangle, |\varphi^+\rangle, |\varphi^-\rangle\}$, respectively. Here, $|\phi^+\rangle$, $|\phi^-\rangle$, $|\varphi^+\rangle$, $|\varphi^-\rangle$ were the four orthogonal states of the Bell state expressed as follows:

$$
\begin{aligned}
|\phi^+\rangle &= \frac{1}{\sqrt{2}}(|00\rangle + |11\rangle) \\
|\phi^-\rangle &= \frac{1}{\sqrt{2}}(|00\rangle - |11\rangle) \\
|\varphi^+\rangle &= \frac{1}{\sqrt{2}}(|01\rangle + |10\rangle) \\
|\varphi^-\rangle &= \frac{1}{\sqrt{2}}(|01\rangle - |10\rangle)
\end{aligned}
\tag{4.11}
$$

There exist 16 types of possible measured results, and the collapsed state of the whole system might be one of the following states:

$$
\begin{aligned}
{}_{a1}\langle\phi^\pm|_{b4}\langle\phi^\pm||sys\rangle &= a_0b_0|000\rangle_{235} \pm^2 a_0b_1|011\rangle_{235} \pm^1 a_1b_0|100\rangle_{235} \pm^1 \pm^2 a_1b_1|111\rangle_{235} \\
{}_{a1}\langle\phi^\pm|_{b4}\langle\varphi^\pm||sys\rangle &= a_0b_0|011\rangle_{235} \pm^2 a_0b_1|000\rangle_{235} \pm^1 a_1b_0|111\rangle_{235} \pm^1 \pm^2 a_1b_1|100\rangle_{235} \\
{}_{a1}\langle\varphi^\pm|_{b4}\langle\phi^\pm||sys\rangle &= a_0b_0|100\rangle_{235} \pm^2 a_0b_1|111\rangle_{235} \pm^1 a_1b_0|000\rangle_{235} \pm^1 \pm^2 a_1b_1|011\rangle_{235} \\
{}_{a1}\langle\varphi^\pm|_{b4}\langle\varphi^\pm||sys\rangle &= a_0b_0|111\rangle_{235} \pm^2 a_0b_1|100\rangle_{235} \pm^1 a_1b_0|011\rangle_{235} \pm^1 \pm^2 a_1b_1|000\rangle_{235}
\end{aligned}
\tag{4.12}
$$

where $\pm^1$ and $\pm^2$ represent the measurement results of Alice and Bob, respectively.

Step 2: Charlie, the controller of the scheme, performed a single-qubit measurement on qubit 5 on the X-basis $\{|+\rangle, |-\rangle\}$. Here, $|+\rangle$ and $|-\rangle$ could be expressed as follows:

$$
\begin{aligned}
|+\rangle &= \frac{1}{\sqrt{2}}(|0\rangle + |1\rangle) \\
|-\rangle &= \frac{1}{\sqrt{2}}(|0\rangle - |1\rangle)
\end{aligned}
\tag{4.13}
$$

Then, the remaining qubits of 2 and 3 might be collapsed into one of the following states:

$$\begin{cases} {}_5\langle \pm|_{a1}\langle \phi^\pm|_{b4}\langle \phi^\pm||sys\rangle \\ = a_0b_0|00\rangle_{23} \pm^2 \pm^3 a_0b_1|01\rangle_{23} \pm^1 a_1b_0|10\rangle_{23} \pm^1 \pm^2 \pm^3 a_1b_1|11\rangle_{23} \\ {}_5\langle \pm|_{a1}\langle \phi^\pm|_{b4}\langle \varphi^\pm||sys\rangle \\ = \pm^3 a_0b_0|01\rangle_{23} \pm^2 a_0b_1|00\rangle_{23} \pm^1 \pm^3 a_1b_0|11\rangle_{23} \pm^1 \pm^2 a_1b_1|10\rangle_{23} \\ {}_5\langle \pm|_{a1}\langle \varphi^\pm|_{b4}\langle \phi^\pm||sys\rangle \\ = a_0b_0|10\rangle_{23} \pm^2 \pm^3 a_0b_1|11\rangle_{23} \pm^1 a_1b_0|00\rangle_{23} \pm^1 \pm^2 \pm^3 a_1b_1|01\rangle_{23} \\ {}_5\langle \pm|_{a1}\langle \varphi^\pm|_{b4}\langle \varphi^\pm||sys\rangle \\ = \pm^3 a_0b_0|11\rangle_{23} \pm^2 a_0b_1|10\rangle_{23} \pm^1 \pm^3 a_1b_0|01\rangle_{23} \pm^1 \pm^2 a_1b_1|00\rangle_{23} \end{cases} \tag{4.14}$$

where $\pm^3$ denotes the measurement result on qubit 5.

Step 3: After performing the measurement operations, three participants announced the outcomes via classical channels. According to the results received from the others, Alice and Bob could reconstruct the desired states by making some appropriate unitary transformations on the qubits in hand.

We assumed that the measurement results of Alice, Bob, and Charlie were $|\phi^+\rangle_{a1}$, $|\varphi^-\rangle_{b4}$, and $|+\rangle_5$, respectively. The collapsed state of qubits 2 and 3 was as follows:

$$|\psi\rangle_{23} = a_0b_0|01\rangle_{23} - a_0b_1|00\rangle_{23} + a_1b_0|11\rangle_{23} - a_1b_1|10\rangle_{23}. \tag{4.15}$$

Step 4: Then, Alice and Bob could reconstruct the original state via unitary transformations I_2 and $\sigma_{3z}\sigma_{3x}$ on qubits 2 and 3, respectively. Here, I, σ^z, and σ^x are three basic unitary transformations used to convert a state to a new one as follows:

$$\begin{array}{l} I = \begin{bmatrix} 1 & 0 \\ 0 & 1 \end{bmatrix}, \sigma^z = \begin{bmatrix} 1 & 0 \\ 0 & -1 \end{bmatrix}, \sigma^x = \begin{bmatrix} 0 & 1 \\ 1 & 0 \end{bmatrix} \\ I|0\rangle \to |0\rangle, I|1\rangle \to |1\rangle \\ \sigma_z|0\rangle \to |0\rangle \ , \sigma_z|1\rangle \to -|1\rangle \\ \sigma_x|0\rangle \to |1\rangle, \ \sigma_x|1\rangle \to |0\rangle \end{array}. \tag{4.16}$$

After the unitary operations, the state $|\vartheta\rangle_{23}$ was changed to the following state:

$$\begin{aligned} |\psi\rangle_{23} &= a_0b_0|00\rangle_{23} + a_0b_1|01\rangle_{23} + a_1b_0|10\rangle_{23} + a_1b_1|11\rangle_{23} \\ &= (a_0|0\rangle + a_1|1\rangle)_2 \otimes (b_0|0\rangle + b_1|1\rangle)_3 \end{aligned}. \tag{4.17}$$

(4.17) implies that Alice obtained the quantum information of the quantum state $|\theta\rangle_b$ on qubit 3. At the same time, Bob received the quantum information of the quantum state $|\theta\rangle_a$ on qubit 2. The schematic of the scheme is illustrated in Fig. 4.1.

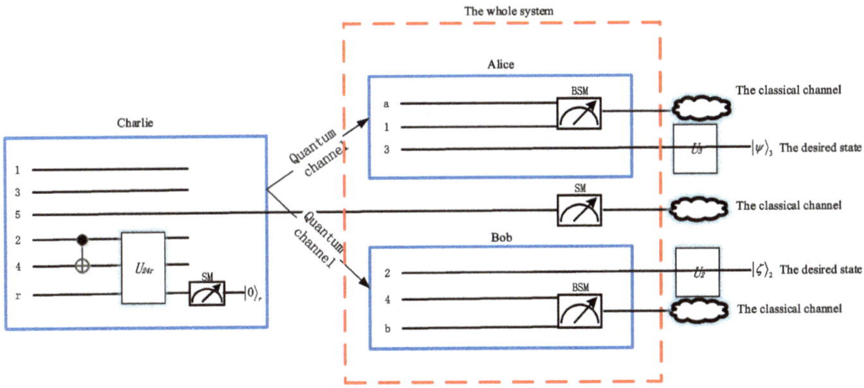

Fig. 4.1 Schematic representation of the scheme

As inferred from the description above, two Bell state measurements and a single-qubit measurement were used in the whole scheme. Therefore, there were 32 types of possible measurement results. For the other measurement results, the collapsed states of qubits 2 and 3 as well as the corresponding unitary transformations to reconstruct the original states are given in Table 4.1.

According to (4.8), if the measurement outcome was $|1\rangle_r$ instead of $|0\rangle_r$, the entangled state collapsed into the following state:

$$\begin{aligned}|\vartheta\rangle_{12345} = & \sqrt{x_0^2 - x_3^2}|00000\rangle_{12345} + \sqrt{x_1^2 - x_3^2}|00111\rangle_{12345} \\ & + \sqrt{x_2^2 - x_3^2}|11000\rangle_{12345} + x_3|11111\rangle_{12345}\end{aligned}. \quad (4.18)$$

State $|\vartheta\rangle_{12345}$ was not appropriate as the quantum channel for performing the quantum teleportation deterministically. However, (4.18) implies that the basis states of state $|\vartheta\rangle_{12345}$ were the same as those of $|\psi\rangle_{12345}$ shown in (4.4). Thus, similar to state $|\psi\rangle_{12345}$, state $|\vartheta\rangle_{12345}$ could be transformed into a new quantum state, which was suitable as the quantum channel for teleporting deterministically. Firstly, we introduced an auxiliary qubit k with the initial state $|0\rangle_r$. Then, a collective unitary transformation was performed on qubits 2, 4, and k under the basis $\{|000\rangle_{24r}, |010\rangle_{24r}, |100\rangle_{24r}, |110\rangle_{24r}, |001\rangle_{24r}, |011\rangle_{24r}, |101\rangle_{24r}, |111\rangle_{24r}\}$. The unitary transformation could be expressed as follows:

$$U'_{24r} = \begin{bmatrix} U'_1 & U'_1 \\ U'_2 & -U'_1 \end{bmatrix}. \quad (4.19)$$

Here, from (4.3), $|x_0| > |x_1| > |x_2| > |x_3|$; therefore, $\sqrt{x_0^2 - x_3^2} > \sqrt{x_1^2 - x_3^2} > \sqrt{x_2^2 - x_3^2}$. If $\sqrt{x_2^2 - x_3^2} > x_3$, then

Table 4.1 Possible measurement results and the corresponding unitary transformations

Alice's results	Bob's results	Charlie's results	Collapsed state of qubits 2 and 3	Unitary transformations (U_2U_3)
$\|\phi^+\rangle_{a1}$	$\|\phi^+\rangle_{b4}$	$\|+\rangle_5$	$a_0b_0\|00\rangle_{23} + a_0b_1\|01\rangle_{23} + a_1b_0\|10\rangle_{23} + a_1b_1\|11\rangle_{23}$	$I_2\ I_3$
$\|\phi^+\rangle_{a1}$	$\|\phi^+\rangle_{b4}$	$\|-\rangle_5$	$a_0b_0\|00\rangle_{23} - a_0b_1\|01\rangle_{23} + a_1b_0\|10\rangle_{23} - a_1b_1\|11\rangle_{23}$	$I_2\ \sigma_{3z}$
$\|\phi^+\rangle_{a1}$	$\|\phi^-\rangle_{b4}$	$\|+\rangle_5$	$a_0b_0\|00\rangle_{23} - a_0b_1\|01\rangle_{23} + a_1b_0\|10\rangle_{23} - a_1b_1\|11\rangle_{23}$	$I_2\ \sigma_{3z}$
$\|\phi^+\rangle_{a1}$	$\|\phi^-\rangle_{b4}$	$\|-\rangle_5$	$a_0b_0\|00\rangle_{23} + a_0b_1\|01\rangle_{23} + a_1b_0\|10\rangle_{23} + a_1b_1\|11\rangle_{23}$	$I_2\ I_3$
$\|\phi^-\rangle_{a1}$	$\|\phi^+\rangle_{b4}$	$\|+\rangle_5$	$a_0b_0\|00\rangle_{23} + a_0b_1\|01\rangle_{23} - a_1b_0\|10\rangle_{23} - a_1b_1\|11\rangle_{23}$	$\sigma_{2z}\ I_3$
$\|\phi^-\rangle_{a1}$	$\|\phi^+\rangle_{b4}$	$\|-\rangle_5$	$a_0b_0\|00\rangle_{23} - a_0b_1\|01\rangle_{23} - a_1b_0\|10\rangle_{23} + a_1b_1\|11\rangle_{23}$	$\sigma_{2z}\ \sigma_{3z}$
$\|\phi^-\rangle_{a1}$	$\|\phi^-\rangle_{b4}$	$\|+\rangle_5$	$a_0b_0\|00\rangle_{23} - a_0b_1\|01\rangle_{23} - a_1b_0\|10\rangle_{23} + a_1b_1\|11\rangle_{23}$	$\sigma_{2z}\ \sigma_{3z}$
$\|\phi^-\rangle_{a1}$	$\|\phi^-\rangle_{b4}$	$\|-\rangle_5$	$a_0b_0\|00\rangle_{23} + a_0b_1\|01\rangle_{23} - a_1b_0\|10\rangle_{23} - a_1b_1\|11\rangle_{23}$	$\sigma_{2z}\ I_3$
$\|\phi^+\rangle_{a1}$	$\|\varphi^+\rangle_{b4}$	$\|+\rangle_5$	$a_0b_0\|01\rangle_{23} + a_0b_1\|00\rangle_{23} + a_1b_0\|11\rangle_{23} + a_1b_1\|10\rangle_{23}$	$I_2\ \sigma_{3x}$
$\|\phi^+\rangle_{a1}$	$\|\varphi^+\rangle_{b4}$	$\|-\rangle_5$	$-a_0b_0\|01\rangle_{23} + a_0b_1\|00\rangle_{23} - a_1b_0\|11\rangle_{23} + a_1b_1\|10\rangle_{23}$	$I_2\ \sigma_{3x}\ \sigma_{3z}$
$\|\phi^+\rangle_{a1}$	$\|\varphi^-\rangle_{b4}$	$\|+\rangle_5$	$a_0b_0\|01\rangle_{23} - a_0b_1\|00\rangle_{23} + a_1b_0\|11\rangle_{23} - a_1b_1\|10\rangle_{23}$	$I_2\ \sigma_{3z}\ \sigma_{3x}$
$\|\phi^+\rangle_{a1}$	$\|\varphi^-\rangle_{b4}$	$\|-\rangle_5$	$a_0b_0\|01\rangle_{23} + a_0b_1\|00\rangle_{23} + a_1b_0\|11\rangle_{23} + a_1b_1\|10\rangle_{23}$	$I_2\ \sigma_{3x}$
$\|\phi^-\rangle_{a1}$	$\|\varphi^+\rangle_{b4}$	$\|+\rangle_5$	$a_0b_0\|01\rangle_{23} + a_0b_1\|00\rangle_{23} - a_1b_0\|11\rangle_{23} - a_1b_1\|10\rangle_{23}$	$\sigma_{2z}\ \sigma_{3x}$
$\|\phi^-\rangle_{a1}$	$\|\varphi^+\rangle_{b4}$	$\|-\rangle_5$	$-a_0b_0\|01\rangle_{23} + a_0b_1\|00\rangle_{23} + a_1b_0\|11\rangle_{23} - a_1b_1\|10\rangle_{23}$	$\sigma_{2z}\ \sigma_{3x}\ \sigma_{3z}$
$\|\phi^-\rangle_{a1}$	$\|\varphi^-\rangle_{b4}$	$\|+\rangle_5$	$a_0b_0\|01\rangle_{23} - a_0b_1\|00\rangle_{23} - a_1b_0\|11\rangle_{23} + a_1b_1\|10\rangle_{23}$	$\sigma_{2z}\ \sigma_{3z}\ \sigma_{3x}$
$\|\phi^-\rangle_{a1}$	$\|\varphi^-\rangle_{b4}$	$\|-\rangle_5$	$-a_0b_0\|01\rangle_{23} - a_0b_1\|00\rangle_{23} + a_1b_0\|11\rangle_{23} + a_1b_1\|10\rangle_{23}$	$\sigma_{2z}\ \sigma_{3x}$
$\|\varphi^+\rangle_{a1}$	$\|\phi^+\rangle_{b4}$	$\|+\rangle_5$	$a_0b_0\|10\rangle_{23} + a_0b_1\|11\rangle_{23} + a_1b_0\|00\rangle_{23} + a_1b_1\|01\rangle_{23}$	$\sigma_{2x}\ I_3$
$\|\varphi^+\rangle_{a1}$	$\|\phi^+\rangle_{b4}$	$\|-\rangle_5$	$a_0b_0\|10\rangle_{23} - a_0b_1\|11\rangle_{23} + a_1b_0\|00\rangle_{23} - a_1b_1\|01\rangle_{23}$	$\sigma_{2x}\ \sigma_{3z}$
$\|\varphi^+\rangle_{a1}$	$\|\phi^-\rangle_{b4}$	$\|+\rangle_5$	$a_0b_0\|10\rangle_{23} - a_0b_1\|11\rangle_{23} + a_1b_0\|00\rangle_{23} - a_1b_1\|01\rangle_{23}$	$\sigma_{2x}\ \sigma_{3z}$
$\|\varphi^+\rangle_{a1}$	$\|\phi^-\rangle_{b4}$	$\|-\rangle_5$	$a_0b_0\|10\rangle_{23} + a_0b_1\|11\rangle_{23} + a_1b_0\|00\rangle_{23} + a_1b_1\|01\rangle_{23}$	$\sigma_{2x}\ I_3$
$\|\varphi^-\rangle_{a1}$	$\|\phi^+\rangle_{b4}$	$\|+\rangle_5$	$a_0b_0\|10\rangle_{23} + a_0b_1\|11\rangle_{23} - a_1b_0\|00\rangle_{23} - a_1b_1\|01\rangle_{23}$	$\sigma_2^z\ \sigma_{2x}$
$\|\varphi^-\rangle_{a1}$	$\|\phi^+\rangle_{b4}$	$\|-\rangle_5$	$a_0b_0\|10\rangle_{23} - a_0b_1\|11\rangle_{23} - a_1b_0\|00\rangle_{23} + a_1b_1\|01\rangle_{23}$	$\sigma_{2z}\ \sigma_{2x}\ \sigma_{3z}$
$\|\varphi^-\rangle_{a1}$	$\|\phi^-\rangle_{b4}$	$\|+\rangle_5$	$a_0b_0\|10\rangle_{23} - a_0b_1\|11\rangle_{23} - a_1b_0\|00\rangle_{23} + a_1b_1\|01\rangle_{23}$	$\sigma_{2z}\ \sigma_{2x}\ \sigma_{3z}$
$\|\varphi^-\rangle_{a1}$	$\|\phi^-\rangle_{b4}$	$\|-\rangle_5$	$a_0b_0\|10\rangle_{23} + a_0b_1\|11\rangle_{23} - a_1b_0\|00\rangle_{23} - a_1b_1\|01\rangle_{23}$	$\sigma_2^z\ \sigma_{2x}$
$\|\varphi^+\rangle_{a1}$	$\|\varphi^+\rangle_{b4}$	$\|+\rangle_5$	$a_0b_0\|11\rangle_{23} + a_0b_1\|10\rangle_{23} + a_1b_0\|01\rangle_{23} + a_1b_1\|00\rangle_{23}$	$\sigma_{2x}\ \sigma_{3x}$
$\|\varphi^+\rangle_{a1}$	$\|\varphi^+\rangle_{b4}$	$\|-\rangle_5$	$-a_0b_0\|11\rangle_{23} + a_0b_1\|10\rangle_{23} - a_1b_0\|01\rangle_{23} + a_1b_1\|00\rangle_{23}$	$\sigma_{2x}\ \sigma_{3x}\ \sigma_{3z}$
$\|\varphi^+\rangle_{a1}$	$\|\varphi^-\rangle_{b4}$	$\|+\rangle_5$	$a_0b_0\|11\rangle_{23} - a_0b_1\|10\rangle_{23} + a_1b_0\|01\rangle_{23} - a_1b_1\|00\rangle_{23}$	$\sigma_{2x}\ \sigma_{3z}\ \sigma_{3x}$
$\|\varphi^+\rangle_{a1}$	$\|\varphi^-\rangle_{b4}$	$\|-\rangle_5$	$-a_0b_0\|11\rangle_{23} - a_0b_1\|10\rangle_{23} - a_1b_0\|01\rangle_{23} - a_1b_1\|00\rangle_{23}$	$\sigma_{2x}\ \sigma_{3x}$
$\|\varphi^-\rangle_{a1}$	$\|\varphi^+\rangle_{b4}$	$\|+\rangle_5$	$a_0b_0\|11\rangle_{23} + a_0b_1\|10\rangle_{23} - a_1b_0\|01\rangle_{23} - a_1b_1\|00\rangle_{23}$	$\sigma_2^z\ \sigma_{2x}\ \sigma_{3x}$
$\|\varphi^-\rangle_{a1}$	$\|\varphi^+\rangle_{b4}$	$\|-\rangle_5$	$-a_0b_0\|11\rangle_{23} + a_0b_1\|10\rangle_{23} + a_1b_0\|01\rangle_{23} - a_1b_1\|00\rangle_{23}$	$\sigma_2^z\ \sigma_{2x}\ \sigma_{3z}\ \sigma_{3x}$
$\|\varphi^-\rangle_{a1}$	$\|\varphi^-\rangle_{b4}$	$\|+\rangle_5$	$a_0b_0\|11\rangle_{23} - a_0b_1\|10\rangle_{23} - a_1b_0\|01\rangle_{23} + a_1b_1\|00\rangle_{23}$	$\sigma_2^z\ \sigma_{2x}\ \sigma_{3z}\ \sigma_{3x}$
$\|\varphi^-\rangle_{a1}$	$\|\varphi^-\rangle_{b4}$	$\|-\rangle_5$	$-a_0b_0\|11\rangle_{23} - a_0b_1\|10\rangle_{23} + a_1b_0\|01\rangle_{23} + a_1b_1\|00\rangle_{23}$	$\sigma_{2x}\ \sigma_2^z\ \sigma_{3x}$

$$
U_1' = \begin{bmatrix} \frac{x_3}{\sqrt{x_0^2 - x_3^2}} & 0 & 0 & 0 \\ 0 & \frac{x_3}{\sqrt{x_1^2 - x_3^2}} & 0 & 0 \\ 0 & 0 & \frac{x_3}{\sqrt{x_2^2 - x_3^2}} & 0 \\ 0 & 0 & 0 & 1 \end{bmatrix},
$$

$$
U_2' = \begin{bmatrix} \sqrt{1 - \frac{x_3^2}{x_0^2 - x_3^2}} & 0 & 0 & 0 \\ 0 & \sqrt{1 - \frac{x_3^2}{x_1^2 - x_3^2}} & 0 & 0 \\ 0 & 0 & \sqrt{1 - \frac{x_3^2}{x_2^2 - x_3^2}} & 0 \\ 0 & 0 & 0 & 1 \end{bmatrix}. \tag{4.20}
$$

Else

$$
U_1' = \begin{bmatrix} \frac{\sqrt{x_2^2 - x_3^2}}{\sqrt{x_0^2 - x_3^2}} & 0 & 0 & 0 \\ 0 & \frac{\sqrt{x_2^2 - x_3^2}}{\sqrt{x_1^2 - x_3^2}} & 0 & 0 \\ 0 & 0 & 1 & 0 \\ 0 & 0 & 0 & \frac{\sqrt{x_2^2 - x_3^2}}{x_3^2} \end{bmatrix},
$$

$$
U_2' = \begin{bmatrix} \sqrt{1 - \frac{x_2^2 - x_3^2}{x_0^2 - x_3^2}} & 0 & 0 & 0 \\ 0 & \sqrt{1 - \frac{x_2^2 - x_3^2}{x_1^2 - x_3^2}} & 0 & 0 \\ 0 & 0 & 1 & 0 \\ 0 & 0 & 0 & \sqrt{1 - \frac{x_2^2 - x_3^2}{x_3^2}} \end{bmatrix}. \tag{4.21}
$$

Compared with other similar schemes, the proposed scheme had the advantage that the non-maximally five-qubit state could be turned into a new quantum state, which was appropriate as the quantum channel to implement teleportation deterministically.

4.2.2 *Teleportation in Noisy Environment*

4.2.2.1 Calculation of Fidelity

In this subsection, we will discuss the influence of the quantum noise on the fidelity of the desired state. We assumed that the qubits to be distributed were influenced by the AD noise, because it is highly relevant for many practical qubit systems, such as spontaneous decay and superconducting qubits by zero-temperature energy relaxation [19]. AD could be expressed using the Kraus operators as follows:

$$E_0 = \begin{bmatrix} 1 & 0 \\ 0 & \sqrt{1-P} \end{bmatrix}, \quad E_1 = \begin{bmatrix} 1 & \sqrt{P} \\ 0 & 0 \end{bmatrix} \tag{4.22}$$

where $P(P \in [0, 1])$ is the decoherence rate of AD. Under the influence of the AD noise, the evolution of a single-qubit state could be described as follows:

$$\begin{aligned} &|0\rangle|0\rangle_E \rightarrow |0\rangle|0\rangle_E \\ &|1\rangle|0\rangle_E \rightarrow \sqrt{1-p}|1\rangle|0\rangle_E + \sqrt{p}|0\rangle|1\rangle_E \end{aligned}. \tag{4.23}$$

(4.23) shows that the E_1 operation changed the $|1\rangle$ state into the $|0\rangle$ state and E_0 left $|0\rangle$ unchanged but reduced the amplitude of the $|1\rangle$ state.

According to the description in Sect. 4.2.1, Charlie had to distribute qubits 1 and 3 to Alice and, at the same time, send qubits 2 and 4 to Bob. We assumed that during the distribution, all of these qubits underwent the AD noise. The pure entangled state in (4.7) evolved into a mixed state as follows:

$$\begin{aligned} |\psi^{AD}\rangle_{12345E} = &(|000000\rangle + (1-p)|001110\rangle + p|000011\rangle \\ &+(1-p)|110000\rangle + p|000001\rangle + \\ &(1-p)^2|111110\rangle + p^2|000011\rangle)_{12345E}. \end{aligned} \tag{4.24}$$

Thus, the quantum state of the whole system could be expressed as follows:

$$|sys^{AD}\rangle = |\theta\rangle_a \otimes |\theta\rangle_b \otimes |\psi^{AD}\rangle_{12345E}. \tag{4.25}$$

Next, Alice, Bob, and Charlie performed quantum measurements on qubits (a, 1), (b, 4), and 5, respectively. We assumed that the outcomes were $|\phi^+\rangle_{a1}$, $|\varphi^+\rangle_{b4}$ and |

$+\rangle_5$, respectively. The collapsed state of qubits 2, 3, and E could be expressed as follows:

$$\begin{aligned}\left|\psi^{AD}\right\rangle_{23E} &= \frac{1}{\sqrt{L_{23E}^{AD}}}\left(a_0b_0|000\rangle + a_0b_0\left(p^2+2p\right)|001\rangle\right.\\ &+a_0b_1(1-p)|010\rangle + a_1b_0(1-p)|100\rangle\\ &\left.+a_1b_1(1-p)^2|110\rangle\right)_{23E}\end{aligned} \tag{4.26}$$

where L_{23E}^{AD} is the normalization factor expressed as follows:

$$L_{23E}^{AD} = a_0^2b_0^2 + a_0^2b_0^2\left(p^2+2p\right)^2 + a_0^2b_1^2(1-p)^2 + a_1^2b_0^2(1-p)^2 + a_1^2b_1^2(1-p)^4. \tag{4.27}$$

Through a partial trace over the noise E in (4.27), the density matrix of $|\psi^{AD}\rangle_{23}$ could be obtained as follows:

$$\begin{aligned}\rho_{23}^{AD} &= \mathrm{Tr}_E\left(\left|\psi^{AD}\right\rangle_{23E}\left\langle\psi^{AD}\right|_{23E}\right)\\ &= \frac{1}{L_{23E}^{AD}}\begin{bmatrix} a_0^2b_0^2+a_0^2b_0^2(p^2+2p)^2 & a_0^2b_0b_1(1-p) & a_0a_1b_0^2(1-p) & a_0a_1b_0b_1(1-p)^2\\ a_0^2b_0b_1(1-p) & a_0^2b_1^2(1-p)^2 & a_0a_1b_0b_1(1-p)^2 & a_0a_1b_1^2(1-p)^3\\ a_0a_1b_0^2(1-p) & a_0a_1b_0b_1(1-p)^2 & a_1^2b_0^2(1-p)^2 & a_1^2b_0b_1(1-p)^3\\ a_0a_1b_0b_1(1-p)^2 & a_0a_1b_1^2(1-p)^3 & a_1^2b_0b_1(1-p)^3 & a_1^2b_1^2(1-p)^4 \end{bmatrix}.\end{aligned} \tag{4.28}$$

The effect of the quantum noise could be visualized by comparing the desired state with the final state in the noisy environment by using the fidelity, which is a distance-based measure corresponding to the closeness of two states. In this scheme, the desired state was $|\theta\rangle_{ab} = |\theta\rangle_a \otimes |\theta\rangle_b$, and the final state in the noisy environment was ρ_{23}^{AD}. Therefore, the fidelity could be expressed as follows:

$$\begin{aligned}F_{AD} &= {}_{ab}\langle\theta|\rho_{23}^{AD}|\theta\rangle_{ab}\\ &= a_0^4b_0^4 + a_0^4b_0^4\left(p^2+2p\right)^2 + a_0^4b_0^2b_1^2(1-p)\\ &+a_0^2a_1^2b_0^4(1-p) + a_0^2a_1^2b_0^2b_1^2(1-p)^2 + a_0^4b_0^2b_1^2(1-p)\\ &+a_0^4b_1^4(1-p)^2 + a_0^2a_1^2b_0^2b_1^2(1-p)^2 + a_0^2a_1^2b_1^4(1-p)^3\\ &+a_0^2a_1^2b_0^4(1-p) + a_0^2a_1^2b_0^2b_1^2(1-p)^2 + a_1^4b_0^4(1-p)^2\\ &+a_1^4b_0^2b_1^2(1-p)^3 + a_0^2a_1^2b_0^2b_1^2(1-p)^2 + a_0^2a_1^2b_1^4(1-p)^3\\ &+a_1^4b_0^2b_1^2(1-p)^3 + a_1^4b_1^4(1-p)^4\end{aligned}. \tag{4.29}$$

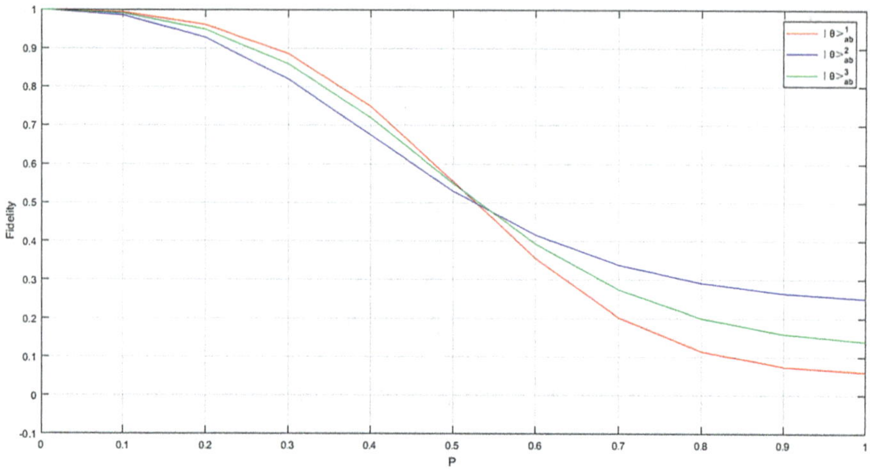

Fig. 4.2 Variation of the fidelity with the AD factor$p$

The fidelity F_{AD} varying with the AD factor p is presented in Fig. 4.2 with three different $|\theta\rangle_{ab}(|\theta\rangle_{ab}^{1}, |\theta\rangle_{ab}^{2}$, and $|\theta\rangle_{ab}^{3})$ expressed as follows:

$$\begin{aligned} |\theta\rangle_{ab}^{1} &= \left(\sqrt{0.2}|0\rangle + \sqrt{0.8}|0\rangle\right) \otimes \left(\sqrt{0.3}|0\rangle + \sqrt{0.7}|0\rangle\right) \\ |\theta\rangle_{ab}^{2} &= \left(\sqrt{0.5}|0\rangle + \sqrt{0.5}|0\rangle\right) \otimes \left(\sqrt{0.5}|0\rangle + \sqrt{0.5}|0\rangle\right). \\ |\theta\rangle_{ab}^{3} &= \left(\sqrt{0.7}|0\rangle + \sqrt{0.3}|0\rangle\right) \otimes \left(\sqrt{0.2}|0\rangle + \sqrt{0.8}|0\rangle\right) \end{aligned} \tag{4.30}$$

This figure shows that the fidelity decreased with an increase in$p$. That is, as the AD noise became more serious, the quality of the teleportation worsened.

4.2.2.2 Weak Measurement and Reversing Measurement

In this subsection, we will describe the scheme to improve the fidelity by making use of the weak measurement and the reversing measurement. Recently, it was demonstrated that the weak measurement and the reversing measurement can effectively suppress the decoherence raised by the AD noise [19]. The central ideal of the approach is that weak measurements are performed on the qubits before they are distributed, and reversing measurements are carried out after they are distributed. This entire process is illustrated in Fig. 4.3.

According to [20], the contribution of the weak measurement and the corresponding reversing measurement on a single qubit can be expressed as follows:

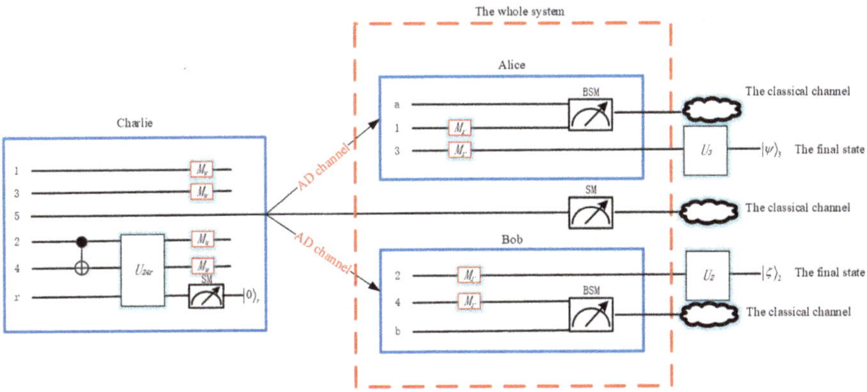

Fig. 4.3 Schematic representation of the scheme used in the case of the AD noise, wherein M_w denotes the weak measurement and M_r denotes the reversing measurement

$$\begin{array}{ll} |0\rangle|0\rangle_{E_w} \to |0\rangle|0\rangle_{E_w}, & |1\rangle|0\rangle_{E_w} \to \sqrt{1-p_W}|1\rangle|0\rangle_{E_w} \\ |0\rangle|0\rangle_{E_r} \to \sqrt{1-p_r}|0\rangle|0\rangle_{E_r}, & |1\rangle|0\rangle_{E_r} \to |1\rangle|0\rangle_{E_r} \end{array} \tag{4.31}$$

where E_w denotes the weak measurement environment, E_r denotes the reversing measurement environment, and p_w and p_r are the strengths of the weak measurement and the reversing measurement, respectively.

Charlie first performed weak measurements on qubits 2, 4, 1, and 3 before they were distributed; then, the quantum state $|\psi\rangle_{12345}$ in (4.7) was converted into the following state:

$$\begin{aligned} |\psi^W\rangle_{12345} &= |00000\rangle_{12345} + (1-p_w)|00111\rangle_{12345} \\ &+ (1-p_w)|11000\rangle_{12345} + (1-p_w)^2|11111\rangle_{12345} \end{aligned}. \tag{4.32}$$

With the help of the weak measurement, the quantum state was less vulnerable to decoherence. After the weak measurements, qubits 2, 4, 1, and 3 were distributed through quantum channels, which were subject to the AD noise. Then, the quantum state $|\psi^W\rangle_{12345}$ was changed into the following:

$$\begin{aligned} |\psi^{W-AD}\rangle_{12345E} &= (|000000\rangle + (1-p_w)(1-p)|001110\rangle \\ &+ (1-p_w)p|000011\rangle + (1-p_w)(1-p)|110000\rangle \\ &+ (1-p_w)p|000001\rangle + (1-p_w)^2(1-p)^2|111110\rangle \\ &+ (1-p_w)^2p^2|000011\rangle)_{12345E}. \end{aligned} \tag{4.33}$$

After receiving the qubits, Alice and Bob carried out the reversing measurements on the qubits. The quantum state $|\psi^{W-AD}\rangle_{12345E}$ evolved into the following state:

$$\left|\psi^{W-AD-R}\right\rangle_{12345E}=\begin{pmatrix} (1-p_r)^2|000000\rangle \\ +(1-p_r)(1-p_w)(1-p)|001110\rangle \\ +(1-p_r)^2(1-p_w)p|000011\rangle \\ +(1-p_r)(1-p_w)(1-p)|110000\rangle \\ +(1-p_r)^2(1-p_w)p|000001\rangle \\ +(1-p_w)^2(1-p)^2|111110\rangle \\ +(1-p_r)^2(1-p_w)^2p^2|000011\rangle \end{pmatrix}_{12345E}. \tag{4.34}$$

Thereafter, the quantum state of the whole system could be expressed as follows:

$$\left|sys^{W-AD-R}\right\rangle=|\theta\rangle_a\otimes|\theta\rangle_b\otimes\left|\psi^{W-AD-R}\right\rangle_{12345E}. \tag{4.35}$$

Next, Alice and Bob performed the Bell state measurements on qubits (a, 1) and (b, 4), and Charlie performed the single-qubit measurement on qubit 5. As the description in Sect. 4.2.2.1, we assumed that the measurement results were $|\phi^+\rangle_{a1}$, $|\varphi^+\rangle_{b4}$, and $|+\rangle_5$. The state of qubits 2, 3, and E collapsed into the following:

$$\left|\psi^{W-AD-R}\right\rangle_{23E}=\frac{1}{\sqrt{L_{23E}^{W-AD-R}}}\begin{pmatrix} a_0b_0(1-p_r)^2|000\rangle \\ +\begin{pmatrix} 2a_0b_0(1-p_r)^2(1-p_w)p \\ +a_0b_0(1-p_r)^2(1-p_w)^2p^2 \end{pmatrix}|001\rangle \\ +a_0b_1(1-p_r)(1-p_w)(1-p)|010\rangle \\ +a_1b_0(1-p_r)(1-p_w)(1-p)|100\rangle \\ +a_1b_1(1-p_w)^2(1-p)^2|110\rangle \end{pmatrix}_{23E} \tag{4.36}$$

where L_{23E}^{W-AD-R} is the normalization factor given as follows:

$$\begin{aligned} L_{23E}^{W-AD-R}=&a_0^2b_0^2(1-p_r)^4+\left(2a_0b_0(1-p_r)^2(1-p_w)p+a_0b_0(1-p_r)^2(1-p_w)^2p^2\right)^2 \\ &+a_0^2b_1^2(1-p_r)^2(1-p_w)^2(1-p)^2+a_1^2b_0^2(1-p_r)^2(1-p_w)^2(1-p)^2 \\ &+a_1^2b_1^2(1-p_w)^4(1-p)^4 \end{aligned}. \tag{4.37}$$

By implementing a partial trace over the noise E in (4.37), we could obtain the density matrix of $|\psi^{W-AD-R}\rangle_{23}$ as follows:

$$\rho_{23}^{W-AD-R} = \mathrm{Tr}_E\left(|\varphi^{W-AD-R}\rangle_{23E}\langle\varphi^{W-AD-R}|_{23E}\right) = \frac{1}{L_{23E}^{W-AD-R}}\begin{bmatrix} \mu_{00} & \mu_{01} & \mu_{02} & \mu_{03} \\ \mu_{10} & \mu_{11} & \mu_{12} & \mu_{13} \\ \mu_{20} & \mu_{21} & \mu_{22} & \mu_{23} \\ \mu_{30} & \mu_{31} & \mu_{32} & \mu_{33} \end{bmatrix} \tag{4.38}$$

where

$$\begin{aligned}
\mu_{00} &= a_0^2b_0^2(1-p_r)^4 + \left(2a_0b_0(1-p_r)^2(1-p_w)p + a_0b_0(1-p_r)^2(1-p_w)^2p^2\right)^2 \\
\mu_{01} &= a_0^2b_0b_1(1-p_r)^3(1-p_w)(1-p), \\
\mu_{02} &= a_0a_1b_0^2(1-p_r)^3(1-p_w)(1-p) \\
\mu_{03} &= a_0a_1b_0b_1(1-p_r)^2(1-p_w)^2(1-p)^2, \\
\mu_{10} &= a_0^2b_0b_1(1-p_r)^3(1-p_w)(1-p) \\
\mu_{11} &= a_0^2b_1^2(1-p_r)^2(1-p_w)^2(1-p)^2, \\
\mu_{12} &= a_0a_1b_0b_1(1-p_r)^2(1-p_w)^2(1-p)^2 \\
\mu_{13} &= a_0a_1b_1^2(1-p_r)(1-p_w)^3(1-p)^3, \\
\mu_{20} &= a_0a_1b_0^2(1-p_r)^3(1-p_w)(1-p) \\
\mu_{21} &= a_0a_1b_0b_1(1-p_r)^2(1-p_w)^2(1-p)^2, \\
\mu_{22} &= a_0^2b_1^2(1-p_r)^2(1-p_w)^2(1-p)^2 \\
\mu_{23} &= a_1^2b_0b_1(1-p_r)(1-p_w)^3(1-p)^3, \\
\mu_{30} &= a_0a_1b_0b_1(1-p_r)^2(1-p_w)^2(1-p)^2 \\
\mu_{31} &= a_0a_1b_1^2(1-p_r)(1-p_w)^3(1-p)^3, \\
\mu_{32} &= a_1^2b_0b_1(1-p_r)(1-p_w)^3(1-p)^3 \\
\mu_{33} &= a_1^2b_1^2(1-p_w)^4(1-p)^4
\end{aligned} \tag{4.39}$$

According to the density matrix ρ_{23}^{W-AD-R}, the fidelity of the desired state $|\theta\rangle_{ab}$ could be calculated as follows:

$$
\begin{aligned}
F_{W-AD-R} = & {}_{ab}\langle\theta|\rho_{23}^{W-AD-R}|\theta\rangle_{ab} \\
= & a_0^4 b_0^4 (1-p_r)^4 + a_0^2 b_0^2 \left(2a_0 b_0 (1-p_r)^2 (1-p_w) p + a_0 b_0 (1-p_r)^2 (1-p_w)^2 p^2\right)^2 \\
& + a_0^4 b_0^2 b_1^2 (1-p_r)^3 (1-p_w)(1-p) + a_0^2 a_1^2 b_0^4 (1-p_r)^3 (1-p_w)(1-p) \\
& + a_0^2 a_1^2 b_0^2 b_1^2 (1-p_r)^2 (1-p_w)^2 (1-p)^2 + a_0^4 b_0^2 b_1^2 (1-p_r)^3 (1-p_w)(1-p) \\
& + a_0^4 b_1^4 (1-p_r)^2 (1-p_w)^2 (1-p)^2 + a_0^2 a_1^2 b_0^2 b_1^2 (1-p_r)^2 (1-p_w)^2 (1-p)^2 \\
& + a_0^2 a_1^2 b_1^4 (1-p_r)(1-p_w)^3 (1-p)^3 + a_0^2 a_1^2 b_0^4 (1-p_r)^3 (1-p_w)(1-p) \\
& + a_0^2 a_1^2 b_0^2 b_1^2 (1-p_r)^2 (1-p_w)^2 (1-p)^2 + a_0^4 b_1^4 (1-p_r)^2 (1-p_w)^2 (1-p)^2 \\
& + a_1^4 b_0^2 b_1^2 (1-p_r)(1-p_w)^3 (1-p)^3 + a_0^2 a_1^2 b_0^2 b_1^2 (1-p_r)^2 (1-p_w)^2 (1-p)^2 \\
& + a_0^2 a_1^2 b_1^4 (1-p_r)(1-p_w)^3 (1-p)^3 + a_1^4 b_0^2 b_1^2 (1-p_r)(1-p_w)^3 (1-p)^3 \\
& + a_1^4 b_1^4 (1-p_w)^4 (1-p)^4.
\end{aligned}
\tag{4.40}
$$

With the same three desired states as those given in (4.30), the comparison between F_{AD} and F_{W-AD-R} is shown in Fig. 4.4. Here, we assumed that $p_w = 0.3$ and $p_r = p_w + p(1-p_w)$. This figure shows that both the fidelities F_{AD} and F_{W-AD-R} decrease with an increase in p. For each desired state ($|\theta\rangle_{ab}^1$, $|\theta\rangle_{ab}^2$, and $|\theta\rangle_{ab}^3$) and a certain p, F_{W-AD-R} was always bigger than F_{AD}, which demonstrated that the weak measurement was a useful approach for increasing the fidelity of the desired state.

The fidelities F_{AD} and F_{W-AD-R} varying with p and p_w are shown in Fig. 4.5. Here, we assumed that $|\theta\rangle_{ab} = \left(\sqrt{0.2}|0\rangle + \sqrt{0.8}|0\rangle\right) \otimes \left(\sqrt{0.2}|0\rangle + \sqrt{0.8}|0\rangle\right)$. This figure shows that for a certain p, F_{W-AD-R} was always bigger than F_{AD} and that F_{W-AD-R} increased with an increase in the weak measurement strength p_w.

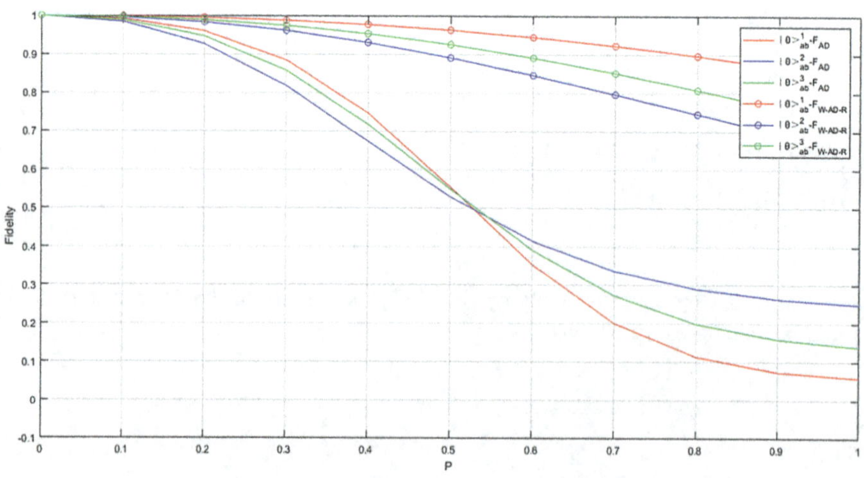

Fig. 4.4 Comparison of the fidelity between with and without the weak measurement

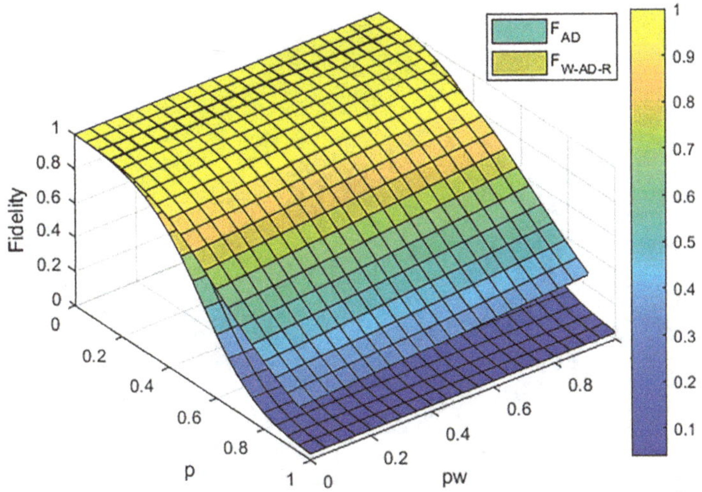

Fig. 4.5 Three-dimensional comparison of the fidelity between with and without the weak measurement

Thus, we could increase the fidelity by using a larger p_w. However, this was not always the best choice. According to [19], for the two-qubit state, the relation between the success probability, the weak measurement strength p_w, and the AD factor p can be expressed as follows:

$$P_s = \frac{1}{4}(1-p)^2(1-p_w)^2[2+p(1-p_w)]^2. \tag{4.41}$$

Based on (4.41), the success probability p_s varying with p_w and p is given in Fig. 4.6. This figure shows that p_s decreased with an increase in p_w and p. That is, although a higher fidelity could be acquired using a largerp_w, we obtained a smaller success probability.

4.2.3 Comparison

In this subsection, the proposed scheme is compared with some of the existing controlled bidirectional teleportation schemes in terms of seven aspects, namely, the number of qubits transmitted (NQT), the quantum resource consumption (QRC), the classical resource consumption (CRC), the quantum channel (QC), the type of the quantum teleportation (TQT), the teleportation environment (TE), and the teleportation efficiency (η), as shown in Table 4.2. Here, the teleportation efficiency was calculated in the following manner, as proposed by Banerjee and Pathak [21]:

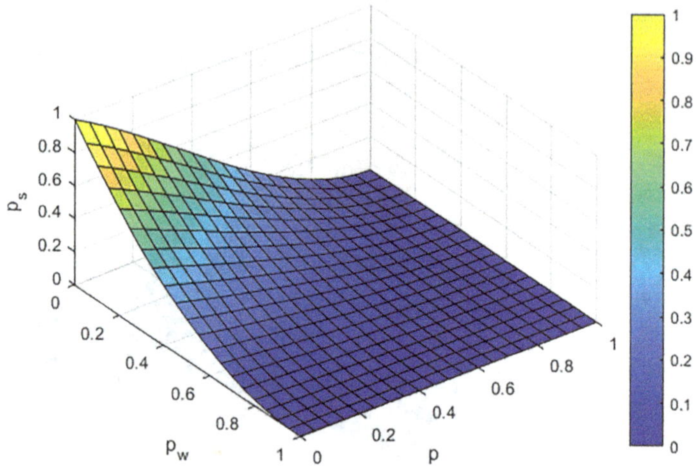

Fig. 4.6 Success probability p_s varying with the weak measurement strength p_w and the AD factor p

Table 4.2 Comparison with other schemes

Scheme	QC	NQT	TQT	TE	QRC	CRC	η
Ref. [22]	Five-qubit cluster state	Two arbitrary single qubits	Deterministic	Ideal	5	5	2/10
Ref. [16]	Two partially double-qubit entangled states	Two arbitrary single qubits	Probabilistic	Ideal	5	4	2/9
Ref. [3]	Six-qubit cluster state	Two arbitrary single qubits	Deterministic	Ideal	6	6	2/12
Ref. [12]	Two non-maximally W-type states	One two-qubit state	Probabilistic	Ideal	7	6	2/13
Ref. [23]	Nine-qubit entangled state	Three-qubit state and a single-qubit state	Deterministic	Noisy	9	9	2/9
Ours	Non-maximally five-qubit cluster state	Two arbitrary single qubits	Deterministic	Noisy	6	5	2/11

$$\eta = \frac{q_s}{q_u + b_t} \tag{4.42}$$

where q_s is the number of qubits to be transmitted, $q_u$denotes the quantum resource consumption, and b_t means the classical resource consumption.

From Table 4.2, the following conclusions were drawn:

(a) Although the teleportation efficiency of the scheme had no advantage over the other schemes, the proposed scheme used a non-maximally entangled state as the quantum channel, which was more in line with the practical environment.
(b) One remarkable advantage of the scheme was that it could improve a probabilistic teleportation to a deterministic teleportation by performing an appropriate unitary transformation on the non-maximally entangled state in advance with the help of an auxiliary qubit.
(c) Another remarkable advantage of the scheme was that we analyzed the influence of the AD noise on the scheme and used the weak measurement and the reversing measurement to eliminate the influence.

Therefore, the scheme could lay the foundation for practical applications in the future.

4.2.4 Summary

In this study, we proposed a deterministic and controlled bidirectional quantum teleportation for an arbitrary single qubit. With Charlie's permission, Alice and Bob could teleport a single-qubit state to each other deterministically. The scheme uses a five-qubit non-maximally entangled state as the quantum channel and has higher practicability because the maximally entangled state is usually difficult to prepare and maintain in practice. In order to change a five-qubit non-maximally entangled state into an appropriate quantum channel for teleporting deterministically, in this scheme, a specific unitary transformation is performed on the non-maximally entangled state in advance with the help of an auxiliary qubit. In addition, unlike the existing schemes that usually discuss only the quantum teleportation in an ideal environment, we investigated the scheme in an environment with AD noise. We analyzed the influence of the AD decoherence on the desired state by calculating the fidelity, which decreased with an increase in the AD factor. In order to battle against the influence, we performed weak measurements and reversing measurements on the qubits before and after their distribution. The results demonstrated that the weak measurement and the reversing measurement could, indeed, be useful approaches for battling against the AD decoherence.

Clearly, many extensions about the work can be considered in future research. For instance, only the AD noise is considered in the work; however, some other quantum noises (phase-damping noise, bit-flip noise, and phase-flip noise) also exist in the surrounding environment. Second, in this work, we showed that a higher weak measurement strength could lead to higher fidelity but a smaller success probability. The solution for the contradiction will be considered in a future work.

4.3 C-BQT Scheme of a Three-Qubit State by Using GHZ States

In this section, we proposed a scheme for implementing C-BQT of a three-qubit state by using GHZ states as quantum channel, and the preparation method of quantum channel is given. The C-BQT task is completed in the following procedure: a three-qubit entangled state is transformed into a two-qubit entangled state by senders that use the Toffoli gate and then a two-qubit entangled state is transferred into a single-qubit superposition state by Bob that uses the controlled-NOT operation. By using Bell state measurements and single-qubit measurement in process, receivers can introduce the appropriate unitary transformation and auxiliary particles to reconstruct the initial state. Finally, a scheme of C-BQT of three-qubit via the control of the supervisor Charlie.

4.3.1 *The Construction of the Quantum Channel*

Implementing this C-BQT scheme by utilizing a six-qubit entangled state generated by two GHZ entangled states as quantum channel [24] is described as (4.43)

$$\begin{aligned} |G\rangle_{a_1b_1b_2a_2c_1b_3} &= \frac{1}{\sqrt{2}}(|000\rangle + |111\rangle)_{a_1b_1b_2} \otimes \frac{1}{\sqrt{2}}(|000\rangle + |111\rangle)_{a_2c_1b_3} \\ &= \frac{1}{2}(|000000\rangle + |000111\rangle + |111000\rangle + |111111\rangle)_{a_1b_1b_2a_2c_1b_3} \end{aligned} \tag{4.43}$$

where qubits a_1, a_2 belong to Alice; qubits b_1, b_2, and b_3 belong to Bob; and qubit c_1 belong to Charlie. Alice, Bob, and Charlie share this six-qubit entangled state.

The process of construction of the quantum channel is given. The proposed quantum channel (a six-qubit entangled state) is not only theoretically existent but also practically feasible. Quantum circuit of quantum channel can be created by using two H gates and four CNOT gates.

The six-qubit product state $|G_0\rangle$ as an input state, where

$$|G_0\rangle = |0\rangle_{a_1} \otimes |0\rangle_{b_1} \otimes |0\rangle_{b_2} \otimes |0\rangle_{a_2} \otimes |0\rangle_{c_1} \otimes |0\rangle_{b_3} \tag{4.44}$$

The construction of the quantum channel is as follows:

1. Executing H operation on qubits a_2 and b_2. $|G_0\rangle$ is transformed into $|G_1\rangle$.

$$|G_1\rangle = |0\rangle_{a_1} \otimes |0\rangle_{b_1} \otimes \left(\frac{|0\rangle + |1\rangle}{\sqrt{2}}\right)_{b_2} \otimes \left(\frac{|0\rangle + |1\rangle}{\sqrt{2}}\right)_{a_2} \otimes |0\rangle_{c_1} \otimes |0\rangle_{b_3}. \tag{4.45}$$

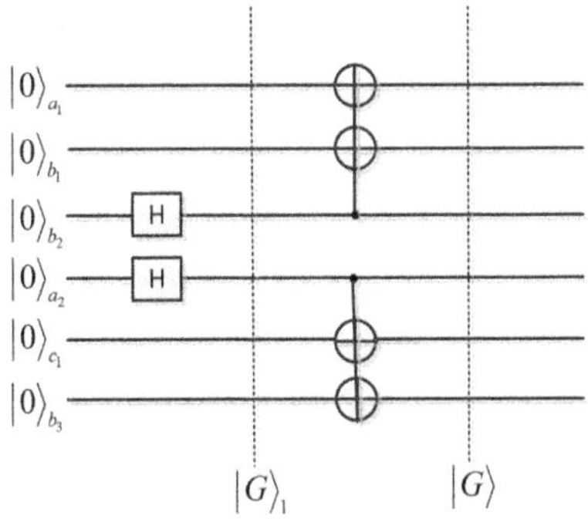

Fig. 4.7 The construction of quantum circuit for generating quantum channel

2. Performing CNOT operations on qubit pairs (b_2, a_1), (b_2, b_1), (a_2, c_1), and (a_2, b_3), respectively, where the first qubit serves as controlled qubit and the second qubit as target qubit. After that, the quantum channel can be constructed as (4.46).

$$|G\rangle_{a_1b_1b_2a_2c_1b_3} = \frac{1}{2}(|000000\rangle + |000111\rangle + |111000\rangle + |111111\rangle)_{a_1b_1b_2a_2c_1b_3}. \tag{4.46}$$

The construction circuit of the quantum channel is shown in Fig. 4.7. Although the generation and maintenance of the multi-qubit entangled state is a difficult task, some experiments [25, 26] have been carried out to study the multi-qubit entangled state. Therefore, we believe that advanced technology can make this task no longer difficult in the future.

4.3.2 C-BQT Scheme Process

This scheme can be described as follows. Alice transmits an unknown three-qubit entangled state of $A_1A_2A_3$ to Bob, and Bob transmits an unknown three-qubit entangled state of $B_1B_2B_3$ to Alice via the control of the supervisor Charlie. The three-qubit entangled state was described as (4.47) severally

$$\begin{aligned} |\Phi\rangle_{A_1A_2A_3} &= \alpha_0|000\rangle + \alpha_1|111\rangle \\ |\Phi\rangle_{B_1B_2B_3} &= \beta_0|000\rangle + \beta_1|111\rangle \end{aligned} \tag{4.47}$$

where $|\alpha_0|^2 + |\alpha_1|^2 = 1$ and $|\beta_0|^2 + |\beta_1|^2 = 1$.

The initial state of the whole system can be expressed as (4.48)

$$|\Psi\rangle_{a_1b_1b_2a_2c_1b_3A_1A_2B_1B_2} = |G\rangle_{a_1b_1b_2a_2c_1b_3} \otimes |\Phi\rangle_{A_1A_2A_3} \otimes |\Phi\rangle_{B_1B_2B_3}. \tag{4.48}$$

The width of the channel is determined; if we want to realize C-BQT, we firstly need to transform a three-qubit entangled state into a two-qubit entangled state. Here, Alice and Bob transform their three-qubit entangled state into the tensor product of a two-qubit entangled state and 0-qubit state by using the Toffoli gate, as follows:

$$|\Phi\rangle_{A_1A_2A_3} = \alpha_0|000\rangle + \alpha_1|111\rangle \xrightarrow{\text{Toffoli Gate}} (\alpha_0|00\rangle + \alpha_1|11\rangle)_{A_1A_2} \otimes |0\rangle_{A_3} = |\Phi\rangle_{A_1A_2} \otimes |0\rangle_{A_3}$$
$$|\Phi\rangle_{B_1B_2B_3} = \beta_0|000\rangle + \beta_1|111\rangle \xrightarrow{\text{Toffoli Gate}} (\beta_0|00\rangle + \beta_1|11\rangle)_{B_1B_2} \otimes |0\rangle_{B_3} = |\Phi\rangle_{B_1B_2} \otimes |0\rangle_{B_3}$$

Then Bob transforms his two-qubit entangled state into the tensor product of a single-qubit superposition state and 0-qubit state by using CNOT operation, as follows:

$$|\Phi\rangle_{B_1B_2} = (\beta_0|00\rangle + \beta_1|11\rangle)_{B_1B_2} \xrightarrow{\text{CNOT operation}} (\beta_0|0\rangle + \beta_1|1\rangle)_{B_1} \otimes |0\rangle_{B_2} = |\Phi\rangle_{B_1} \otimes |0\rangle_{B_2}$$

The circuit of C-BQT is accomplished as illustrated in Fig. 4.8.

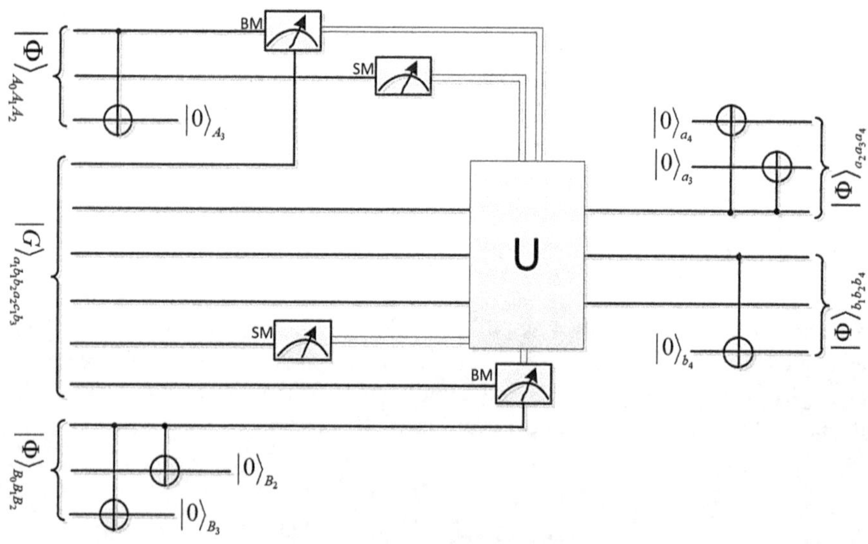

Fig. 4.8 The schematic diagram for C-BQT. BM corresponds to the Bell state measurement, and SM corresponds to the single measurement. The line "–" carries classical bits, which corresponds to classical channel of transmitting the measurement outcomes from Alice, Bob, and Charlie, which enables Alice and Bob to apply unitary operation U. The line "==" carries quantum bits. The circuit of the right is used to accomplish task of teleporting a two-qubit entangled state A_1A_2 and a single-qubit superposition state B_1 and recovering the initial three-qubit state. Finally, Alice receives $|\Phi\rangle_{a_2a_3a_4}$, and Bob receives $|\Phi\rangle_{b_1b_2a_4}$

The actual teleportation becomes asymmetric C-BQT by using a six-qubit entangled state as quantum channel, so the state of the whole system can be expressed as (4.49)

$$\begin{aligned}|\Psi\rangle_{a_1b_1b_2a_2a_3b_3A_1B_1} &= |G\rangle_{a_1b_1b_2a_2a_3b_3} \otimes |\Phi\rangle_{A_1A_2} \otimes |\Phi\rangle_{B_1} \\ &= \frac{1}{2}(|000000\rangle + |000111\rangle + |111000\rangle + |111111\rangle)_{a_1b_1b_2a_2a_3b_3} \\ &\quad \otimes(\alpha_0|00\rangle + \alpha_1|11\rangle)_{A_1A_2} \otimes (\beta_0|0\rangle + \beta_1|1\rangle)_{B_1}\end{aligned} \tag{4.49}$$

where $|\Phi^{\pm}\rangle = \frac{1}{\sqrt{2}}(|00\rangle \pm |11\rangle)$ and $|\Psi^{\pm}\rangle = \frac{1}{\sqrt{2}}(|01\rangle \pm |10\rangle)$ are Bell states.

Secondly, to achieve the purpose of C-BQT task, Alice and Bob perform a Bell state measurement on qubit pairs (A_1, a_1) and (B_1, b_3), respectively. Alice and Bob may obtain one of the 16 kinds of possible measured results with equal probability, and the remaining qubits may collapse into one of the 16 states after their measurements. The outcome of the measurement performed by Alice and Bob and the corresponding collapsed state of qubits A_2, b_1, b_2, a_2, c_1 are shown in Table 4.3.

Then Alice takes the X-basis measurement of $\{|\pm\rangle = (|0\rangle \pm |1\rangle)/\sqrt{2}\}$ on qubit A_2. Alice (Bob) tells the result to Bob (Alice) and Charlie via classical channel. Whether it is possible for Alice and Bob to reconstruct the single-qubit superposition state and the two-qubit entangled state, respectively, with local operations depends on the controller Charlie. If Charlie allows Bob and Alice to reconstruct the

Table 4.3 The measurement results of users and the corresponding collapsed state

Alice's results	Bob's results	The collapsed state of qubits A_2, b_1, b_2, a_2, and c_1
$\vert\Phi^+\rangle_{A_1a_1}$	$\vert\Phi^+\rangle_{B_1b_3}$	$\alpha_0\beta_0\vert00000\rangle + \alpha_0\beta_1\vert00011\rangle + \alpha_1\beta_0\vert11100\rangle + \alpha_1\beta_1\vert11111\rangle$
$\vert\Phi^+\rangle_{A_1a_1}$	$\vert\Phi^-\rangle_{B_1b_3}$	$\alpha_0\beta_0\vert00000\rangle - \alpha_0\beta_1\vert00011\rangle + \alpha_1\beta_0\vert11100\rangle - \alpha_1\beta_1\vert11111\rangle$
$\vert\Phi^-\rangle_{A_1a_1}$	$\vert\Phi^+\rangle_{B_1b_3}$	$\alpha_0\beta_0\vert00000\rangle + \alpha_0\beta_1\vert00011\rangle - \alpha_1\beta_0\vert11100\rangle - \alpha_1\beta_1\vert11111\rangle$
$\vert\Phi^-\rangle_{A_1a_1}$	$\vert\Phi^-\rangle_{B_1b_3}$	$\alpha_0\beta_0\vert00000\rangle - \alpha_0\beta_1\vert00011\rangle - \alpha_1\beta_0\vert11100\rangle + \alpha_1\beta_1\vert11111\rangle$
$\vert\Phi^+\rangle_{A_1a_1}$	$\vert\psi^+\rangle_{B_1b_3}$	$\alpha_0\beta_0\vert00011\rangle + \alpha_0\beta_1\vert00000\rangle + \alpha_1\beta_0\vert11111\rangle + \alpha_1\beta_1\vert11100\rangle$
$\vert\Phi^+\rangle_{A_1a_1}$	$\vert\psi^-\rangle_{B_1b_3}$	$\alpha_0\beta_0\vert00011\rangle - \alpha_0\beta_1\vert00000\rangle + \alpha_1\beta_0\vert11111\rangle - \alpha_1\beta_1\vert11100\rangle$
$\vert\Phi^-\rangle_{A_1a_1}$	$\vert\psi^+\rangle_{B_1b_3}$	$\alpha_0\beta_0\vert00011\rangle + \alpha_0\beta_1\vert00000\rangle - \alpha_1\beta_0\vert11111\rangle - \alpha_1\beta_1\vert11100\rangle$
$\vert\Phi^-\rangle_{A_1a_1}$	$\vert\psi^-\rangle_{B_1b_3}$	$\alpha_0\beta_0\vert00011\rangle - \alpha_0\beta_1\vert00000\rangle - \alpha_1\beta_0\vert11111\rangle + \alpha_1\beta_1\vert11100\rangle$
$\vert\Psi^+\rangle_{A_1a_1}$	$\vert\Phi^+\rangle_{B_1b_3}$	$\alpha_0\beta_0\vert01100\rangle + \alpha_0\beta_1\vert01111\rangle + \alpha_1\beta_0\vert10000\rangle + \alpha_1\beta_1\vert10011\rangle$
$\vert\Psi^+\rangle_{A_1a_1}$	$\vert\Phi^-\rangle_{B_1b_3}$	$\alpha_0\beta_0\vert01100\rangle - \alpha_0\beta_1\vert01111\rangle + \alpha_1\beta_0\vert10000\rangle - \alpha_1\beta_1\vert10011\rangle$
$\vert\Psi^-\rangle_{A_1a_1}$	$\vert\Phi^+\rangle_{B_1b_3}$	$\alpha_0\beta_0\vert01100\rangle + \alpha_0\beta_1\vert01111\rangle - \alpha_1\beta_0\vert10000\rangle - \alpha_1\beta_1\vert10011\rangle$
$\vert\Psi^-\rangle_{A_1a_1}$	$\vert\Phi^-\rangle_{B_1b_3}$	$\alpha_0\beta_0\vert01100\rangle - \alpha_0\beta_1\vert01111\rangle - \alpha_1\beta_0\vert10000\rangle + \alpha_1\beta_1\vert10011\rangle$
$\vert\Psi^+\rangle_{A_1a_1}$	$\vert\psi^+\rangle_{B_1b_3}$	$\alpha_0\beta_0\vert01111\rangle + \alpha_0\beta_1\vert01100\rangle + \alpha_1\beta_0\vert10011\rangle + \alpha_1\beta_1\vert10000\rangle$
$\vert\Psi^+\rangle_{A_1a_1}$	$\vert\psi^-\rangle_{B_1b_3}$	$\alpha_0\beta_0\vert01111\rangle - \alpha_0\beta_1\vert01100\rangle + \alpha_1\beta_0\vert10011\rangle - \alpha_1\beta_1\vert10000\rangle$
$\vert\Psi^-\rangle_{A_1a_1}$	$\vert\psi^+\rangle_{B_1b_3}$	$\alpha_0\beta_0\vert01111\rangle + \alpha_0\beta_1\vert01100\rangle - \alpha_1\beta_0\vert10011\rangle - \alpha_1\beta_1\vert10000\rangle$
$\vert\Psi^-\rangle_{A_1a_1}$	$\vert\psi^-\rangle_{B_1b_3}$	$\alpha_0\beta_0\vert01111\rangle - \alpha_0\beta_1\vert01100\rangle - \alpha_1\beta_0\vert10011\rangle + \alpha_1\beta_1\vert10000\rangle$

unknown state, he needs to carry out the X-basis measurement on qubit c_1 and tells receivers his result via classical channel. By combining information from Alice, Bob, and the controller Charlie, Alice and Bob can reconstruct the transferred single-qubit superposition state and the transferred two-qubit entangled state, respectively, with an appropriate unitary transformation.

The last and important steps are that Alice performs CNOT operation to reconstruct the two-qubit entangled state at first, where the qubit a_2 serves as controlled particle and ancillary qubit $|0\rangle_{a_3}$ as target particle. And then Alice performs the Toffoli gate operation to reconstruct the three-qubit entangled state, where the qubits a_2a_3 serve as controlled particles and the ancillary qubit $|0\rangle_{a_4}$ serves as target particle. Bob performs the Toffoli gate operation to reconstruct the three-qubit entangled state, where the qubits b_1b_2 serve as controlled particles and the ancillary qubit $|0\rangle_{b_4}$ serves as target particle. Finally, Alice and Bob successfully recover the initial three-qubit state. This C-BQT scheme of a three-qubit state by using GHZ states as channel is completed.

Now, we take an example to demonstrate the principle of this C-BQT scheme of a three-qubit state by using GHZ states via the control of the supervisor Charlie. Suppose Alice's Bell state measurement result is $|\Phi^+\rangle_{A_1a_1}$ and Bob's Bell state measurement result is $|\Phi^+\rangle_{B_1b_3}$; so the corresponding collapse result of qubits A_2, c_1, b_1, b_2, and a_2 are given as

$$|\Omega\rangle_{A_2c_1b_1b_2a_2}=(\alpha_0\beta_0|00000\rangle+\alpha_0\beta_1|01001\rangle+\alpha_1\beta_0|10110\rangle+\alpha_1\beta_1|11111\rangle)_{A_2c_1b_1b_2a_2}. \tag{4.50}$$

Then Alice takes X-basis measurement on qubit A_2; if Alice's measurement result is $|+\rangle_{A_2}$, the remaining qubits collapse into the state (4.51)

$$\begin{aligned}|\Omega\rangle_{c_1b_1b_2a_2} &= (\alpha_0\beta_0|0000\rangle+\alpha_0\beta_1|1001\rangle+\alpha_1\beta_0|0110\rangle+\alpha_1\beta_1|1111\rangle)_{c_1b_1b_2a_2}\\ &=\frac{1}{2}\Big[|+\rangle_{c_1}(\alpha_0\beta_0|000\rangle+\alpha_0\beta_1|001\rangle+\alpha_1\beta_0|110\rangle+\alpha_1\beta_1|111\rangle)_{b_1b_2a_2}+\\ &\quad|-\rangle_{c_1}\Big(\alpha_0\beta_0|000\rangle-\alpha_0\beta_1|001\rangle+\alpha_1\beta_0|110\rangle-\alpha_1\beta_1|111\rangle)_{b_1b_2a_2}\Big]\\ &=\frac{1}{2}\Big[|+\rangle_{c_1}(\alpha_0|00\rangle+\alpha_1|11\rangle)_{b_1b_2}(\beta_0|0\rangle+\beta_1|1\rangle)_{a_2}+\\ &\quad|-\rangle_{c_1}(\alpha_0|00\rangle+\alpha_1|11\rangle)_{b_1b_2}(\beta_0|0\rangle-\beta_1|1\rangle)_{a_2}\Big]\end{aligned}. \tag{4.51}$$

If Charlie allows Bob and Alice to reconstruct the unknown state, he needs to make X-basis measurement on qubit c_1, and then he sends the results of his measurement to Bob and Alice. Charlie's measurement result, Bob's operation, and Alice's operation are described in Table 4.4. So Alice and Bob can reconstruct the transferred single-qubit superposition state and the two-qubit entangled state, respectively, by applying appropriate unitary operation.

Table 4.4 Relation between the measurement results and appropriate unitary operations

Charlie's measurement result	Bob's operation	Alice's operation
$\|+\rangle_{c_1}$	$I_{b_1} \otimes I_{b_2}$	I_{a_2}
$\|-\rangle_{c_1}$	$\sigma^z_{b_1} \otimes \sigma^z_{b_2}$	I_{a_2}

Table 4.5 Comparison between the proposed quantum teleportation protocol and previous works

Scheme	Type of protocol	Quantum channel	Teleported quantum states	Efficiency (η%)
Protocol [3]	C-BQT	$Cluster_6$	Single-qubit state	33.33
Protocol [35]	QT	$Cluster_6$	Three-qubit state	50
Protocol [25]	C-BQT	$Cluster_6$	A single-qubit state and a two-qubit state	50
Protocol [36]	QT	$Cluster_6$	Four-qubit state	66.66
Protocol [24]	BQT	GHZ_6	Pure EPR state	66.66
Proposed scheme	C-BQT	GHZ_6	Three-qubit state	66.66

Lastly, Alice firstly performs CNOT operation to reconstruct the two-qubit entangled state $(\beta_0|00\rangle + \beta_1|11\rangle)_{a_2a_3}$, where the qubit a_2 serves as controlled particle and ancillary qubit $|0\rangle_{a_3}$ as target particle. And then Alice performs the Toffoli gate operation to reconstruct the three-qubit entangled state $(\beta_0|000\rangle + \beta_1|111\rangle)_{a_2a_3a_4}$, where the qubits a_2a_3 serve as controlled particles and the ancillary qubit $|0\rangle_{a_4}$ serves as target particle. Bob performs the Toffoli gate operation to reconstruct the three-qubit entangled state $(\alpha_0|000\rangle + \alpha_1|111\rangle)_{b_1b_2b_4}$, where the qubits b_1b_2 serve as controlled particles and the ancillary qubit $|0\rangle_{b_4}$ serves as target particle [27]. Finally, Alice and Bob successfully recover the initial three-qubit state. Thus, C-BQT scheme of a three-qubit state by using GHZ states as channel is completed.

4.3.3 Comparison

One of the most important advantages of this scheme is efficiency; efficiency [28–34] is also the main factor that can be utilized for comparing the previous protocols. In quantum communication protocols, it is defined as (4.52)

$$\eta = \frac{c}{q} \tag{4.52}$$

where c indicates the number of quantum states that the senders teleport in the quantum channel and q implies to the number of total qubits. The details of comparison between the proposed protocol and previous works are given in Table 4.5.

4.3.4 Summary

In Sect. 4.3, we propose a scheme of controlled bidirectional quantum teleportation of a three-qubit state by using GHZ states via the control of the supervisor Charlie. Alice transmits an unknown three-qubit entangled state to Bob, and Bob transmits an unknown three-qubit entangled state to Alice via the control of the supervisor Charlie.

The advantage of the proposed scheme is that the preparation method of quantum channel is also given in Sect. 4.3.2. The proposed protocol is that two users can simultaneously teleport pure three qubits to each other via the control of the supervisor Charlie. We can see from Sect. 4.3.3 that the efficiency also has obvious advantages over other schemes. Compared with previous schemes by using a six-qubit entangled state as quantum channel, it realizes bidirectional, controlled, and more than two particles' teleportation.

This scheme is completed by using the Toffoli gate, CNOT operation, and three auxiliary qubits. So, it achieves high efficiency through quantum transformation but also slightly increases the waste of resources. We hope the proposed C-BQT scheme can be achieved experimentally in the future.

4.4 C-BQT Scheme of a 3-Qubit State by a New Entangled 11-Qubit State

In this section, an improved asymmetric bidirectional controlled quantum teleportation (C-ABQT) scheme is investigated, in which an entangled 11-qubit state acts as the quantum channel, and two participants can transmit the special 3-qubit state to each other. Note that the entangled 11-qubit quantum channel is formed by the tensor products of two Bell states, two GHZ states, and one single-qubit state. The teleportation of quantum state is realized by first implementing the GHZ state measurement, Bell state measurement, and single-qubit measurement, followed by implementing specific local unitary transformation.

4.4.1 C-BQT Scheme

The procedures of the proposed scheme are explained in detail within two parts. The first part provides a basic introduction, including the illustration of the unknown three-qubit states held by Alice and Bob, respectively, and the specific entangled eleven-qubit state that acts as the quantum channel shared among Alice, Bob, and the controller Charlie. The second part introduces the transmission process of the proposed protocol in detail; in particular, there are four steps in this part, including two GHZ state measurements, two Bell state measurements, one single-qubit state

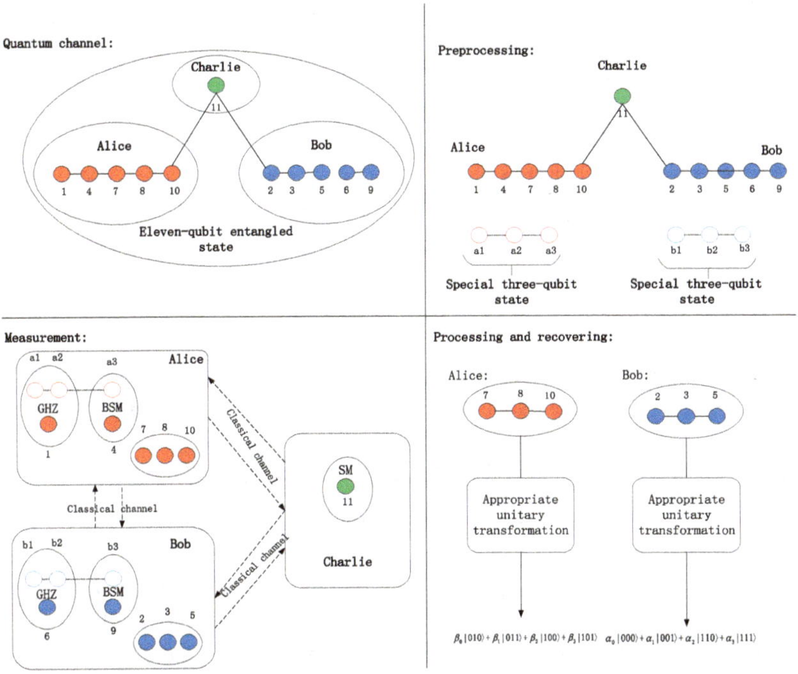

Fig. 4.9 The schematic diagram of C-BQT scheme. Herein, (**a**) the solid circle represents the qubits on the quantum channel; (**b**) the hollow circle represents the qubits to be transmitted between Alice and Bob; (**c**) the numbers 1, 4, 7, 8, and 10 correspond to the letters A_1, A_2, A_3, A_4, and A_5, and the numbers 2, 3, 5, 6, and 9 correspond to the letters B_3, B_4, B_5, B_1, and B_2; and (**d**) BSM denotes Bell state measurement, GHZ denotes GHZ state measurement, and SM denotes single-qubit state measurement

measurement, and unitary transformations. The schematic of the proposed scheme is illustrated in Fig. 4.9.

4.4.1.1 The Form of Qubit State of Alice and bob and the Quantum Channel

Suppose there are two legitimate participants, Alice and Bob, over a long distance. Alice has a special entangled three-qubit state $|\Phi\rangle$ in her qubits a_1, a_2, and a_3, and Bob has a special entangled three-qubit state $|\chi\rangle$ in his qubits b_1, b_2, and b_3. In the ideal circumstance, let's assume that the generalized forms of three-qubit states $|\Phi\rangle_{a_1a_2a_3}$ and $|\chi\rangle_{b_1b_2b_3}$ are written as

$$|\Phi\rangle_{a_1a_2a_3} = \alpha_0|000\rangle + \alpha_1|001\rangle + \alpha_2|110\rangle + \alpha_3|111\rangle \tag{4.53}$$

$$|\chi\rangle_{b_1b_2b_3} = \beta_0|010\rangle + \beta_1|011\rangle + \beta_2|100\rangle + \beta_3|101\rangle \tag{4.54}$$

where parameters $\alpha_0, \alpha_1, \alpha_2, \alpha_3, \beta_0, \beta_1, \beta_2, \beta_3$ are eight independent complex numbers and satisfy that $|\alpha_0|^2 + |\alpha_1|^2 + |\alpha_2|^2 + |\alpha_3|^2 = 1$ and $|\beta_0|^2 + |\beta_1|^2 + |\beta_2|^2 + |\beta_3|^2 = 1$.

Suppose that under a third controller party, Charlie, Alice intends to teleport her three qubits a_1, a_2, and a_3 to Bob, and at the same time, Bob intends to teleport his three qubits b_1, b_2, and b_3 to Alice. To achieve this aim, Alice, Bob, and Charlie need to share the entangled 11-qubit state $|\Omega\rangle_{1,2,3,4,5,6,7,8,9,10,11}$ in advance, which is employed as the quantum channel in the proposed protocol, written as

$$\begin{aligned}|\Omega\rangle_{1,2,3,4,5,6,7,8,9,10,11} = &|\phi^+\rangle_{123}|\varepsilon^+\rangle_{45}|\phi^+\rangle_{678}|\varepsilon^+\rangle_{910}|0\rangle_{11}+\\ &|\gamma^+\rangle_{123}|\nu^+\rangle_{45}|\gamma^+\rangle_{678}|\nu^+\rangle_{910}|1\rangle_{11}\end{aligned} \tag{4.55}$$

where qubits {1,4,7,8,10} belong to Alice, qubits {2,3,5,6,9} belong to Bob, and qubit 11 belongs to Charlie. Hence, the quantum channel can be further written as

$$\begin{aligned}|\Omega\rangle_{A_1B_3B_4A_2B_5B_1A_3A_4B_2A_5C} = &|\phi^+\rangle_{A_1B_3B_4}|\varepsilon^+\rangle_{A_2B_5}|\phi^+\rangle_{B_1A_3A_4}|\varepsilon^+\rangle_{B_2A_5}|0\rangle_C\\ &+|\gamma^+\rangle_{A_1B_3B_4}|\nu^+\rangle_{A_2B_5}|\gamma^+\rangle_{B_1A_3A_4}|\nu^+\rangle_{B_2A_5}|1\rangle_C\end{aligned}. \tag{4.56}$$

Apparently, the quantum channel is composed of two GHZ states, two Bell states, and one single-qubit state. After taking the Alice's and Bob's three-qubit states into consideration, the whole quantum system can be written into the tensor product as

$$\left|\sum\right\rangle_{a_1a_2a_3b_1b_2b_3A_1B_3B_4A_2B_5B_1A_3A_4B_2A_5C} = |\Phi\rangle_{a_1a_2a_3} \otimes |\chi\rangle_{b_1b_2b_3} \otimes |\Omega\rangle_{A_1B_3B_4A_2B_5B_1A_3A_4B_2A_5C}. \tag{4.57}$$

To better understand our choice of measurement basis, we put the position of the particles to be measured first in the system state, which does not change any information of the system, so we rewrite the system state, shown in (4.58).

$$\begin{aligned}&\left|\sum\right\rangle_{a_1a_2A_1B_3B_4a_3A_2B_5b_1b_2B_1A_3A_4b_3B_2A_5C} =\\ &\frac{1}{2}\begin{bmatrix}|000\rangle\otimes(\alpha_0|000\rangle+\alpha_1|001\rangle)+|111\rangle\otimes(\alpha_2|110\rangle+\alpha_3|111\rangle)\\ +|001\rangle\otimes(\alpha_0|110\rangle+\alpha_1|111\rangle)+|110\rangle\otimes(\alpha_2|000\rangle+\alpha_3|001\rangle)\end{bmatrix}_{a_1a_2A_1B_3B_4a_3}\otimes|\varepsilon^+\rangle_{A_2B_5}\\ &\otimes\begin{bmatrix}|010\rangle\otimes(\beta_0|000\rangle+\beta_1|001\rangle)+|101\rangle\otimes(\beta_2|110\rangle+\beta_3|111\rangle)\\ +|011\rangle\otimes(\beta_0|110\rangle+\beta_1|111\rangle)+|100\rangle\otimes(\beta_2|000\rangle+\beta_3|001\rangle)\end{bmatrix}_{b_1b_2B_1A_3A_4b_3}\otimes|\varepsilon^+\rangle_{B_2A_5}\otimes|0\rangle_C.\\ &+\frac{1}{2}\begin{bmatrix}|000\rangle\otimes(\alpha_0|100\rangle+\alpha_1|101\rangle)+|111\rangle\otimes(\alpha_2|010\rangle+\alpha_3|011\rangle)\\ +|001\rangle\otimes(\alpha_0|010\rangle+\alpha_1|011\rangle)+|110\rangle\otimes(\alpha_2|100\rangle+\alpha_3|101\rangle)\end{bmatrix}_{a_1a_2A_1B_3B_4a_3}\otimes|\nu^+\rangle_{A_2B_5}\\ &\otimes\begin{bmatrix}|010\rangle\otimes(\beta_0|100\rangle+\beta_1|101\rangle)+|101\rangle\otimes(\beta_2|010\rangle+\beta_3|011\rangle)\\ +|011\rangle\otimes(\beta_0|010\rangle+\beta_1|011\rangle)+|100\rangle\otimes(\beta_2|000\rangle+\beta_3|101\rangle)\end{bmatrix}_{b_1b_2B_1A_3A_4b_3}\otimes|\nu^+\rangle_{B_2A_5}\otimes|1\rangle_C\end{aligned} \tag{4.58}$$

4.4.1.2 The Transmission Process in Detail

The proposed C-BQT scheme can be described within the following four steps in order:

Step 1. GHZ state measurement operations.

According to the diagram shown in Fig. 4.9, at the beginning, Alice performs a GHZ state measurement on qubits (a_1, a_2, A_1) and conveys her measurement results to Bob, where she can choose randomly one of the four GHZ states $|\phi^+\rangle_{a_1a_2A_1}$, $|\phi^-\rangle_{a_1a_2A_1}$, $|\eta^+\rangle_{a_1a_2A_1}$, and $|\eta^-\rangle_{a_1a_2A_1}$ as the measurement basis. The four different GHZ states are given as

$$\begin{aligned} |\phi^+\rangle_{a_1a_2A_1} &= \frac{1}{\sqrt{2}}(|000\rangle + |111\rangle), \ |\phi^-\rangle_{a_1a_2A_1} = \frac{1}{\sqrt{2}}(|000\rangle - |111\rangle) \\ |\eta^+\rangle_{a_1a_2A_1} &= \frac{1}{\sqrt{2}}(|001\rangle + |110\rangle), |\eta^-\rangle_{a_1a_2A_1} = \frac{1}{\sqrt{2}}(|001\rangle - |110\rangle) \end{aligned}. \quad (4.59)$$

Following that, Bob performs a GHZ state measurement on qubits (b_1, b_2, B_1), where he can choose one of the four GHZ states $|\gamma^+\rangle_{b_1b_2B_1}$, $|\gamma^-\rangle_{b_1b_2B_1}$, $|\varsigma^+\rangle_{b_1b_2B_1}$, and $|\varsigma^-\rangle_{b_1b_2B_1}$ randomly as the measurement basis. The four different GHZ states are given as

$$\begin{aligned} |\gamma^+\rangle_{b_1b_2B_1} &= \frac{1}{\sqrt{2}}(|010\rangle + |101\rangle), |\gamma^-\rangle_{b_1b_2B_1} = \frac{1}{\sqrt{2}}(|010\rangle - |101\rangle) \\ |\varsigma^+\rangle_{b_1b_2B_1} &= \frac{1}{\sqrt{2}}(|011\rangle + |100\rangle), \ |\varsigma^-\rangle_{b_1b_2B_1} = \frac{1}{\sqrt{2}}(|011\rangle - |100\rangle) \end{aligned}. \quad (4.60)$$

Because both Alice and Bob have four different choices of measurement basis, the remaining qubits $\{B_3, B_4, a_3, A_2, B_5, A_3, A_4, b_3, B_2, A_5, C\}$ would be collapsed into 16 different qubit states with equal probability. Table 4.6 in Appendix shows the measurement results of Alice and Bob and the corresponding collapsed state.

To simplify the analysis in here, let's assume that Alice chose the measurement basis $|\phi^+\rangle_{a_1a_2A_1}$ and Bob chose the measurement basis $|\gamma^+\rangle_{b_1b_2B_1}$ as an example; the remaining 11 qubits collapse to the value shown in (4.61).

$$\begin{aligned} &{}_{a_1a_2A_1}\langle\phi^+|_{b_1b_2B_1}\langle\gamma^+|\Big|\sum\Big\rangle_{a_1a_2A_1B_3B_4a_3A_2B_5b_1b_2B_1A_3A_4b_3B_2A_5C} \\ &= \frac{1}{4}\begin{bmatrix} (\alpha_0|000\rangle + \alpha_1|001\rangle + \alpha_2|110\rangle + \alpha_3|111\rangle)_{B_3B_4a_3} \otimes |\varepsilon^+\rangle_{A_2B_5} \\ \otimes(\beta_0|000\rangle + \beta_1|001\rangle + \beta_2|110\rangle + \beta_3|111\rangle)_{A_3A_4b_3} \otimes |\varepsilon^+\rangle_{B_2A_5}|0\rangle_C \\ +(\alpha_0|100\rangle + \alpha_1|101\rangle + \alpha_2|010\rangle + \alpha_3|011\rangle)_{B_3B_4a_3} \otimes |\nu^+\rangle_{A_2B_5} \\ \otimes(\beta_0|100\rangle + \beta_1|101\rangle + \beta_2|010\rangle + \beta_3|011\rangle)_{A_3A_4b_3} \otimes |\nu^+\rangle_{B_2A_5}|1\rangle_C \end{bmatrix}. \end{aligned} \quad (4.61)$$

Table 4.6 The collapsed states of qubits {B_3, B_4, a_3, A_2, B_5, A_3, A_4, b_3, B_2, A_5, C} under Alice's and Bob's GHZ state measurement results

Alice's measurement results	Bob's measurement results	The collapsed states of qubits {B_3, B_4, a_3, A_2, B_5, A_3, A_4, b_3, B_2, A_5, C}
$\vert\phi^+\rangle_{a_1a_2A_1}$	$\vert\gamma^+\rangle_{b_1b_2B_1}$	$(\alpha_0\vert 000\rangle+\alpha_1\vert 001\rangle+\alpha_2\vert 110\rangle+\alpha_3\vert 111\rangle)_{B_3B_4a_3}\otimes\vert\varepsilon^+\rangle_{A_2B_5}$ $\otimes(\beta_0\vert 000\rangle+\beta_1\vert 001\rangle+\beta_2\vert 110\rangle+\beta_3\vert 111\rangle)_{A_3A_4b_3}\otimes\vert\varepsilon^+\rangle_{B_2A_5}\vert 0\rangle_C$ $+(\alpha_0\vert 100\rangle+\alpha_1\vert 101\rangle+\alpha_2\vert 010\rangle+\alpha_3\vert 011\rangle)_{B_3B_4a_3}\otimes\vert\nu^+\rangle_{A_2B_5}$ $\otimes(\beta_0\vert 100\rangle+\beta_1\vert 101\rangle+\beta_2\vert 010\rangle+\beta_3\vert 011\rangle)_{A_3A_4b_3}\otimes\vert\nu^+\rangle_{B_2A_5}\vert 1\rangle_C$
$\vert\phi^+\rangle_{a_1a_2A_1}$	$\vert\gamma^-\rangle_{b_1b_2B_1}$	$(\alpha_0\vert 000\rangle+\alpha_1\vert 001\rangle+\alpha_2\vert 110\rangle+\alpha_3\vert 111\rangle)_{B_3B_4a_3}\otimes\vert\varepsilon^+\rangle_{A_2B_5}$ $\otimes(\beta_0\vert 000\rangle+\beta_1\vert 001\rangle-\beta_2\vert 110\rangle-\beta_3\vert 111\rangle)_{A_3A_4b_3}\otimes\vert\varepsilon^+\rangle_{B_2A_5}\vert 0\rangle_C$ $+(\alpha_0\vert 100\rangle+\alpha_1\vert 101\rangle+\alpha_2\vert 010\rangle+\alpha_3\vert 011\rangle)_{B_3B_4a_3}\otimes\vert\nu^+\rangle_{A_2B_5}$ $\otimes(\beta_0\vert 100\rangle+\beta_1\vert 101\rangle-\beta_2\vert 010\rangle-\beta_3\vert 011\rangle)_{A_3A_4b_3}\otimes\vert\nu^+\rangle_{A_2B_5}\vert 1\rangle_C$
$\vert\phi^+\rangle_{a_1a_2A_1}$	$\vert\varsigma^+\rangle_{b_1b_2B_1}$	$(\alpha_0\vert 000\rangle-\alpha_1\vert 001\rangle+\alpha_2\vert 110\rangle-\alpha_3\vert 111\rangle)_{B_3B_4a_3}\otimes\vert\varepsilon^+\rangle_{A_2B_5}$ $\otimes(\beta_0\vert 110\rangle+\beta_1\vert 111\rangle+\beta_2\vert 000\rangle+\beta_3\vert 001\rangle)_{A_3A_4b_3}\otimes\vert\varepsilon^+\rangle_{B_2A_5}\vert 0\rangle_C$ $+(\alpha_0\vert 100\rangle+\alpha_1\vert 101\rangle-\alpha_2\vert 010\rangle-\alpha_3\vert 011\rangle)_{B_3B_4a_3}\otimes\vert\nu^+\rangle_{A_2B_5}$ $\otimes(\beta_0\vert 010\rangle+\beta_1\vert 011\rangle+\beta_2\vert 100\rangle+\beta_3\vert 101\rangle)_{A_3A_4b_3}\otimes\vert\nu^+\rangle_{B_2A_5}\vert 1\rangle_C$
$\vert\phi^+\rangle_{a_1a_2A_1}$	$\vert\varsigma^-\rangle_{b_1b_2B_1}$	$(\alpha_0\vert 000\rangle+\alpha_1\vert 001\rangle+\alpha_2\vert 110\rangle+\alpha_3\vert 111\rangle)_{B_3B_4a_3}\otimes\vert\varepsilon^+\rangle_{A_2B_5}$ $\otimes(\beta_0\vert 110\rangle+\beta_1\vert 111\rangle-\beta_2\vert 000\rangle-\beta_3\vert 001\rangle)_{A_3A_4b_3}\otimes\vert\varepsilon^+\rangle_{B_2A_5}\vert 0\rangle_C$ $+(\alpha_0\vert 100\rangle+\alpha_1\vert 101\rangle+\alpha_2\vert 010\rangle+\alpha_3\vert 011\rangle)_{B_3B_4a_3}\otimes\vert\nu^+\rangle_{A_2B_5}$ $\otimes(\beta_0\vert 010\rangle+\beta_1\vert 011\rangle-\beta_2\vert 100\rangle-\beta_3\vert 101\rangle)_{A_3A_4b_3}\otimes\vert\nu^+\rangle_{B_2A_5}\vert 1\rangle_C$
$\vert\phi^-\rangle_{a_1a_2A_1}$	$\vert\gamma^+\rangle_{b_1b_2B_1}$	$(\alpha_0\vert 000\rangle-\alpha_1\vert 001\rangle+\alpha_2\vert 110\rangle-\alpha_3\vert 111\rangle)_{B_3B_4a_3}\otimes\vert\varepsilon^+\rangle_{A_2B_5}$ $\otimes(\beta_0\vert 000\rangle+\beta_1\vert 001\rangle+\beta_2\vert 110\rangle+\beta_3\vert 111\rangle)_{A_3A_4b_3}\otimes\vert\varepsilon^+\rangle_{B_2A_5}\vert 0\rangle_C$ $+(\alpha_0\vert 100\rangle+\alpha_1\vert 101\rangle-\alpha_2\vert 010\rangle-\alpha_3\vert 011\rangle)_{B_3B_4a_3}\otimes\vert\nu^+\rangle_{A_2B_5}$ $\otimes(\beta_0\vert 100\rangle+\beta_1\vert 101\rangle+\beta_2\vert 010\rangle+\beta_3\vert 011\rangle)_{A_3A_4b_3}\otimes\vert\nu^+\rangle_{B_2A_5}\vert 1\rangle_C$

<table>
<tr><td>$\|\phi^-\rangle_{a_1a_2A_1}$</td><td>$\|\gamma^-\rangle_{b_1b_2B_1}$</td><td>$(\alpha_0\|000\rangle - \alpha_1\|001\rangle + \alpha_2\|110\rangle - \alpha_3\|111\rangle)_{B_3B_4a_3} \otimes \|\varepsilon^+\rangle_{A_2B_5}$
$\otimes(\beta_0\|000\rangle + \beta_1\|001\rangle - \beta_2\|110\rangle - \beta_3\|111\rangle)_{A_3A_4b_3} \otimes \|\varepsilon^+\rangle_{B_2A_5}\|0\rangle_C$
$+(\alpha_0\|100\rangle + \alpha_1\|101\rangle - \alpha_2\|010\rangle - \alpha_3\|011\rangle)_{B_3B_4a_3} \otimes \|\nu^+\rangle_{A_2B_5}$
$\otimes(\beta_0\|100\rangle + \beta_1\|101\rangle - \beta_2\|010\rangle - \beta_3\|011\rangle)_{A_3A_4b_3} \otimes \|\nu^+\rangle_{B_2A_5}\|1\rangle_C$</td></tr>
<tr><td>$\|\phi^-\rangle_{a_1a_2A_1}$</td><td>$\|\varsigma^+\rangle_{b_1b_2B_1}$</td><td>$(\alpha_0\|000\rangle - \alpha_1\|001\rangle + \alpha_2\|110\rangle - \alpha_3\|111\rangle)_{B_3B_4a_3} \otimes \|\varepsilon^+\rangle_{A_2B_5}$
$\otimes(\beta_0\|110\rangle + \beta_1\|111\rangle + \beta_2\|000\rangle + \beta_3\|001\rangle)_{A_3A_4b_3} \otimes \|\varepsilon^+\rangle_{B_2A_5}\|0\rangle_C$
$+(\alpha_0\|100\rangle + \alpha_1\|101\rangle - \alpha_2\|010\rangle - \alpha_3\|011\rangle)_{B_3B_4a_3} \otimes \|\nu^+\rangle_{A_2B_5}$
$\otimes(\beta_0\|010\rangle + \beta_1\|011\rangle + \beta_2\|100\rangle + \beta_3\|101\rangle)_{A_3A_4b_3} \otimes \|\nu^+\rangle_{B_2A_5}\|1\rangle_C$</td></tr>
<tr><td>$\|\phi^-\rangle_{a_1a_2A_1}$</td><td>$\|\varsigma^-\rangle_{b_1b_2B_1}$</td><td>$(\alpha_0\|000\rangle - \alpha_1\|001\rangle + \alpha_2\|110\rangle - \alpha_3\|111\rangle)_{B_3B_4a_3} \otimes \|\varepsilon^+\rangle_{A_2B_5}$
$\otimes(\beta_0\|110\rangle + \beta_1\|111\rangle - \beta_2\|000\rangle - \beta_3\|001\rangle)_{A_3A_4b_3} \otimes \|\varepsilon^+\rangle_{B_2A_5}\|0\rangle_C$
$+(\alpha_0\|100\rangle + \alpha_1\|101\rangle - \alpha_2\|010\rangle - \alpha_3\|011\rangle)_{B_3B_4a_3} \otimes \|\nu^+\rangle_{A_2B_5}$
$\otimes(\beta_0\|010\rangle + \beta_1\|011\rangle - \beta_2\|100\rangle - \beta_3\|101\rangle)_{A_3A_4b_3} \otimes \|\nu^+\rangle_{B_2A_5}\|1\rangle_C$</td></tr>
<tr><td>$\|\eta^+\rangle_{a_1a_2A_1}$</td><td>$\|\gamma^+\rangle_{b_1b_2B_1}$</td><td>$(\alpha_0\|110\rangle + \alpha_1\|111\rangle + \alpha_2\|000\rangle + \alpha_3\|001\rangle)_{B_3B_4a_3} \otimes \|\varepsilon^+\rangle_{A_2B_5}$
$\otimes(\beta_0\|000\rangle + \beta_1\|001\rangle + \beta_2\|110\rangle + \beta_3\|111\rangle)_{A_3A_4b_3} \otimes \|\varepsilon^+\rangle_{B_2A_5}\|0\rangle_C$
$+(\alpha_0\|010\rangle + \alpha_1\|011\rangle + \alpha_2\|100\rangle + \alpha_3\|101\rangle)_{B_3B_4a_3} \otimes \|\nu^+\rangle_{A_2B_5}$
$\otimes(\beta_0\|100\rangle + \beta_1\|101\rangle + \beta_2\|010\rangle + \beta_3\|011\rangle)_{A_3A_4b_3} \otimes \|\nu^+\rangle_{B_2A_5}\|1\rangle_C$</td></tr>
<tr><td>$\|\eta^+\rangle_{a_1a_2A_1}$</td><td>$\|\gamma^-\rangle_{b_1b_2B_1}$</td><td>$(\alpha_0\|110\rangle + \alpha_1\|111\rangle + \alpha_2\|000\rangle + \alpha_3\|001\rangle)_{B_3B_4a_3} \otimes \|\varepsilon^+\rangle_{A_2B_5}$
$\otimes(\beta_0\|000\rangle + \beta_1\|001\rangle - \beta_2\|110\rangle - \beta_3\|111\rangle)_{A_3A_4b_3} \otimes \|\varepsilon^+\rangle_{B_2A_5}\|0\rangle_C$
$+(\alpha_0\|010\rangle + \alpha_1\|011\rangle + \alpha_2\|100\rangle + \alpha_3\|101\rangle)_{B_3B_4a_3} \otimes \|\nu^+\rangle_{A_2B_5}$
$\otimes(\beta_0\|100\rangle + \beta_1\|101\rangle - \beta_2\|010\rangle - \beta_3\|011\rangle)_{A_3A_4b_3} \otimes \|\nu^+\rangle_{B_2A_5}\|1\rangle_C$</td></tr>
<tr><td>$\|\eta^+\rangle_{a_1a_2A_1}$</td><td>$\|\varsigma^+\rangle_{b_1b_2B_1}$</td><td>$(\alpha_0\|110\rangle + \alpha_1\|111\rangle + \alpha_2\|000\rangle + \alpha_3\|001\rangle)_{B_3B_4a_3} \otimes \|\varepsilon^+\rangle_{A_2B_5}$
$\otimes(\beta_0\|000\rangle + \beta_1\|001\rangle + \beta_2\|110\rangle + \beta_3\|111\rangle)_{A_3A_4b_3} \otimes \|\varepsilon^+\rangle_{B_2A_5}\|0\rangle_C$
$+(\alpha_0\|010\rangle + \alpha_1\|011\rangle + \alpha_2\|100\rangle + \alpha_3\|101\rangle)_{B_3B_4a_3} \otimes \|\nu^+\rangle_{A_2B_5}$
$\otimes(\beta_0\|100\rangle + \beta_1\|101\rangle + \beta_2\|010\rangle + \beta_3\|011\rangle)_{A_3A_4b_3} \otimes \|\nu^+\rangle_{B_2A_5}\|1\rangle_C$</td></tr>
</table>

(continued)

Table 4.6 (continued)

Alice's measurement results	Bob's measurement results	The collapsed states of qubits {B$_3$, B$_4$, a$_3$, A$_2$, B$_5$, A$_3$, A$_4$, b$_3$, B$_2$, A$_5$, C}
$\|\eta^+\rangle_{a_1a_2A_1}$	$\|\varsigma^-\rangle_{b_1b_2B_1}$	$(\alpha_0\|110\rangle+\alpha_1\|111\rangle+\alpha_2\|000\rangle+\alpha_3\|001\rangle)_{B_3B_4a_3}\otimes\|\varepsilon^+\rangle_{A_2B_5}$ $\otimes(\beta_0\|000\rangle+\beta_1\|001\rangle-\beta_2\|110\rangle-\beta_3\|111\rangle)_{A_3A_4b_3}\otimes\|\varepsilon^+\rangle_{B_2A_5}\|0\rangle_C$ $+(\alpha_0\|010\rangle+\alpha_1\|011\rangle+\alpha_2\|100\rangle+\alpha_3\|101\rangle)_{B_3B_4a_3}\otimes\|\nu^+\rangle_{A_2B_5}$ $\otimes(\beta_0\|100\rangle+\beta_1\|101\rangle-\beta_2\|010\rangle-\beta_3\|011\rangle)_{A_3A_4b_3}\otimes\|\nu^+\rangle_{B_2A_5}\|1\rangle_C$
$\|\eta^-\rangle_{a_1a_2A_1}$	$\|\gamma^+\rangle_{b_1b_2B_1}$	$(\alpha_0\|110\rangle+\alpha_1\|111\rangle-\alpha_2\|000\rangle-\alpha_3\|001\rangle)_{B_3B_4a_3}\otimes\|\varepsilon^+\rangle_{A_2B_5}$ $\otimes(\beta_0\|000\rangle+\beta_1\|001\rangle+\beta_2\|110\rangle+\beta_3\|111\rangle)_{A_3A_4b_3}\otimes\|\varepsilon^+\rangle_{B_2A_5}\|0\rangle_C$ $+(\alpha_0\|010\rangle+\alpha_1\|011\rangle-\alpha_2\|100\rangle-\alpha_3\|101\rangle)_{B_3B_4a_3}\otimes\|\nu^+\rangle_{A_2B_5}$ $\otimes(\beta_0\|100\rangle+\beta_1\|101\rangle+\beta_2\|010\rangle+\beta_3\|011\rangle)_{A_3A_4b_3}\otimes\|\nu^+\rangle_{B_2A_5}\|1\rangle_C$
$\|\eta^-\rangle_{a_1a_2A_1}$	$\|\gamma^-\rangle_{b_1b_2B_1}$	$(\alpha_0\|110\rangle+\alpha_1\|111\rangle-\alpha_2\|000\rangle-\alpha_3\|001\rangle)_{B_3B_4a_3}\otimes\|\varepsilon^+\rangle_{A_2B_5}$ $\otimes(\beta_0\|000\rangle+\beta_1\|001\rangle-\beta_2\|110\rangle-\beta_3\|111\rangle)_{A_3A_4b_3}\otimes\|\varepsilon^+\rangle_{B_2A_5}\|0\rangle_C$ $+(\alpha_0\|010\rangle+\alpha_1\|011\rangle-\alpha_2\|100\rangle-\alpha_3\|101\rangle)_{B_3B_4a_3}\otimes\|\nu^+\rangle_{A_2B_5}$ $\otimes(\beta_0\|100\rangle+\beta_1\|101\rangle-\beta_2\|010\rangle-\beta_3\|011\rangle)_{A_3A_4b_3}\otimes\|\nu^+\rangle_{B_2A_5}\|1\rangle_C$
$\|\eta^-\rangle_{a_1a_2A_1}$	$\|\varsigma^+\rangle_{b_1b_2B_1}$	$(\alpha_0\|110\rangle+\alpha_1\|111\rangle-\alpha_2\|000\rangle-\alpha_3\|001\rangle)_{B_3B_4a_3}\otimes\|\varepsilon^+\rangle_{A_2B_5}$ $\otimes(\beta_0\|000\rangle+\beta_1\|001\rangle+\beta_2\|110\rangle+\beta_3\|111\rangle)_{A_3A_4b_3}\otimes\|\varepsilon^+\rangle_{B_2A_5}\|0\rangle_C$ $+(\alpha_0\|010\rangle+\alpha_1\|011\rangle-\alpha_2\|100\rangle-\alpha_3\|101\rangle)_{B_3B_4a_3}\otimes\|\nu^+\rangle_{A_2B_5}$ $\otimes(\beta_0\|100\rangle+\beta_1\|101\rangle+\beta_2\|010\rangle+\beta_3\|011\rangle)_{A_3A_4b_3}\otimes\|\nu^+\rangle_{B_2A_5}\|1\rangle_C$
$\|\eta^-\rangle_{a_1a_2A_1}$	$\|\varsigma^-\rangle_{b_1b_2B_1}$	$(\alpha_0\|110\rangle+\alpha_1\|111\rangle-\alpha_2\|000\rangle-\alpha_3\|001\rangle)_{B_3B_4a_3}\otimes\|\varepsilon^+\rangle_{A_2B_5}$ $\otimes(\beta_0\|000\rangle+\beta_1\|001\rangle-\beta_2\|110\rangle-\beta_3\|111\rangle)_{A_3A_4b_3}\otimes\|\varepsilon^+\rangle_{B_2A_5}\|0\rangle_C$ $+(\alpha_0\|010\rangle+\alpha_1\|011\rangle-\alpha_2\|100\rangle-\alpha_3\|101\rangle)_{B_3B_4a_3}\otimes\|\nu^+\rangle_{A_2B_5}$ $\otimes(\beta_0\|100\rangle+\beta_1\|101\rangle-\beta_2\|010\rangle-\beta_3\|011\rangle)_{A_3A_4b_3}\otimes\|\nu^+\rangle_{B_2A_5}\|1\rangle_C$

Step 2. Bell state measurement operations.

After the above two GHZ state measurements, Alice then implements a Bell state measurement on the quantum pair (a_3, A_2), and at the same time, Bob also implements a Bell state measurement on the quantum pair (b_3, B_2). Herein, Alice chooses one of the four different Bell states $\left\{|\varepsilon^+\rangle_{\mathrm{a3A_2}}, |\varepsilon^-\rangle_{\mathrm{a3A_2}}, |\nu^+\rangle_{\mathrm{a3A_2}}, |\nu^-\rangle_{\mathrm{a3A_2}}\right\}$ as the measurement basis. These four Bell states are given as

$$\begin{array}{l} |\varepsilon^+\rangle_{\mathrm{a3A_2}} = \frac{1}{\sqrt{2}}(|00\rangle + |11\rangle),\ |\varepsilon^-\rangle_{\mathrm{a3A_2}} = \frac{1}{\sqrt{2}}(|00\rangle - |11\rangle) \\ |\nu^+\rangle_{\mathrm{a3A_2}} = \frac{1}{\sqrt{2}}(|01\rangle + |10\rangle),\ |\nu^-\rangle_{\mathrm{a3A_2}} = \frac{1}{\sqrt{2}}(|01\rangle - |10\rangle) \end{array}. \tag{4.62}$$

Bob can choose one of the four Bell states $\left\{|\varepsilon^+\rangle_{b3\mathrm{B_2}}, |\varepsilon^-\rangle_{b3\mathrm{B_2}}, |\nu^+\rangle_{b3\mathrm{B_2}}, |\nu^-\rangle_{b3\mathrm{B_2}}\right\}$ as his measurement basis. These four Bell states are given as

$$\begin{array}{l} |\varepsilon^+\rangle_{b3\mathrm{B_2}} = \frac{1}{\sqrt{2}}(|00\rangle + |11\rangle),\ |\varepsilon^-\rangle_{b3\mathrm{B_2}} = \frac{1}{\sqrt{2}}(|00\rangle - |11\rangle) \\ |\nu^+\rangle_{b3\mathrm{B_2}} = \frac{1}{\sqrt{2}}(|01\rangle + |10\rangle),\ |\nu^+\rangle_{b3\mathrm{B_2}} = \frac{1}{\sqrt{2}}(|01\rangle - |10\rangle) \end{array}. \tag{4.63}$$

If Alice and Bob choose $|\varepsilon^+\rangle_{\mathrm{a3A_2}}$ and $|\varepsilon^+\rangle_{b3\mathrm{B_2}}$ as the measurement result, on the basis of (4.61), the remaining particles $\{B_3, B_4, B_5, A_3, A_4, A_5, C\}$ are collapsed into following state:

$$\begin{array}{l} {}_{\mathrm{a_1a_2A_1}}\langle\phi^+|_{\mathrm{b_1b_2B_1}}\langle\gamma^+|_{\mathrm{a3A_2}}\langle\varepsilon^+|_{b3\mathrm{B_2}}\langle\varepsilon^+|\left|\sum\right\rangle_{\mathrm{a_1a_2A_1B_3B_4a_3A_2B_5}b_1\mathrm{b_2B_1A_3A_4}b_3\mathrm{B_2A_5C}} \\ = \frac{1}{8}\left[\begin{array}{l} (\alpha_0|000\rangle + \alpha_1|001\rangle + \alpha_2|110\rangle + \alpha_3|111\rangle)_{\mathrm{B_3B_4B_5}} \\ \otimes(\beta_0|000\rangle + \beta_1|001\rangle + \beta_2|110\rangle + \beta_3|111\rangle)_{\mathrm{A_3A_4A_5}} \otimes |0\rangle_{\mathrm{C}} \\ +(\alpha_0|101\rangle + \alpha_1|100\rangle + \alpha_2|011\rangle + \alpha_3|010\rangle)_{\mathrm{B_3B_4B_5}} \\ \otimes(\beta_0|101\rangle + \beta_1|100\rangle + \beta_2|011\rangle + \beta_3|011\rangle)_{\mathrm{A_3A_4A_5}} \otimes |1\rangle_{\mathrm{C}} \end{array}\right]. \end{array} \tag{4.64}$$

In this process, Alice and Bob will tell the measurement result to each other through classical channel. And Bob will tell the measurement result to Charlie via classical channel. On the basis of Alice's and Bob's GHZ state measurement results $|\phi^+\rangle_{\mathrm{a_1a_2A_1}}$ and $|\gamma^+\rangle_{\mathrm{b_1b_2B_1}}$, all the Bell state measurement choices and corresponding 16 different collapsed states of qubits $\{B3, B4, B5, A3, A4, A5, C\}$, with the equal probability, are shown in Table 4.7.

Table 4.7 On the basis of Alice's and Bob's GHZ state measurement results $|\phi^+\rangle_{a_1a_2A_1}$ and $|\gamma^+\rangle_{b_1b_2B_1}$ of Step 1, the collapsed states of qubits {B₃, B₄, B₅, A₃, A₄, A₅, C} correspond to Alice's and Bob's Bell state measurement results of Step 2

Alice's results	Bob's results	The collapsed state of qubits $\{B_3, B_4, B_5, A_3, A_4, A_5, C\}$
$\|\varepsilon^+\rangle_{a3A_2}$	$\|\varepsilon^+\rangle_{b3B_2}$	$(\alpha_0\|000\rangle+\alpha_1\|001\rangle+\alpha_2\|110\rangle+\alpha_3\|111\rangle)_{B_3B_4B_5}$ $\otimes(\beta_0\|000\rangle+\beta_1\|001\rangle+\beta_2\|110\rangle+\beta_3\|111\rangle)_{A_3A_4A_5}\otimes\|0\rangle_C$ $+(\alpha_0\|101\rangle+\alpha_1\|100\rangle+\alpha_2\|011\rangle+\alpha_3\|010\rangle)_{B_3B_4B_5}$ $\otimes(\beta_0\|101\rangle+\beta_1\|100\rangle+\beta_2\|011\rangle+\beta_3\|010\rangle)_{A_3A_4A_5}\otimes\|1\rangle_C$
$\|\varepsilon^+\rangle_{a3A_2}$	$\|\varepsilon^-\rangle_{b3B_2}$	$(\alpha_0\|000\rangle+\alpha_1\|001\rangle+\alpha_2\|110\rangle+\alpha_3\|111\rangle)_{B_3B_4B_5}$ $\otimes(\beta_0\|000\rangle-\beta_1\|001\rangle+\beta_2\|110\rangle-\beta_3\|111\rangle)_{A_3A_4A_5}\otimes\|0\rangle_C$ $+(\alpha_0\|101\rangle+\alpha_1\|100\rangle+\alpha_2\|011\rangle+\alpha_3\|010\rangle)_{B_3B_4B_5}$ $\otimes(\beta_0\|101\rangle-\beta_1\|100\rangle+\beta_2\|011\rangle-\beta_3\|010\rangle)_{A_3A_4A_5}\otimes\|1\rangle_C$
$\|\varepsilon^+\rangle_{a3A_2}$	$\|\nu^+\rangle_{b3B_2}$	$(\alpha_0\|000\rangle+\alpha_1\|001\rangle+\alpha_2\|110\rangle+\alpha_3\|111\rangle)_{B_3B_4B_5}$ $\otimes(\beta_0\|001\rangle+\beta_1\|000\rangle+\beta_2\|111\rangle+\beta_3\|110\rangle)_{A_3A_4A_5}\otimes\|0\rangle_C$ $+(\alpha_0\|101\rangle+\alpha_1\|100\rangle+\alpha_2\|011\rangle+\alpha_3\|010\rangle)_{B_3B_4B_5}$ $\otimes(\beta_0\|100\rangle+\beta_1\|101\rangle+\beta_2\|010\rangle+\beta_3\|011\rangle)_{A_3A_4A_5}\otimes\|1\rangle_C$
$\|\varepsilon^+\rangle_{a3A_2}$	$\|\nu^-\rangle_{b3B_2}$	$(\alpha_0\|000\rangle+\alpha_1\|001\rangle+\alpha_2\|110\rangle+\alpha_3\|111\rangle)_{B_3B_4B_5}$ $\otimes(\beta_0\|001\rangle-\beta_1\|000\rangle+\beta_2\|111\rangle-\beta_3\|110\rangle)_{A_3A_4A_5}\otimes\|0\rangle_C$ $+(\alpha_0\|101\rangle+\alpha_1\|100\rangle+\alpha_2\|011\rangle+\alpha_3\|010\rangle)_{B_3B_4B_5}$ $\otimes(\beta_0\|100\rangle-\beta_1\|101\rangle+\beta_2\|010\rangle-\beta_3\|011\rangle)_{A_3A_4A_5}\otimes\|1\rangle_C$
$\|\varepsilon^-\rangle_{a3A_2}$	$\|\varepsilon^+\rangle_{b3B_2}$	$(\alpha_0\|000\rangle-\alpha_1\|001\rangle+\alpha_2\|110\rangle-\alpha_3\|111\rangle)_{B_3B_4B_5}$ $\otimes(\beta_0\|000\rangle+\beta_1\|001\rangle+\beta_2\|110\rangle+\beta_3\|111\rangle)_{A_3A_4A_5}\otimes\|0\rangle_C$ $+(\alpha_0\|101\rangle-\alpha_1\|100\rangle+\alpha_2\|011\rangle-\alpha_3\|010\rangle)_{B_3B_4B_5}$ $\otimes(\beta_0\|101\rangle+\beta_1\|100\rangle+\beta_2\|011\rangle+\beta_3\|010\rangle)_{A_3A_4A_5}\otimes\|1\rangle_C$
$\|\varepsilon^-\rangle_{a3A_2}$	$\|\varepsilon^-\rangle_{b3B_2}$	$(\alpha_0\|000\rangle-\alpha_1\|001\rangle+\alpha_2\|110\rangle-\alpha_3\|111\rangle)_{B_3B_4B_5}$ $\otimes(\beta_0\|000\rangle-\beta_1\|001\rangle+\beta_2\|110\rangle-\beta_3\|111\rangle)_{A_3A_4A_5}\otimes\|0\rangle_C$ $+(\alpha_0\|101\rangle-\alpha_1\|100\rangle+\alpha_2\|011\rangle-\alpha_3\|010\rangle)_{B_3B_4B_5}$ $\otimes(\beta_0\|101\rangle-\beta_1\|100\rangle+\beta_2\|011\rangle-\beta_3\|010\rangle)_{A_3A_4A_5}\otimes\|1\rangle_C$
$\|\varepsilon^-\rangle_{a3A_2}$	$\|\nu^+\rangle_{b3B_2}$	$(\alpha_0\|000\rangle-\alpha_1\|001\rangle+\alpha_2\|110\rangle-\alpha_3\|111\rangle)_{B_3B_4B_5}$ $\otimes(\beta_0\|001\rangle+\beta_1\|000\rangle+\beta_2\|111\rangle+\beta_3\|110\rangle)_{A_3A_4A_5}\otimes\|0\rangle_C$ $+(\alpha_0\|101\rangle-\alpha_1\|100\rangle+\alpha_2\|011\rangle-\alpha_3\|010\rangle)_{B_3B_4B_5}$ $\otimes(\beta_0\|100\rangle+\beta_1\|101\rangle+\beta_2\|010\rangle+\beta_3\|011\rangle)_{A_3A_4A_5}\otimes\|1\rangle_C$
$\|\varepsilon^-\rangle_{a3A_2}$	$\|\nu^-\rangle_{b3B_2}$	$(\alpha_0\|000\rangle-\alpha_1\|001\rangle+\alpha_2\|110\rangle-\alpha_3\|111\rangle)_{B_3B_4B_5}$ $\otimes(\beta_0\|001\rangle-\beta_1\|000\rangle+\beta_2\|111\rangle-\beta_3\|110\rangle)_{A_3A_4A_5}\otimes\|0\rangle_C$ $+(\alpha_0\|101\rangle-\alpha_1\|100\rangle+\alpha_2\|011\rangle-\alpha_3\|010\rangle)_{B_3B_4B_5}$ $\otimes(\beta_0\|100\rangle-\beta_1\|101\rangle+\beta_2\|010\rangle-\beta_3\|011\rangle)_{A_3A_4A_5}\otimes\|1\rangle_C$
$\|\nu^+\rangle_{a3A_2}$	$\|\varepsilon^+\rangle_{b3B_2}$	$(\alpha_0\|001\rangle+\alpha_1\|000\rangle+\alpha_2\|111\rangle+\alpha_3\|110\rangle)_{B_3B_4B_5}$ $\otimes(\beta_0\|000\rangle+\beta_1\|001\rangle+\beta_2\|110\rangle+\beta_3\|111\rangle)_{A_3A_4A_5}\otimes\|0\rangle_C$ $+(\alpha_0\|100\rangle+\alpha_1\|101\rangle+\alpha_2\|010\rangle+\alpha_3\|011\rangle)_{B_3B_4B_5}$ $\otimes(\beta_0\|101\rangle+\beta_1\|100\rangle+\beta_2\|011\rangle+\beta_3\|010\rangle)_{A_3A_4A_5}\otimes\|1\rangle_C$

(continued)

Table 4.7 (continued)

Alice's results	Bob's results	The collapsed state of qubits {B₃, B₄, B₅, A₃, A₄, A₅, C}
$\vert\nu^{+}\rangle_{a3A_2}$	$\vert\varepsilon^{-}\rangle_{b3B_2}$	$(\alpha_0\vert 001\rangle+\alpha_1\vert 000\rangle+\alpha_2\vert 111\rangle+\alpha_3\vert 110\rangle)_{B_3B_4B_5}$ $\otimes(\beta_0\vert 000\rangle-\beta_1\vert 001\rangle+\beta_2\vert 110\rangle-\beta_3\vert 111\rangle)_{A_3A_4A_5}\otimes\vert 0\rangle_C$ $+(\alpha_0\vert 100\rangle+\alpha_1\vert 101\rangle+\alpha_2\vert 010\rangle+\alpha_3\vert 011\rangle)_{B_3B_4B_5}$ $\otimes(\beta_0\vert 101\rangle-\beta_1\vert 100\rangle+\beta_2\vert 011\rangle-\beta_3\vert 010\rangle)_{A_3A_4A_5}\otimes\vert 1\rangle_C$
$\vert\nu^{+}\rangle_{a3A_2}$	$\vert\nu^{+}\rangle_{b3B_2}$	$(\alpha_0\vert 001\rangle+\alpha_1\vert 000\rangle+\alpha_2\vert 111\rangle+\alpha_3\vert 110\rangle)_{B_3B_4B_5}$ $\otimes(\beta_0\vert 001\rangle+\beta_1\vert 000\rangle+\beta_2\vert 111\rangle+\beta_3\vert 110\rangle)_{A_3A_4A_5}\otimes\vert 0\rangle_C$ $+(\alpha_0\vert 100\rangle+\alpha_1\vert 101\rangle+\alpha_2\vert 010\rangle+\alpha_3\vert 011\rangle)_{B_3B_4B_5}$ $\otimes(\beta_0\vert 100\rangle+\beta_1\vert 101\rangle+\beta_2\vert 010\rangle+\beta_3\vert 011\rangle)_{A_3A_4A_5}\otimes\vert 1\rangle_C$
$\vert\nu^{+}\rangle_{a3A_2}$	$\vert\nu^{-}\rangle_{b3B_2}$	$(\alpha_0\vert 001\rangle+\alpha_1\vert 000\rangle+\alpha_2\vert 111\rangle+\alpha_3\vert 110\rangle)_{B_3B_4B_5}$ $\otimes(\beta_0\vert 001\rangle-\beta_1\vert 000\rangle+\beta_2\vert 111\rangle-\beta_3\vert 110\rangle)_{A_3A_4A_5}\otimes\vert 0\rangle_C$ $+(\alpha_0\vert 100\rangle+\alpha_1\vert 101\rangle+\alpha_2\vert 010\rangle+\alpha_3\vert 011\rangle)_{B_3B_4B_5}$ $\otimes(\beta_0\vert 100\rangle-\beta_1\vert 101\rangle+\beta_2\vert 010\rangle-\beta_3\vert 011\rangle)_{A_3A_4A_5}\otimes\vert 1\rangle_C$
$\vert\nu^{-}\rangle_{a3A_2}$	$\vert\varepsilon^{+}\rangle_{b3B_2}$	$(\alpha_0\vert 001\rangle-\alpha_1\vert 000\rangle+\alpha_2\vert 111\rangle-\alpha_3\vert 110\rangle)_{B_3B_4B_5}$ $\otimes(\beta_0\vert 000\rangle+\beta_1\vert 001\rangle+\beta_2\vert 110\rangle+\beta_3\vert 111\rangle)_{A_3A_4A_5}\otimes\vert 0\rangle_C$ $+(\alpha_0\vert 100\rangle-\alpha_1\vert 101\rangle+\alpha_2\vert 010\rangle-\alpha_3\vert 011\rangle)_{B_3B_4B_5}$ $\otimes(\beta_0\vert 101\rangle+\beta_1\vert 100\rangle+\beta_2\vert 011\rangle+\beta_3\vert 010\rangle)_{A_3A_4A_5}\otimes\vert 1\rangle_C$
$\vert\nu^{-}\rangle_{a3A_2}$	$\vert\varepsilon^{-}\rangle_{b3B_2}$	$(\alpha_0\vert 001\rangle-\alpha_1\vert 000\rangle+\alpha_2\vert 111\rangle-\alpha_3\vert 110\rangle)_{B_3B_4B_5}$ $\otimes(\beta_0\vert 000\rangle-\beta_1\vert 001\rangle+\beta_2\vert 110\rangle-\beta_3\vert 111\rangle)_{A_3A_4A_5}\otimes\vert 0\rangle_C$ $+(\alpha_0\vert 100\rangle-\alpha_1\vert 101\rangle+\alpha_2\vert 010\rangle-\alpha_3\vert 011\rangle)_{B_3B_4B_5}$ $\otimes(\beta_0\vert 101\rangle-\beta_1\vert 100\rangle+\beta_2\vert 011\rangle-\beta_3\vert 010\rangle)_{A_3A_4A_5}\otimes\vert 1\rangle_C$
$\vert\nu^{-}\rangle_{a3A_2}$	$\vert\nu^{+}\rangle_{b3B_2}$	$(\alpha_0\vert 001\rangle-\alpha_1\vert 000\rangle+\alpha_2\vert 111\rangle-\alpha_3\vert 110\rangle)_{B_3B_4B_5}$ $\otimes(\beta_0\vert 001\rangle+\beta_1\vert 000\rangle+\beta_2\vert 111\rangle+\beta_3\vert 110\rangle)_{A_3A_4A_5}\otimes\vert 0\rangle_C$ $+(\alpha_0\vert 100\rangle-\alpha_1\vert 101\rangle+\alpha_2\vert 010\rangle-\alpha_3\vert 011\rangle)_{B_3B_4B_5}$ $\otimes(\beta_0\vert 100\rangle+\beta_1\vert 101\rangle+\beta_2\vert 010\rangle+\beta_3\vert 011\rangle)_{A_3A_4A_5}\otimes\vert 1\rangle_C$
$\vert\nu^{-}\rangle_{a3A_2}$	$\vert\nu^{-}\rangle_{b3B_2}$	$(\alpha_0\vert 001\rangle-\alpha_1\vert 000\rangle+\alpha_2\vert 111\rangle-\alpha_3\vert 110\rangle)_{B_3B_4B_5}$ $\otimes(\beta_0\vert 001\rangle-\beta_1\vert 000\rangle+\beta_2\vert 111\rangle-\beta_3\vert 110\rangle)_{A_3A_4A_5}\otimes\vert 0\rangle_C$ $+(\alpha_0\vert 100\rangle-\alpha_1\vert 101\rangle+\alpha_2\vert 010\rangle-\alpha_3\vert 011\rangle)_{B_3B_4B_5}$ $\otimes(\beta_0\vert 100\rangle-\beta_1\vert 101\rangle+\beta_2\vert 010\rangle-\beta_3\vert 011\rangle)_{A_3A_4A_5}\otimes\vert 1\rangle_C$

Step 3. Single-qubit state measurement operation

After two GHZ state measurements and two Bell state measurements, the initial system has become a seven-qubit entangled state of qubits $\{B_3, B_4, B_5, A_3, A_4, A_5, C\}$, shown in Table 4.7. For the smooth implementation of our protocol, Charlie needs to measure the single-qubit C in Z-basis $\{|0\rangle, |1\rangle\}$. Charlie announces his measurement results to both Alice and Bob. Based on Charlie's measurement result, Alice and Bob utilize appropriate unitary transformation for reconstructing the initial state.

Let's take the collapsed state of seven qubits $\{B_3, B_4, B_5, A_3, A_4, A_5, C\}$ in above (4.64) as an example: if Charlie's measurement result is $|0\rangle_C$, then the six qubits $\{B_3, B_4, B_5, A_3, A_4, A_5\}$ collapsed into

$$|M\rangle_{B_3B_4B_5A_3A_4A_5} = {}_{a_1a_2A_1}\langle\phi^+|_{b_1b_2B_1}\langle\gamma^+|_{a3A_2}\langle\varepsilon^+|_{b3B_2}\langle\varepsilon^+|_C\langle 0|\Big|\sum\Big\rangle_{a_1a_2A_1B_3B_4a_3A_2B_5b_1b_2B_1A_3A_4b_3B_2A_5C} = \frac{1}{8}\begin{bmatrix}(\alpha_0|000\rangle + \alpha_1|001\rangle + \alpha_2|110\rangle + \alpha_3|111\rangle)_{B_3B_4B_5} \\ \otimes(\beta_0|000\rangle + \beta_1|001\rangle + \beta_2|110\rangle + \beta_3|111\rangle)_{A_3A_4A_5}\end{bmatrix}. \tag{4.65}$$

Otherwise, the six qubits $\{B_3, B_4, B_5, A_3, A_4, A_5\}$ collapsed into

$$|M\rangle_{B_3B_4B_5A_3A_4A_5} = {}_{a_1a_2A_1}\langle\phi^+|_{b_1b_2B_1}\langle\gamma^+|_{a3A_2}\langle\varepsilon^+|_{b3B_2}\langle\varepsilon^+|_C\langle 1|\Big|\sum\Big\rangle_{a_1a_2A_1B_3B_4a_3A_2B_5b_1b_2B_1A_3A_4b_3B_2A_5C} = \frac{1}{8}\begin{bmatrix}+(\alpha_0|101\rangle + \alpha_1|100\rangle + \alpha_2|011\rangle + \alpha_3|010\rangle)_{B_3B_4B_5} \\ \otimes(\beta_0|101\rangle + \beta_1|100\rangle + \beta_2|011\rangle + \beta_3|011\rangle)_{A_3A_4A_5}\end{bmatrix}. \tag{4.66}$$

Step 4. Unitary transformations

According to Charlie's results, Alice and Bob can utilize the four different unitary operators $\{I, X, iY, Z\}$ to reconstruct the transmitted state by each other. Four unitary operators are given as

$$\begin{aligned} I &= |0\rangle\langle 0| + |1\rangle\langle 1|, \quad X = |0\rangle\langle 1| + |1\rangle\langle 0|, \\ iY &= |1\rangle\langle 0| - |0\rangle\langle 1|, \quad Z = |0\rangle\langle 0| - |1\rangle\langle 1|, \end{aligned} \tag{4.67}$$

Through implementing the appropriate unitary transformations, three qubits in Alice's side and Bob's side can, respectively, transform into following states:

$$\begin{aligned} |\Phi\rangle_{B_3B_4B_5} &= \alpha_0|000\rangle + \alpha_1|001\rangle + \alpha_2|110\rangle + \alpha_3|111\rangle \\ |\chi\rangle_{A_3A_4A_5} &= \beta_0|010\rangle + \beta_1|011\rangle + \beta_2|100\rangle + \beta_3|101\rangle \end{aligned} \tag{4.68}$$

On the basis of both Alice's and Bob's GHZ state measurement results $|\phi^+\rangle_{a_1a_2A_1}$ and $|\gamma^+\rangle_{b_1b_2B_1}$, the specific Bell state measurement results of Alice and Bob, and Charlie's measurement results, Table 4.8 gives the specific unitary operators respectively implemented by Alice and Bob to transform their three-qubit state into the desired states shown in (4.68).

Table 4.8 On the basis of Alice's and Bob's GHZ state measurement results $|\phi^+\rangle_{a_1a_2A_1}$ and $|\gamma^+\rangle_{b_1b_2B_1}$ of Step 1, the specific unitary transformation employed to transmit collapsed states of qubits $\{B_3, B_4, B_5, A_3, A_4, A_5\}$ into the desired state shown in (4.68). corresponds to Alice's and Bob's Bell state measurement results and Charlie's single-qubit measurement results

Alice's measurement results	Bob's measurement results	Charlie's measurement results	Unitary transformation on qubits $\{B_3, B_4, B_5, A_3, A_4, A_5\}$
$\|\varepsilon^+\rangle_{a3A_2}$	$\|\varepsilon^+\rangle_{b3B_2}$	$\|0\rangle$	$I\otimes I\otimes I\otimes I\otimes I\otimes I$ $I\otimes I\otimes I\otimes I\otimes X\otimes I$
$\|\varepsilon^+\rangle_{a3A_2}$	$\|\varepsilon^+\rangle_{b3B_2}$	$\|1\rangle$	$X\otimes I\otimes X\otimes X\otimes X\otimes X$
$\|\varepsilon^+\rangle_{a3A_2}$	$\|\varepsilon^-\rangle_{b3B_2}$	$\|0\rangle$	$I\otimes I\otimes I\otimes I\otimes X\otimes Z$
$\|\varepsilon^+\rangle_{a3A_2}$	$\|\varepsilon^-\rangle_{b3B_2}$	$\|1\rangle$	$X\otimes I\otimes X\otimes X\otimes X(X\otimes Z)$
$\|\varepsilon^+\rangle_{a3A_2}$	$\|\nu^+\rangle_{b3B_2}$	$\|0\rangle$	$I\otimes I\otimes I\otimes I\otimes X\otimes X$
$\|\varepsilon^+\rangle_{a3A_2}$	$\|\nu^+\rangle_{b3B_2}$	$\|1\rangle$	$X\otimes I\otimes X\otimes X\otimes X\otimes I$
$\|\varepsilon^+\rangle_{a3A_2}$	$\|\nu^-\rangle_{b3B_2}$	$\|0\rangle$	$I\otimes I\otimes I\otimes X\otimes X\otimes (Z\otimes X)$
$\|\varepsilon^+\rangle_{a3A_2}$	$\|\nu^-\rangle_{b3B_2}$	$\|1\rangle$	$X\otimes I\otimes X\otimes X\otimes X\otimes Z$
$\|\varepsilon^-\rangle_{a3A_2}$	$\|\varepsilon^+\rangle_{b3B_2}$	$\|0\rangle$	$Z\otimes Z\otimes Z\otimes I\otimes I\otimes I$
$\|\varepsilon^-\rangle_{a3A_2}$	$\|\varepsilon^+\rangle_{b3B_2}$	$\|1\rangle$	$X\otimes X(X\otimes Z)\otimes X\otimes X\otimes X$
$\|\varepsilon^-\rangle_{a3A_2}$	$\|\varepsilon^-\rangle_{b3B_2}$	$\|0\rangle$	$Z\otimes Z\otimes Z\otimes \quad \otimes I\otimes X\otimes Z$
$\|\varepsilon^-\rangle_{a3A_2}$	$\|\varepsilon^-\rangle_{b3B_2}$	$\|1\rangle$	$X\otimes X\otimes (X\otimes Z)\otimes X\otimes X(X\otimes Z)$
$\|\varepsilon^-\rangle_{a3A_2}$	$\|\nu^+\rangle_{b3B_2}$	$\|0\rangle$	$Z\otimes Z\otimes Z\otimes I\otimes X\otimes X$
$\|\varepsilon^-\rangle_{a3A_2}$	$\|\nu^+\rangle_{b3B_2}$	$\|1\rangle$	$X\otimes X(X\otimes Z)\otimes X\otimes X\otimes I$
$\|\varepsilon^-\rangle_{a3A_2}$	$\|\nu^-\rangle_{b3B_2}$	$\|0\rangle$	$Z\otimes Z\otimes Z\otimes X\otimes X\otimes (Z\otimes X)$
$\|\varepsilon^-\rangle_{a3A_2}$	$\|\nu^-\rangle_{b3B_2}$	$\|1\rangle$	$X\otimes X(X\otimes Z)\otimes X\otimes X\otimes Z$
$\|\nu^+\rangle_{a3A_2}$	$\|\varepsilon^+\rangle_{b3B_2}$	$\|0\rangle$	$I\otimes I\otimes X\otimes I\otimes I\otimes I$
$\|\nu^+\rangle_{a3A_2}$	$\|\varepsilon^+\rangle_{b3B_2}$	$\|1\rangle$	$X\otimes I\otimes I\otimes X\otimes X\otimes X$
$\|\nu^+\rangle_{a3A_2}$	$\|\varepsilon^-\rangle_{b3B_2}$	$\|0\rangle$	$I\otimes I\otimes X\otimes I\otimes X\otimes Z$
$\|\nu^+\rangle_{a3A_2}$	$\|\varepsilon^-\rangle_{b3B_2}$	$\|1\rangle$	$X\otimes I\otimes I\otimes X\otimes X(X\otimes Z)$
$\|\nu^+\rangle_{a3A_2}$	$\|\nu^+\rangle_{b3B_2}$	$\|0\rangle$	$I\otimes I\otimes X\otimes I\otimes X\otimes X$
$\|\nu^+\rangle_{a3A_2}$	$\|\nu^+\rangle_{b3B_2}$	$\|1\rangle$	$X\otimes I\otimes I\otimes X\otimes X\otimes \mathrm{I}$
$\|\nu^+\rangle_{a3A_2}$	$\|\nu^-\rangle_{b3B_2}$	$\|0\rangle$	$I\otimes I\otimes X\otimes X\otimes X\otimes (Z\otimes X)$
$\|\nu^+\rangle_{a3A_2}$	$\|\nu^-\rangle_{b3B_2}$	$\|1\rangle$	$X\otimes I\otimes I\otimes X\otimes X\otimes Z$
$\|\nu^-\rangle_{a3A_2}$	$\|\varepsilon^+\rangle_{b3B_2}$	$\|0\rangle$	$I\otimes I\otimes (X\otimes Z)\otimes I\otimes I\otimes I$
$\|\nu^-\rangle_{a3A_2}$	$\|\varepsilon^+\rangle_{b3B_2}$	$\|1\rangle$	$X\otimes I\otimes Z\otimes X\otimes X\otimes X$
$\|\nu^-\rangle_{a3A_2}$	$\|\varepsilon^-\rangle_{b3B_2}$	$\|0\rangle$	$I\otimes I\otimes (X\otimes Z)\otimes \mathrm{I}\otimes X\otimes Z$
$\|\nu^-\rangle_{a3A_2}$	$\|\varepsilon^-\rangle_{b3B_2}$	$\|1\rangle$	$X\otimes I\otimes Z\otimes X\otimes X(X\otimes Z)$
$\|\nu^-\rangle_{a3A_2}$	$\|\nu^+\rangle_{b3B_2}$	$\|0\rangle$	$I\otimes I\otimes (X\otimes Z)\otimes I\otimes X\otimes X$
$\|\nu^-\rangle_{a3A_2}$	$\|\nu^+\rangle_{b3B_2}$	$\|1\rangle$	$X\otimes I\otimes Z\otimes X\otimes X\otimes I$
$\|\nu^-\rangle_{a3A_2}$	$\|\nu^-\rangle_{b3B_2}$	$\|0\rangle$	$I\otimes I\otimes (X\otimes Z)\otimes X\otimes X\otimes (Z\otimes X)$
$\|\nu^-\rangle_{a3A_2}$	$\|\nu^-\rangle_{b3B_2}$	$\|1\rangle$	$X\otimes I\otimes Z\otimes X\otimes X\otimes Z$

4.4.2 Efficiency Analysis

On the basis of an 11-qubit entangled state acting as quantum channel, the C-BQT protocol of transmitting two specific three-qubit states proposed in this paper is realized through two GHZ state measurements, two Bell state measurements, single-qubit measurement, and some exchanged classical information. Hence, the protocol efficiency of the presented protocol is defined as

$$\kappa = \frac{q_t}{q_c + b_t} \tag{4.69}$$

where q_t indicates the total number of qubits that Alice and Bob want to teleport, q_c indicates the number of qubits of the quantum channel, and b_t indicates the number of classical bits transmitted.

Consider that the GHZ state measurement respectively performed by Alice and Bob only four different results, thus, their measurement results can be mapped into 2 bits information. Hence, the efficiency of the proposed protocol is given as

$$\kappa = \frac{3+3}{11+4+4+1} = 30\%. \tag{4.70}$$

The comparisons between our protocol and previous works are illustrated in Table 4.9. Therein, letter "*a*" corresponds to the arbitrary unknown states, letter "*p*" corresponds to the unknown pure states, and letter "*s*" corresponds to special unknown states presented in this paper, i.e., three qubits $|\Phi\rangle_{a_1a_2a_3}$ and $|\chi\rangle_{b_1b_2b_3}$ shown in previous (4.53) and (4.54).

For arbitrary single-qubit, two-qubit, and three-qubit states in refs. [37, 38], the general form can be written as

Table 4.9 Comparison of different protocols with our protocols in efficiency

Schemes	The type of schemes	The number of transmitted qubits and the type of states	The number of qubits that are used as quantum channel	Efficiency	The number of different measurement operations: GHZ states	Bell states	Single qubit
[37]	B-CQT	(a)1 and (a)3	9	22.2%	0	0	9
[38]	B-CQT	(a)2 and (a)3	11	23.8%	0	5	0
[39]	B-CQT	(p)2 and (p)3	9	29.4%	2	1	0
Ours	C-ABQT	(s)3 and (s)3	11	30%	2	2	1

$$\begin{aligned}|A\rangle_1 &= a_0|0\rangle + a_1|1\rangle \\ |A\rangle_{12} &= b_0|00\rangle + b_1|01\rangle + b_2|10\rangle + b_3|11\rangle \\ |A\rangle_{123} &= c_0|000\rangle + c_1|001\rangle + c_2|010\rangle + c_3|011\rangle \\ &\quad + c_4|100\rangle + c_5|101\rangle + c_6|110\rangle + c_7|111\rangle\end{aligned} \tag{4.71}$$

where all parameters in (4.71) are complex numbers and satisfy the normalization condition expressed as $|a_0|^2 + |a_1|^2 = 1$, $\sum_{i=0}^{3}|b_i|^2 = 1$, and $\sum_{i=0}^{7}|c_i|^2 = 1$.

For the two-qubit pure states in ref. [35], it can be expressed as

$$|P\rangle_{12} = a|00\rangle + b|11\rangle, |P\rangle_{123} = a|000\rangle + b|111\rangle \tag{4.72}$$

where the complex numbers a and b also satisfy the normalization condition $|a|^2 + |b|^2 = 1$.

Hence, on the basis of comparisons illustrated in Table 4.9, our proposed scheme is quite different to other schemes [37–39]. Firstly, this scheme can transmit two totally different three-qubit states simultaneously. Secondly, even the number of channel qubits becomes larger, and the proposed scheme has better efficiency than other three schemes.

4.4.3 Summary

In summary, a novel C-BQT scheme is proposed in this paper, which can teleport two different unknown three-qubit states at the same time. To achieve this aim, an 11-qubit entangled state is employed as the quantum channel shared among two participants, Alice and Bob, and a controller, Charlie, in advance. Then, Alice and Bob implement GHZ state measurement and Bell state measurement orderly; Charlie finally implements single-qubit measurement. On the basis of all measurement results, both Alice and Bob can transmit a three-qubit state and receive a three-qubit state in the meantime. Compared to the former existing research, our protocol poses the following two distinct advantages: (1) our proposed C-BQT protocol can transmit two totally different three-qubit states simultaneously with higher scheme efficiency and (2) the quantum channel employed in here can be generated from single-qubit, two-qubit Bell states, and three-qubit GHZ states.

References

1. Li YH, Li XL, Sang MH, Nie YY, Wang ZS (2013) Bidirectional controlled quantum teleportation and secure direct communication using five-qubit entangled state. Quantum Inf Process 12:3835–3844

2. Sun X, Zha X (2013) A scheme of bidirectional quantum controlled teleportation via six-qubit maximally entangled state. Acta Photonica Sin 42:1052
3. Yan A (2013) Bidirectional controlled teleportation via six-qubit cluster state. Int J Theor Phys 52:3870–3873
4. Hong W (2016) Qin: asymmetric bidirectional controlled teleportation by using a seven-qubit entangled state. Int J Theor Phys 55:384–387
5. Jiang S-X, Zhou R-G, Luo G, Liang X, Fan P (2020) Controlled bidirectional quantum teleportation of arbitrary single qubit via a non-maximally entangled state. Int J Theor Phys 59(9):2966–2983
6. Zhou R-G, Zhang Y-N (2019) Bidirectional quantum controlled teleportation of three-qubit state by using GHZ states. Int J Theor Phys 58(10):3594–3601
7. Ya-Li J, Ri-Gui Z, Hao D-Y, WenWen H (2021) Bidirectional controlled quantum teleportation of three-qubit state by a new entangled eleven-qubit state. Int J Theor Phys 60(9):3618–3630
8. Linden N, Massar S, Popescu S (1998) Purifying noisy entanglement requires collective measurements. Phys Rev Lett 81:3279–3282
9. Li WL, Li CF, Guo GC (2000) Probabilistic teleportation and entanglement matching. Phys Rev A – At Mol Opt Phys 61:3
10. Lu H (2001) Probabilistic teleportation of the three-particle entangled state via entanglement swapping. Chin Phys Lett 18:1004–1006
11. Agrawal P, Pati AK (2002) Probabilistic quantum teleportation. Phys Lett Sect A Gen At Solid State Phys 305:12–17
12. Dai HY, Chen PX, Li CZ (2004) Probabilistic teleportation of an arbitrary two-particle state by two partial three-particle entangled W states. J Opt B Quantum Semiclassical Opt 6:106–109
13. Dai HY, Chen PX, Li CZ (2004) Probabilistic teleportation of an arbitrary two-particle state by a partially entangled three-particle GHZ state and W state. Opt Commun 231:281–287
14. Han LF, Yuan H (2008) Probabilistic teleportation of an arbitrary two-qubit state via one-dimensional four-qubit cluster-class state. Int J Quantum Inf 6:1093–1099
15. Choudhury BS, Dhara A (2016) Probabilistically teleporting arbitrary two-qubit states. Quantum Inf Process 15:5063–5071
16. Gou YT, Shi HL, Wang XH, Liu SY (2017) Probabilistic resumable bidirectional quantum teleportation. Quantum Inf Process 16:1–13
17. Wei J, Shi L, Han C, Xu Z, Zhu Y, Wang G, Wu H (2018) Probabilistic teleportation via multi-parameter measurements and partially entangled states. Quantum Inf Process 17:1–8
18. Yu LZ, Wu T (2013) Probabilistic teleportation of three-qubit entangled state via five-qubit cluster state. Int J Theor Phys 52:1461–1465
19. Kim YS, Lee JC, Kwon O, Kim YH (2012) Protecting entanglement from decoherence using weak measurement and quantum measurement reversal. Nat Phys 8:117
20. Lee JC, Jeong YC, Kim YS, Kim YH (2011) Experimental demonstration of decoherence suppression by quantum measurement reversal. In: Optics InfoBase Conference Papers
21. Banerjee A, Pathak A (2012) Maximally efficient protocols for direct secure quantum communication. Phys Lett Sect A Gen At Solid State Phys 376:2944–2950
22. Zha XW, Zou ZC, Qi JX, Song HY (2013) Bidirectional quantum controlled teleportation via five-qubit cluster state. Int J Theor Phys 52:1740–1744
23. Zhou RG, Zhang YN, Xu R, Qian C, Ian H (2019) Asymmetric bidirectional controlled teleportation by using nine-qubit entangled state in noisy environment. IEEE Access 7: 75247–75264
24. Hassanpour S, Houshmand M (2016) Bidirectional teleportation of a pure EPR state by using GHZ states. Quantum Inf Process 15:905–912
25. Li YH, Nie LP, Li XL, Sang MH (2016) Asymmetric bidirectional controlled teleportation by using six-qubit cluster state. Int J Theor Phys 55:3008–3016
26. Zhou R, Qian C, Ian H (2018) Bidirectional quantum teleportation of two-qubit state via four-qubit cluster state. Int J Theor Phys 58:150

27. Sisodia M, Pathak A (2018) Comment on "quantum teleportation of eight-qubit state via six-qubit cluster state". Int J Theor Phys 57:2213–2217
28. Joy D, Sabir M (2018) Efficient schemes for the quantum teleportation of a sub-class of tripartite entangled states. Quantum Inf Process 17:1–11
29. Zadeh MSS, Houshmand M, Aghababa H (2018) Bidirectional quantum teleportation of a class of n-qubit states by using (2n + 2)-qubit entangled states as quantum channel. Int J Theor Phys 57:175–183
30. Cai T, Jiang M (2018) Improving the teleportation scheme of three-qubit state with a four-qubit quantum channel. Int J Theor Phys 57:131–137
31. Frey MR (2016) Efficiency of quantum energy teleportation within particle pairs. Quantum Inf Process 15:1103–1116
32. Li W, Qi XZJ (2016) Tripartite quantum controlled teleportation via seven-qubit cluster state. Int J Theor Phys 9:3927–3933
33. Runhua SHI, Liusheng H, Wei Y, Hong Z (2011) Controlled quantum perfect teleportation of multiple arbitrary multi-qubit states. Sci CHINA Physics, Mech Astron 54:2208–2216
34. Yang Y, Jiang M, Zhou L (2018) Improving the teleportation scheme of five-qubit state with a seven-qubit quantum channel. Int J Theor Phys 57:3485–3491
35. Yu LZ, Sun SX, Teleportation of three-qubit state via six-qubit cluster state (2015) Int J Theor Phys 54:1600–1603
36. Li YH, Sang MH, Wang XP, Nie Y (2016) You: quantum teleportation of a four-qubit state by using six-qubit cluster state. Int J Theor Phys 55:3547–3550
37. Jiang SX, Zhou RG, Xu R, Luo G (2019) Cyclic hybrid double-channel quantum communication via bell-state and GHZ-state in noisy environments. IEEE Access 7:80530–80541
38. Huo G, Zhang T, Zha X, Zhang X, Zhang M (2021) Controlled asymmetric bidirectional quantum teleportation of two- and three-qubit states. Quantum Inf Process 20:1–11
39. Choudhury BS, Samanta S (2017) Asymmetric bidirectional 3 $\Leftrightarrow$ 2 qubit teleportation protocol between Alice and bob via 9-qubit cluster state. Int J Theor Phys 56:3285–3296

Chapter 5
Cyclic Quantum Teleportation

Abstract Cyclic quantum teleportation represents a cutting-edge research direction in the fields of quantum communication and quantum computation. By enabling the cyclic transmission of quantum states among multiple locations, it offers new possibilities for constructing complex quantum networks and achieving efficient quantum computation. However, realizing cyclic quantum teleportation and its widespread practical application necessitates overcoming a series of technical and practical challenges. With the continuous development and refinement of quantum technologies, it is believed that cyclic quantum teleportation will play a significant role in the future.

5.1 Introduction

The theoretical research of quantum teleportation has developed from single-particle transmission to asymmetric quantum teleportation, controlled quantum teleportation, bidirectional quantum teleportation, and so on. The previous teleportation scheme and the C-BQT scheme are all information transfer between two parties. The cyclic quantum teleportation, a brand-new model of transmission, has been proposed lately.

In 2017, Chen [1] proposed a scheme of cyclic quantum teleportation using a six-qubit maximally entangled state as a quantum channel. In the presented scheme, there are three observers Alice, Bob, and Charlie, where Alice can transfer an arbitrary single-qubit state to Bob, Bob can transfer an arbitrary single-qubit state to Charlie, and Charlie can also transfer an arbitrary single-qubit state to Alice.

Then, Sang [2] presented a scheme that realized cyclic controlled quantum teleportation using a seven-qubit entangled state as a quantum channel. Recently, Sisodia [3] proposed an optimal scheme for teleportation of an n-qubit quantum state, and Toy [4] presented an efficient scheme for the quantum teleportation of a subclass of tripartite entangled states; both of them have designed the corresponding quantum circuit, and the latter has done the simulation using IBM's real quantum processor.

In this chapter, two cyclic quantum teleportation schemes are proposed, which uses pseudo-multi-qubit states and nine-qubit entangled state.

R.-G. Zhou et al., *Design of Quantum Teleportation Schemes*,
https://doi.org/10.1007/978-3-031-82725-9_5

In Sect. 5.2, we demonstrate a novel and efficient scheme on the basis of combination between cyclic model and traditional bidirectional model of teleportation [5], where quantum channel is a pseudo-multi-qubit state and generated with the help of controlled-NOT (CNOT). Alice, Bob, and Charlie, the three legitimate users, hold two-qubit entangled states, respectively, and teleport their qubits to each other; the scheme can be prepared well by appropriate unitary operation. Compared with the schemes above, this is a more efficient idea.

In Sect. 5.3, we proposed an asymmetric cyclic controlled quantum teleportation (C-ACQT) scheme to complete the task of transferring an arbitrary two-qubit state from Alice to Bob, an arbitrary two-qubit state from Bob to Charlie, and an arbitrary single-qubit state from Charlie to Alice by the nine-qubit entangled state with the cooperation of controller David [6]. The senders Alice, Bob, and Charlie perform a GHZ state measurement and Bell state measurement, respectively, and the success probability of our scheme is 100%.

5.2 C-BQT Scheme via Pseudo-multi-qubit States

To teleport more particles among three particles and realize a both original and efficient scheme, here, we proposed a C-BQT scheme, which realizes the cyclic and bidirectional teleportation via pseudo-multi-qubit entangled states. To realize the teleportation completely, it utilizes controlled-NOT operation to obtain quantum channel whose number of qubits reaches 12. In this scheme, Alice can teleport her qubits to Bob and Charlie, and Bob can teleport his qubits to Alice and Charlie, while Charlie teleports his qubits to Alice and Bob, respectively; they reconstruct the original state and make the scheme perfectly by performing Bell state measurement and appropriate unitary operation.

5.2.1 C-BQT Scheme for Tripartite Teleportation of Two-Qubit Entangled States

a) Generation of the pseudo-multi-qubit state

It is necessary to prepare the multi-qubit states with enough entangled particles to achieve this cyclic bidirectional teleportation. Here, we prepare a six-qubit entangled state firstly, and it can be described as follows:

$$
\begin{aligned}
|\psi\rangle_6 = \frac{1}{8}(&|000000\rangle + |000001\rangle + |000010\rangle \\
&+|000011\rangle + |000100\rangle + |000101\rangle + |000110\rangle \\
&+|000111\rangle + \ldots + |100000\rangle + \ldots + |111111\rangle)_{123456}.
\end{aligned} \tag{5.1}
$$

Then we can obtain the pseudo-multi-qubit states by "copying" every qubit. In $|\psi\rangle_6$, the specific way is doing CNOT operations on the six-qubit cluster state and auxiliary qubits $|0\rangle_t$, where the six qubits and auxiliary qubits work as control qubits and target qubits, respectively; the action of the operation can be represented by

$$|1\rangle|2\rangle|3\rangle|4\rangle|5\rangle|6\rangle \rightarrow |1\rangle|t\oplus 1\rangle|2\rangle|t\oplus 2\rangle|3\rangle|t\oplus 3\rangle|4\rangle|t\oplus 4\rangle|5\rangle|t\oplus 5\rangle|6\rangle|t\oplus 6\rangle. \tag{5.2}$$

After the operation, the whole state will be changed into the following state:

$$\begin{aligned}|\psi\rangle_{12} = \frac{1}{8}(&|0000000000000\rangle + |000000000011\rangle + |000000001100\rangle \\ &+ |000000001111\rangle + |000000110000\rangle + |000000110011\rangle \\ &+ |000000111100\rangle + |000000111111\rangle + \ldots + |110000000000\rangle \\ &+ \ldots + |111111111111\rangle)_{1,2,3,4,5,6,7,8,9,10,11,12}.\end{aligned} \tag{5.3}$$

The state has been expanded through the CNOT, and the state $|\psi\rangle_{12}$ above will be utilized as the quantum channel for implementing the bidirectional teleportation among three legitimate participants in this scheme.

b) Description of the C-BQT scheme

In this scheme, the three users, Alice, Bob, and Charlie, have an arbitrary two-qubit entangled state, respectively, which are given by

$$\begin{aligned}|\psi\rangle_\alpha &= a_0|00\rangle_{\alpha_1\alpha_2} + a_1|11\rangle_{\alpha_1\alpha_2} \\ |\psi\rangle_\beta &= b_0|00\rangle_{\beta_1\beta_2} + b_1|11\rangle_{\beta_1\beta_2} \\ |\psi\rangle_\gamma &= c_0|00\rangle_{\gamma_1\gamma_2} + c_1|11\rangle_{\gamma_1\gamma_2}.\end{aligned} \tag{5.4}$$

Here, assume that Alice holds qubit α and she wants to transfer α_1 and α_2 to Bob and Charlie, respectively. Bob owns qubit β, and he wants to transfer β_1 to Charlie and β_2 to Alice, while Charlie possesses qubit γ, and he will transfer qubits γ_1 and γ_2 to Alice and Bob, respectively. In addition, a_0, a_1, b_0, b_1, c_0, c_1 are amplitudes satisfying normalization that $|a_0|^2 + |a_1|^2 = 1$, $|b_0|^2 + |b_1|^2 = 1$, and $|c_0|^2 + |c_1|^2 = 1$.

The schematic demonstration and circuit are illustrated in Figures 5.1 and 5.2.

The scheme consists of the following steps:

Step 1. Suppose that three legitimate users Alice, Bob, and Charlie share the multi-qubit entangled state above as (5.2), where qubits 3, 6, 9, and 12 belong to Alice; qubits 1, 4, 7, and 10 belong to Bob; and qubits 2, 5, 8, and 11 belong to Charlie, respectively, so the multi-qubit entangled state can be rewritten as (5.5)

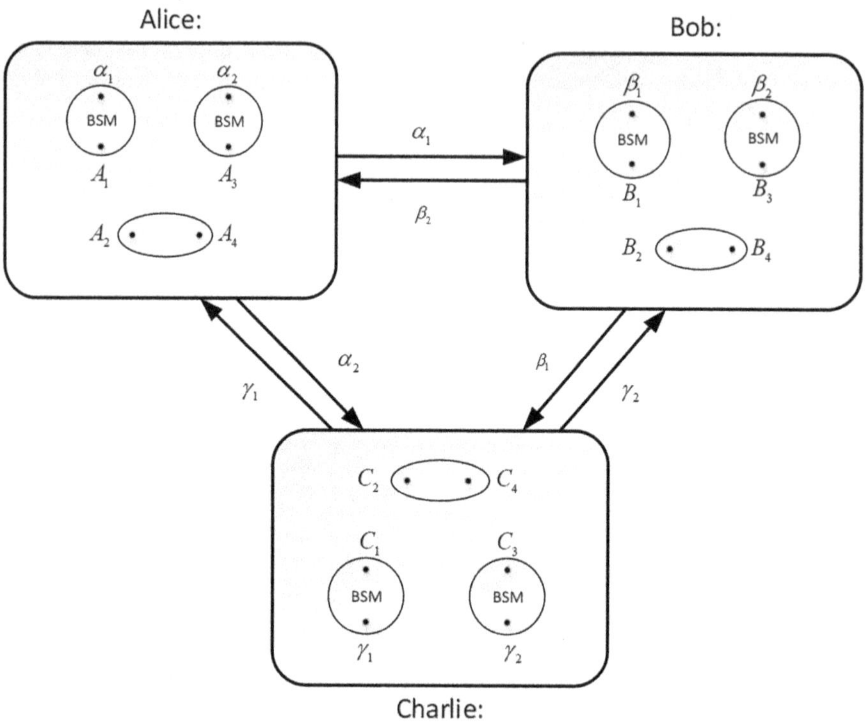

Fig. 5.1 The schematic of C-ABQT scheme

$$
\begin{aligned}
|\psi\rangle_{12} = \frac{1}{8}(&|0000000000000\rangle + |000000000011\rangle + |000000001100\rangle \\
&+ |000000001111\rangle + |000000110000\rangle + |000000110011\rangle \\
&+ |000000111100\rangle + |000000111111\rangle + \ldots + |110000000000\rangle \\
&+ \ldots + |111111111111\rangle)_{B_1C_2A_1B_2C_1A_2B_3C_4A_3B_4C_3A_4}.
\end{aligned} \tag{5.5}
$$

All the qubits are in pure product states, and the whole system here can be expressed by

$$
|\Psi\rangle_{\alpha_1\alpha_2\beta_1\beta_2\gamma_1\gamma_2B_1C_2A_1B_2C_1A_2B_3C_4A_3B_4C_3A_4} = |\psi\rangle_\alpha \otimes |\psi\rangle_\beta \otimes |\psi\rangle_\gamma \otimes |\psi\rangle_{12}. \tag{5.6}
$$

Step 2. In order to realize the cyclic and bidirectional teleportation, Alice must perform a complete Bell state measurement (BSM) on her qubits α_1 and A_1; the Bell state measurement basics are given by

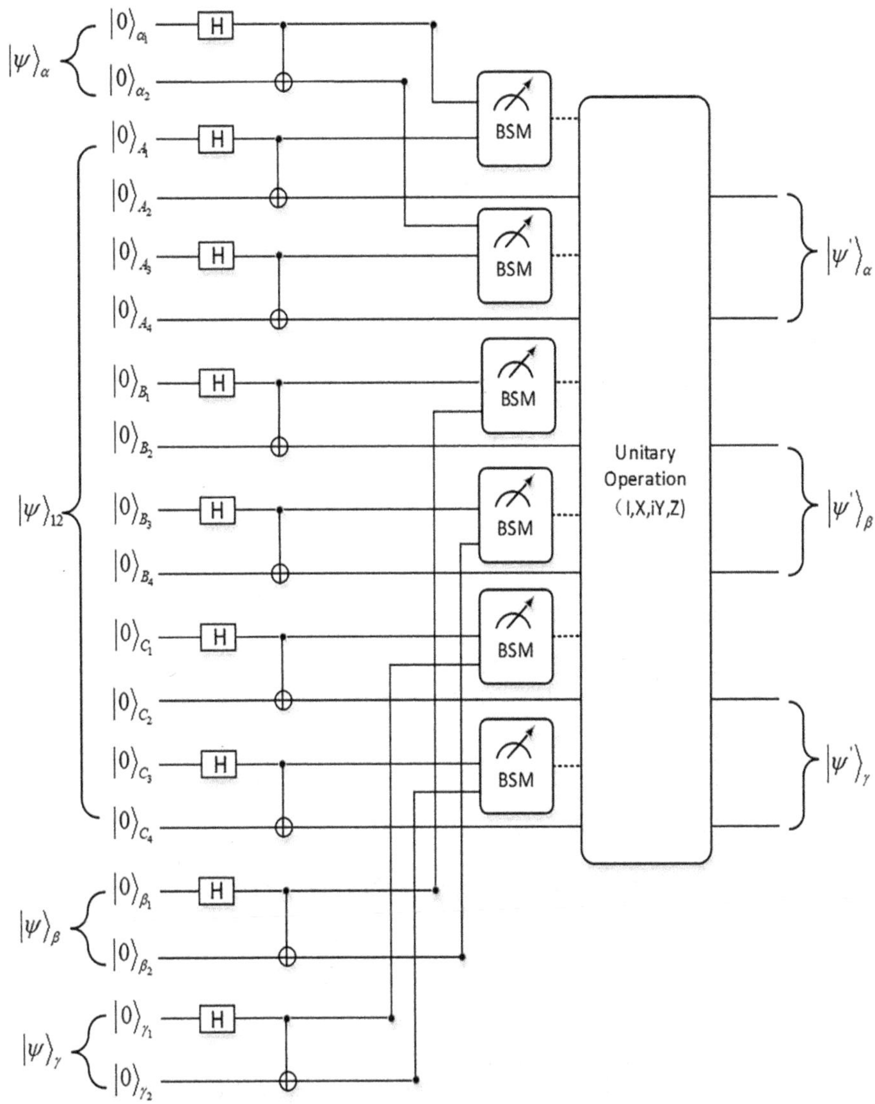

Fig. 5.2 The circuit for C-BQT scheme of a two-qubit entangled state using pseudo-multi-qubit state

$$
\begin{aligned}
|\Phi^+\rangle &= \frac{1}{\sqrt{2}}(|00\rangle + |11\rangle) \\
|\Phi^-\rangle &= \frac{1}{\sqrt{2}}(|00\rangle - |11\rangle) \\
|\Psi^+\rangle &= \frac{1}{\sqrt{2}}(|01\rangle + |10\rangle) \\
|\Psi^-\rangle &= \frac{1}{\sqrt{2}}(|01\rangle - |10\rangle).
\end{aligned} \tag{5.7}
$$

If Alice's BSM result was $|\Phi^+\rangle_{\alpha_1 A_1}$, the remaining particles might collapse into the product state

$$
\begin{aligned}
|\psi\rangle_{16} = &\left(b_0|00\rangle_{\beta_1\beta_2} + b_1|11\rangle_{\beta_1\beta_2}\right) \otimes \left(c_0|00\rangle_{\gamma_1\gamma_2} + c_1|11\rangle_{\gamma_1\gamma_2}\right) \\
&\otimes \frac{\sqrt{2}}{8}(a_0|0000\rangle + a_0|0110\rangle + a_1|1001\rangle + a_1|1111\rangle)_{\alpha_2 B_1 C_2 B_2} \\
&\otimes (|0000\rangle + |0011\rangle + |1100\rangle + |1111\rangle)_{C_1 A_2 B_3 C_4} \\
&\otimes (|0000\rangle + |0011\rangle + |1100\rangle + |1111\rangle)_{A_3 B_4 C_3 A_4}.
\end{aligned} \tag{5.8}
$$

Step 3. Subsequently, Bob carries out a BSM on his own qubits β_2 and β_3; the basics are the same as (5.9) above. If the result was $|\Phi^+\rangle_{\beta_2 B_3}$, the rest of particles would collapse into a new state

$$
\begin{aligned}
|\psi\rangle_{14} = &\left(c_0|00\rangle_{\gamma_1\gamma_2} + c_1|11\rangle_{\gamma_1\gamma_2}\right) \\
&\otimes \frac{1}{4}(a_0|0000\rangle + a_0|0110\rangle + a_1|1001\rangle + a_1|1111\rangle)_{\alpha_2 B_1 C_2 A_2}. \\
&\otimes (b_0|0000\rangle + b_1|0011\rangle + b_0|1100\rangle + b_1|1111\rangle)_{\beta_1 C_1 A_2 C_4} \\
&\otimes (|0000\rangle + |0011\rangle + |1100\rangle + |1111\rangle)_{A_3 B_4 C_3 A_4}
\end{aligned} \tag{5.9}
$$

Step 4. At the same time, Charlie measures his qubits γ_1 and C_1 based on the same basis; if the result was $|\Phi^+\rangle_{\gamma_1 C_1}$, the other qubits would collapse into the following state:

$$
\begin{aligned}
|\psi\rangle_{12} = &\frac{\sqrt{2}}{4}(a_0|0000\rangle + a_0|0110\rangle + a_1|1001\rangle + a_1|1111\rangle)_{\alpha_2 B_1 C_2 B_2} \\
&\otimes (c_0 b_0|0000\rangle + c_0 b_1|0101\rangle + c_1 b_0|1010\rangle + c_1 b_1|1111\rangle)_{\gamma_2 \beta_1 A_2 C_4}. \\
&\otimes (|0000\rangle + |0011\rangle + |1100\rangle + |1111\rangle)_{A_3 B_4 C_3 A_4}
\end{aligned} \tag{5.10}
$$

Step 5. It will realize the reversely circulatory teleportation in this step. Firstly, Alice sticks at measuring her qubits α_2 and A_3; when the result of measurement is $|\Phi^+\rangle_{\alpha_2 A_3}$, the remaining qubits are collapsed into the state

$$\begin{aligned}|\psi\rangle_{10} = & \frac{1}{2}(a_0|000000\rangle + a_0|000011\rangle + a_0|110000\rangle + a_0|110011\rangle \\ & + a_1|001100\rangle + a_1|001111\rangle + a_1|111100\rangle + a_1|111111\rangle)_{B_1C_2A_2B_4C_3A_4} \\ & \otimes (c_0b_0|0000\rangle + c_0b_1|0101\rangle + c_1b_0|1010\rangle + c_1b_1|1111\rangle)_{\gamma_2\beta_1A_2C_4}\end{aligned} \quad (5.11)$$

Step 6. Charlie performs the BSM on qubits γ_2 and C_1, and if the result was $|\Phi^+\rangle_{\gamma_2 C_1}$, the other qubits might collapse to the following state:

$$\begin{aligned}|\psi\rangle_8 = & \frac{\sqrt{2}}{2}\Big[(a_0|00000\rangle + a_1|00110\rangle + a_0|11000\rangle + a_1|11110\rangle)_{B_1C_2B_2B_4A_4} \\ & \otimes (c_0b_0|000\rangle + c_0b_1|101\rangle)_{\beta_1A_2C_4} + (a_0|00001\rangle + a_0|11001\rangle \\ & + a_1|00111\rangle + a_1|11111\rangle)_{B_1C_2B_2B_4A_4} \otimes (c_1b_0|010\rangle + c_1b_1|111\rangle)_{\beta_1A_2C_4}\Big]\end{aligned} \quad (5.12)$$

Step 7. Finally, Bob performs the BSM on his qubits β_1 and B_1; when the result is $|\Phi^+\rangle_{\gamma_1 B_1}$, the remaining particles are collapsed into the state shown in (5.13):

$$\begin{aligned}|\psi\rangle_6 = & a_0c_0b_0|000000\rangle + a_1c_0b_0|011000\rangle + a_0c_0b_1|100001\rangle \\ & + a_1c_0b_1|111001\rangle + a_0c_1b_0|000110\rangle + a_1c_1b_0|011110\rangle \\ & + a_0c_1b_1|100111\rangle + a_1c_1b_1|111111\rangle_{C_2B_2B_4A_4A_2C_4} \\ = & a_0b_0c_0|000000\rangle + a_0b_0c_1|000011\rangle + a_0b_1c_0|001100\rangle \\ & + a_0b_1c_1|001111\rangle + a_1b_0c_0|110000\rangle + a_1b_0c_1|110011\rangle \\ & + a_1b_1c_0|111100\rangle + a_1b_1c_1|111111\rangle_{B_2B_4C_2C_4A_4A_2}\end{aligned} \quad (5.13)$$

The formula above can be rewritten as follows:

$$\begin{aligned}|\psi\rangle_6 = & (c_0|00\rangle + c_1|11\rangle)_{A_2A_4} \\ & \otimes (a_0|00\rangle + a_1|11\rangle)_{B_2B_4} \\ & \otimes (b_0|00\rangle + b_1|11\rangle)_{C_2C_4}\end{aligned} \quad (5.14)$$

As is shown in (5.14) clearly, Alice, Bob, and Charlie can reconstruct the desired state after sharing their results of measurement via classical communication.

One of the possible outcomes of measurements and appropriate unitary operations are presented in Table 5.1 (the possible outcomes are so many but legitimate and verifiable, so we do not show them all).

There are 4^6 possible results of measurement in this scheme; according to which, the three legitimate participants can obtain the original target state by applying the appropriate unitary operation on their own qubits. Thus, the cyclic and bidirectional teleportation based on pseudo-multi-qubit states is successfully achieved.

5.2.2 *Efficiency Analysis and Comparison*

The most significant innovation of the C-BQT scheme reflected in realizing the tripartite teleportation among senders (receivers) Alice, Bob, and Charlie via multi-qubit states, which is not only cyclic but also bidirectional. Moreover, efficiency is a main factor utilized for weighing the superiority of schemes [7], and it is defined as (5.15)

$$\eta = \frac{c}{q}. \tag{5.15}$$

Here, the efficiency equals to the ratio of c (the number of qubits teleported in quantum channel) and q (the total number of qubits).

The C-BQT scheme realizes the teleportation of 2-qubit entangled states by using 12-qubit states as quantum channel among Alice, Bob, and Charlie, so the qubit efficiency of C-BQT scheme can reach 50%. Comparing with the scheme [1, 2], it is obvious that the C-BQT scheme is high-efficiency according to the date. In addition, the qubit efficiency can be calculated in another manner proposed by Banerjee and Pathak [8], where the classical qubits are put into consideration, and it can be described as (5.16)

$$\eta = \frac{q_s}{q_u + b_t}. \tag{5.16}$$

Here, b_t represents the number of qubits used as the classical channel. Because of performing the same BSM for six times, the qubit efficiency will be 25% in the second way, but it is still a respectable number.

5.2.3 *Summary*

In this section, it is the first time to demonstrate the scheme for cyclic and bidirectional teleportation, where the three legitimate users Alice, Bob, and Charlie are all in possession of two-qubit entangled states; Alice transmits her qubits to Bob and

Table 5.1 One of the possible outcomes of measurement and corresponding unitary operations

Alice's	Bob's	Charlie's	Alice's	Bob's	Charlie's	Unitary operation		
BSM 1	BSM 1	BSM 1	BSM 2	BSM 2	BSM 2	Alice	Bob	Charlie
$\|\Phi^+\rangle_{\alpha_1 A_1}$	$\|\Phi^+\rangle_{\beta_2 B_3}$	$\|\Phi^+\rangle_{\gamma_1 C_1}$	$\|\Phi^+\rangle_{\alpha_2 A_3}$	$\|\Phi^+\rangle_{\gamma_2 C_1}$	$\|\Phi^+\rangle_{\beta_1 B_1}$	$I_{A_2} \otimes I_{A_4}$	$I_{B_2} \otimes I_{B_4}$	$I_{C_2} \otimes I_{C_4}$
					$\|\Phi^-\rangle_{\beta_1 B_1}$	$I_{A_2} \otimes I_{A_4}$	$I_{B_2} \otimes I_{B_4}$	$\sigma^Z_{C_2} \otimes I_{C_4}$
					$\|\Psi^+\rangle_{\beta_1 B_1}$	$I_{A_2} \otimes I_{A_4}$	$I_{B_2} \otimes I_{B_4}$	$\sigma^X_{C_2} \otimes I_{C_4}$
					$\|\Psi^-\rangle_{\beta_1 B_1}$	$I_{A_2} \otimes I_{A_4}$	$I_{B_2} \otimes I_{B_4}$	$\sigma^{iY}_{C_2} \otimes I_{C_4}$

Charlie, respectively, and Bob transmits his qubits to Alice and Charlie; meanwhile, Charlie transmits his qubits to the other two users. In C-BQT scheme, it can get the pseudo-multi-qubit states by the CNOT and tensor product of six Bell states, which has been demonstrated experimentally. As we all know, Bell states can be prepared by the CNOT on superposition of single-qubit states, so the pseudo-multi-qubit states can be easily realized experimentally by performing the same idea. In the process, the three legitimate users choose the Bell state measurement basis independently. Eventually, they make the teleportation come true by implementing the appropriate unitary operation after getting the measurement results that are transmitted from others via classical channel. What's more, we suggest the efficiency could be further increased if the senders employed more suitable quantum channel and measurement basis; we also hope that the C-BQT will be implemented into experiment.

5.3 C-ACQT Scheme by Using a Nine-Qubit Entangled State

In this section, we propose a novel scheme of asymmetric cyclic controlled quantum teleportation (C-ACQT), using a maximally nine-qubit entangled state as the quantum channel. In C-ACQT scheme, we can simultaneously transmit two arbitrary two-qubit entangled states and an arbitrary single-qubit state. That is to say, Alice can transmit an arbitrary two-qubit state to Bob, Bob can transmit an arbitrary two-qubit state to Charlie, and Charlie can transmit an arbitrary single-qubit state to Alice. The scheme requires not only the respective measurement information of Alice, Bob, and Charlie but also the single-qubit measurement information of the controller David; otherwise, the transmission of quantum information cannot be completed. We also compared with other related schemes, and the proposed C-ACQT scheme possesses higher transmission efficiency.

5.3.1 C-ACQT Scheme

Now let us present asymmetric cyclic controlled quantum teleportation scheme in detail. C-ACQT scheme can be described as follows. Suppose that Alice wants to transmit an arbitrary two-qubit entangled state to Bob and Bob wants to transmit an arbitrary two-qubit entangled state to Charlie, which are given by

$$|\varphi\rangle_{A_1A_2} = a_0|00\rangle + a_1|11\rangle \tag{5.17}$$

$$|\varphi\rangle_{B_1B_2} = b_0|00\rangle + b_1|11\rangle \tag{5.18}$$

where a_0, a_1, b_0, b_1 are complex coefficients with

$$|a_0|^2 + |a_1|^2 = |b_0|^2 + |b_1|^2 = 1. \tag{5.19}$$

At the same time, Charlie wants to transmit an arbitrary single-qubit to Alice, which is given by

$$|\varphi\rangle_C = c_0|0\rangle + c_1|1\rangle \tag{5.20}$$

where c_0, c_1 are complex coefficients with

$$|c_0|^2 + |c_1|^2 = 1. \tag{5.21}$$

Assume that Alice, Bob, Charlie, and the controller David share a maximally nine-qubit entangled state, which has the form

$$\begin{aligned} |\varphi\rangle_{123456789} = \frac{1}{2\sqrt{2}}(&|000001100\rangle + |100111100\rangle + |010000000\rangle + |110110000\rangle \\ &+ |001001111\rangle + |101111111\rangle + |011000011\rangle + |111110011\rangle)_{123456789} \end{aligned} \tag{5.22}$$

where qubits 1 and 8 belong to Alice; qubits 2, 4, and 5 belong to Bob; qubits 3, 6, and 7 belong to Charlie; and qubit 9 belongs to David, respectively. Therefore, the initial state of the total system can be defined as

$$|\varphi\rangle_{123456789A_1A_2B_1B_2C} = |\varphi\rangle_{123456789} \otimes |\varphi\rangle_{A_1A_2} \otimes |\varphi\rangle_{B_1B_2} \otimes |\varphi\rangle_C. \tag{5.23}$$

In order to implement the controlled cyclic asymmetric quantum teleportation scheme, Charlie firstly performs a Bell state measurement on qubits C and 3, and the measurement result is given by

$$|\phi^{\pm}\rangle = 1/\sqrt{2}(|00\rangle \pm |11\rangle)_{c3}. \tag{5.24}$$

$$|\psi^{\pm}\rangle = 1/\sqrt{2}(|01\rangle \pm |10\rangle)_{c3}. \tag{5.25}$$

If Charlie's Bell state measurement result is $|\phi^-\rangle C_3$, the other qubits A_1, A_2, B_1, B_2, 1, 2, 4, 5, 6, 7, 8, and 9 are collapsed into the following product state:

$$\begin{aligned} |\varphi\rangle_{12456789A_1A_2B_1B_2} = 1/2(&c_0|00001100\rangle + c_0|10111100\rangle + c_0|01000000\rangle \\ &+ c_0|11110000\rangle - c_1|00001111\rangle - c_1|10111111\rangle \\ &- c_1|01000011\rangle - c_1|11110011\rangle)_{12456789} \\ &\otimes (a_0|00\rangle + a_1|11\rangle)_{A_1A_2} \otimes (b_0|00\rangle + b_1|11\rangle)_{B_1B_2} \end{aligned}. \tag{5.26}$$

Subsequently, Alice and Bob perform a three-qubit measurement on $(A_1, A_2, 1)$ and $(B_1, B_2, 2)$, respectively. And the measurement result is given by

$$|\eta^1\rangle = 1/\sqrt{2}(|000\rangle + |111\rangle). \tag{5.27}$$

$$|\eta^2\rangle = 1/\sqrt{2}(|000\rangle - |111\rangle). \tag{5.28}$$

$$|\eta^3\rangle = 1/\sqrt{2}(|001\rangle + |110\rangle). \tag{5.29}$$

$$|\eta^4\rangle = 1/\sqrt{2}(|001\rangle - |110\rangle). \tag{5.30}$$

If Alice's three-qubit measurement result is $|\eta^1\rangle_{A_1A_21}$ and Bob's three-qubit measurement result is $|\eta^3\rangle_{B_1B_23}$, then the other qubits 4, 5, 6, 7, 8, and 9 are collapsed into the following entangled state:

$$\begin{aligned}|\varphi\rangle_{456789} = &(a_0b_0c_0|000000\rangle - a_0b_0c_1|000011\rangle - a_1b_0c_0|110000\rangle \\ &+a_1b_0c_1|110011\rangle + a_0b_1c_0|001100\rangle - a_0b_1c_1|001111\rangle . \\ &- a_1b_1c_0|111100\rangle + a_1b_1c_1|11110011\rangle)_{456789}\end{aligned} \tag{5.31}$$

Finally, the controller David has to perform a single-qubit von Neumann measurement on his qubit 9, and the measurement result is given by

$$|+\rangle_9 = 1/\sqrt{2}(|0\rangle + |1\rangle)_9. \tag{5.32}$$

$$|-\rangle_9 = 1/\sqrt{2}(|0\rangle - |1\rangle)_9. \tag{5.33}$$

If David's single-qubit measurement result is $|+\rangle_9$, then qubits 4, 5, 6, 7, and 8 will collapse into the following state:

$$\begin{aligned}|\varphi\rangle_{45678} = &(a_0b_0c_0|00000\rangle + a_1b_0c_0|11000\rangle + a_0b_1c_0|00110\rangle \\ &+a_1b_1c_0|11110\rangle - a_0b_0c_1|00001\rangle - a_1b_0c_1|11001\rangle \\ &- a_0b_1c_1|00111\rangle - a_1b_1c_1|11111\rangle)_{45678}\end{aligned} \tag{5.34}$$

Formula (5.35) can be rewritten as

$$|\varphi\rangle_{45678} = (c_0|0\rangle - c_1|1\rangle)_8 \otimes (a_0|00\rangle + a_1|11\rangle)_{45} \otimes (b_0|00\rangle + b_1|11\rangle)_{67}. \tag{5.35}$$

Then if Alice, Bob, and Charlie perform the local unitary operation $(\sigma_z)_8 \otimes I_4 \otimes I_5 \otimes I_6 \otimes I_7$, then the desired state can be reconstructed. Therefore, the controlled cyclic asymmetric teleportation is successfully realized. Theoretically, after all the measurement operations, there will be 128 measurement results. We will show all possible collapsed states after measurement in Table 5.2 and list the corresponding unitary operations in Table 5.3.

Table 5.2 All measurement results and corresponding collapsed states (in the first column, from left to right, are the measurement results of Charlie, Alice, Bob, and David)

Measurement Result	The corresponding collapsed state
$\vert\phi^{+}\rangle\vert\eta^{1}\rangle\vert\eta^{1}\rangle\vert\pm\rangle$	$(c_0\vert 0\rangle \pm c_1\vert 1\rangle)_8(a_0\vert 00\rangle + a_1\vert 11\rangle)_{45}(b_0\vert 11\rangle + b_1\vert 00\rangle)_{67}$
$\vert\phi^{+}\rangle\vert\eta^{1}\rangle\vert\eta^{2}\rangle\vert\pm\rangle$	$(c_0\vert 0\rangle \pm c_1\vert 1\rangle)_8(a_0\vert 00\rangle + a_1\vert 11\rangle)_{45}(b_0\vert 11\rangle - b_1\vert 00\rangle)_{67}$
$\vert\phi^{+}\rangle\vert\eta^{1}\rangle\vert\eta^{3}\rangle\vert\pm\rangle$	$(c_0\vert 0\rangle \pm c_1\vert 1\rangle)_8(a_0\vert 00\rangle + a_1\vert 11\rangle)_{45}(b_0\vert 00\rangle + b_1\vert 11\rangle)_{67}$
$\vert\phi^{+}\rangle\vert\eta^{1}\rangle\vert\eta^{4}\rangle\vert\pm\rangle$	$(c_0\vert 0\rangle \pm c_1\vert 1\rangle)_8(a_0\vert 00\rangle + a_1\vert 11\rangle)_{45}(b_0\vert 00\rangle - b_1\vert 11\rangle)_{67}$
$\vert\phi^{+}\rangle\vert\eta^{2}\rangle\vert\eta^{1}\rangle\vert\pm\rangle$	$(c_0\vert 0\rangle \pm c_1\vert 1\rangle)_8(a_0\vert 00\rangle - a_1\vert 11\rangle)_{45}(b_0\vert 11\rangle + b_1\vert 00\rangle)_{67}$
$\vert\phi^{+}\rangle\vert\eta^{2}\rangle\vert\eta^{2}\rangle\vert\pm\rangle$	$(c_0\vert 0\rangle \pm c_1\vert 1\rangle)_8(a_1\vert 11\rangle - a_0\vert 00\rangle)_{45}(b_1\vert 00\rangle - b_1\vert 11\rangle)_{67}$
$\vert\phi^{+}\rangle\vert\eta^{2}\rangle\vert\eta^{3}\rangle\vert\pm\rangle$	$(c_0\vert 0\rangle \pm c_1\vert 1\rangle)_8(a_0\vert 00\rangle - a_1\vert 11\rangle)_{45}(b_0\vert 00\rangle + b_1\vert 11\rangle)_{67}$
$\vert\phi^{+}\rangle\vert\eta^{2}\rangle\vert\eta^{4}\rangle\vert\pm\rangle$	$(c_0\vert 0\rangle \pm c_1\vert 1\rangle)_8(a_0\vert 00\rangle - a_1\vert 11\rangle)_{45}(b_0\vert 00\rangle - b_1\vert 11\rangle)_{67}$
$\vert\phi^{+}\rangle\vert\eta^{3}\rangle\vert\eta^{1}\rangle\vert\pm\rangle$	$(c_0\vert 0\rangle \pm c_1\vert 1\rangle)_8(a_0\vert 11\rangle + a_1\vert 00\rangle)_{45}(b_1\vert 00\rangle + b_0\vert 11\rangle)_{67}$
$\vert\phi^{+}\rangle\vert\eta^{3}\rangle\vert\eta^{2}\rangle\vert\pm\rangle$	$(c_0\vert 0\rangle \pm c_1\vert 1\rangle)_8(a_0\vert 11\rangle + a_1\vert 00\rangle)_{45}(b_0\vert 11\rangle - b_1\vert 00\rangle)_{67}$
$\vert\phi^{+}\rangle\vert\eta^{3}\rangle\vert\eta^{3}\rangle\vert\pm\rangle$	$(c_0\vert 0\rangle \pm c_1\vert 1\rangle)_8(a_0\vert 11\rangle + a_1\vert 00\rangle)_{45}(b_0\vert 00\rangle + b_1\vert 11\rangle)_{67}$
$\vert\phi^{+}\rangle\vert\eta^{3}\rangle\vert\eta^{4}\rangle\vert\pm\rangle$	$(c_0\vert 0\rangle \pm c_1\vert 1\rangle)_8(a_0\vert 11\rangle + a_1\vert 00\rangle)_{45}(b_0\vert 00\rangle - b_1\vert 11\rangle)_{67}$
$\vert\phi^{+}\rangle\vert\eta^{4}\rangle\vert\eta^{1}\rangle\vert\pm\rangle$	$(c_0\vert 0\rangle \pm c_1\vert 1\rangle)_8(a_0\vert 11\rangle - a_1\vert 00\rangle)_{45}(b_0\vert 11\rangle + b_1\vert 00\rangle)_{67}$
$\vert\phi^{+}\rangle\vert\eta^{4}\rangle\vert\eta^{2}\rangle\vert\pm\rangle$	$(c_0\vert 0\rangle \pm c_1\vert 1\rangle)_8(a_0\vert 11\rangle - a_1\vert 00\rangle)_{45}(b_0\vert 11\rangle - b_1\vert 00\rangle)_{67}$
$\vert\phi^{+}\rangle\vert\eta^{4}\rangle\vert\eta^{3}\rangle\vert\pm\rangle$	$(c_0\vert 0\rangle \pm c_1\vert 1\rangle)_8(a_0\vert 11\rangle - a_1\vert 00\rangle)_{45}(b_0\vert 00\rangle + b_1\vert 11\rangle)_{67}$
$\vert\phi^{+}\rangle\vert\eta^{4}\rangle\vert\eta^{4}\rangle\vert\pm\rangle$	$(c_0\vert 0\rangle \pm c_1\vert 1\rangle)_8(a_0\vert 11\rangle - a_1\vert 00\rangle)_{45}(b_0\vert 00\rangle - b_1\vert 11\rangle)_{67}$
$\vert\phi^{-}\rangle\vert\eta^{1}\rangle\vert\eta^{1}\rangle\vert\pm\rangle$	$(c_0\vert 0\rangle \mp c_1\vert 1\rangle)_8(a_0\vert 00\rangle + a_1\vert 11\rangle)_{45}(b_0\vert 11\rangle + b_1\vert 00\rangle)_{67}$
$\vert\phi^{-}\rangle\vert\eta^{1}\rangle\vert\eta^{2}\rangle\vert\pm\rangle$	$(c_0\vert 0\rangle \mp c_1\vert 1\rangle)_8(a_0\vert 00\rangle + a_1\vert 11\rangle)_{45}(b_0\vert 11\rangle - b_1\vert 00\rangle)_{67}$
$\vert\phi^{-}\rangle\vert\eta^{1}\rangle\vert\eta^{3}\rangle\vert\pm\rangle$	$(c_0\vert 0\rangle \mp c_1\vert 1\rangle)_8(a_0\vert 00\rangle + a_1\vert 11\rangle)_{45}(b_0\vert 00\rangle + b_1\vert 11\rangle)_{67}$
$\vert\phi^{-}\rangle\vert\eta^{1}\rangle\vert\eta^{4}\rangle\vert\pm\rangle$	$(c_0\vert 0\rangle \mp c_1\vert 1\rangle)_8(a_0\vert 00\rangle + a_1\vert 11\rangle)_{45}(b_0\vert 00\rangle - b_1\vert 11\rangle)_{67}$
$\vert\phi^{-}\rangle\vert\eta^{2}\rangle\vert\eta^{1}\rangle\vert\pm\rangle$	$(c_0\vert 0\rangle \mp c_1\vert 1\rangle)_8(a_0\vert 00\rangle - a_1\vert 11\rangle)_{45}(b_0\vert 11\rangle - b_1\vert 00\rangle)_{67}$
$\vert\phi^{-}\rangle\vert\eta^{2}\rangle\vert\eta^{2}\rangle\vert\pm\rangle$	$(c_0\vert 0\rangle \mp c_1\vert 1\rangle)_8(a_0\vert 00\rangle - a_1\vert 11\rangle)_{45}(b_0\vert 11\rangle - b_1\vert 00\rangle)_{67}$
$\vert\phi^{-}\rangle\vert\eta^{2}\rangle\vert\eta^{3}\rangle\vert\pm\rangle$	$(c_0\vert 0\rangle \mp c_1\vert 1\rangle)_8(a_0\vert 00\rangle - a_1\vert 11\rangle)_{45}(b_0\vert 00\rangle + b_1\vert 11\rangle)_{67}$
$\vert\phi^{-}\rangle\vert\eta^{2}\rangle\vert\eta^{4}\rangle\vert\pm\rangle$	$(c_0\vert 0\rangle \mp c_1\vert 1\rangle)_8(a_0\vert 00\rangle - a_1\vert 11\rangle)_{45}(b_0\vert 00\rangle - b_1\vert 11\rangle)_{67}$
$\vert\phi^{-}\rangle\vert\eta^{3}\rangle\vert\eta^{1}\rangle\vert\pm\rangle$	$(c_0\vert 0\rangle \mp c_1\vert 1\rangle)_8(a_0\vert 11\rangle + a_1\vert 00\rangle)_{45}(b_0\vert 11\rangle + b_1\vert 00\rangle)_{67}$
$\vert\phi^{-}\rangle\vert\eta^{3}\rangle\vert\eta^{2}\rangle\vert\pm\rangle$	$(c_0\vert 0\rangle \mp c_1\vert 1\rangle)_8(a_0\vert 11\rangle + a_1\vert 00\rangle)_{45}(b_0\vert 11\rangle - b_1\vert 00\rangle)_{67}$
$\vert\phi^{-}\rangle\vert\eta^{3}\rangle\vert\eta^{3}\rangle\vert\pm\rangle$	$(c_0\vert 0\rangle \mp c_1\vert 1\rangle)_8(a_0\vert 11\rangle + a_1\vert 00\rangle)_{45}(b_0\vert 00\rangle + b_1\vert 11\rangle)_{67}$
$\vert\phi^{-}\rangle\vert\eta^{3}\rangle\vert\eta^{4}\rangle\vert\pm\rangle$	$(c_0\vert 0\rangle \mp c_1\vert 1\rangle)_8(a_0\vert 11\rangle + a_1\vert 00\rangle)_{45}(b_0\vert 00\rangle - b_1\vert 11\rangle)_{67}$
$\vert\phi^{-}\rangle\vert\eta^{4}\rangle\vert\eta^{1}\rangle\vert\pm\rangle$	$(c_0\vert 0\rangle \mp c_1\vert 1\rangle)_8(a_0\vert 11\rangle - a_1\vert 00\rangle)_{45}(b_0\vert 11\rangle + b_1\vert 00\rangle)_{67}$
$\vert\phi^{-}\rangle\vert\eta^{4}\rangle\vert\eta^{2}\rangle\vert\pm\rangle$	$(c_0\vert 0\rangle \mp c_1\vert 1\rangle)_8(a_0\vert 11\rangle - a_1\vert 00\rangle)_{45}(b_0\vert 11\rangle - b_1\vert 00\rangle)_{67}$
$\vert\phi^{-}\rangle\vert\eta^{4}\rangle\vert\eta^{3}\rangle\vert\pm\rangle$	$(c_0\vert 0\rangle \mp c_1\vert 1\rangle)_8(a_0\vert 11\rangle - a_1\vert 00\rangle)_{45}(b_0\vert 00\rangle + b_1\vert 11\rangle)_{67}$
$\vert\phi^{-}\rangle\vert\eta^{4}\rangle\vert\eta^{4}\rangle\vert\pm\rangle$	$(c_0\vert 0\rangle \mp c_1\vert 1\rangle)_8(a_0\vert 11\rangle - a_1\vert 00\rangle)_{45}(b_0\vert 00\rangle - b_1\vert 11\rangle)_{67}$
$\vert\psi^{+}\rangle\vert\eta^{1}\rangle\vert\eta^{1}\rangle\vert\pm\rangle$	$(c_1\vert 0\rangle \pm c_0\vert 1\rangle)_8(a_0\vert 00\rangle + a_1\vert 11\rangle)_{45}(b_0\vert 11\rangle + b_1\vert 00\rangle)_{67}$
$\vert\psi^{+}\rangle\vert\eta^{1}\rangle\vert\eta^{2}\rangle\vert\pm\rangle$	$(c_1\vert 0\rangle \pm c_0\vert 1\rangle)_8(a_0\vert 00\rangle + a_1\vert 11\rangle)_{45}(b_0\vert 11\rangle - b_1\vert 00\rangle)_{67}$
$\vert\psi^{+}\rangle\vert\eta^{1}\rangle\vert\eta^{3}\rangle\vert\pm\rangle$	$(c_1\vert 0\rangle \pm c_0\vert 1\rangle)_8(a_0\vert 00\rangle + a_1\vert 11\rangle)_{45}(b_0\vert 00\rangle + b_1\vert 11\rangle)_{67}$
$\vert\psi^{+}\rangle\vert\eta^{1}\rangle\vert\eta^{4}\rangle\vert\pm\rangle$	$(c_1\vert 0\rangle \pm c_0\vert 1\rangle)_8(a_0\vert 00\rangle + a_1\vert 11\rangle)_{45}(b_0\vert 00\rangle - b_1\vert 11\rangle)_{67}$
$\vert\psi^{+}\rangle\vert\eta^{2}\rangle\vert\eta^{1}\rangle\vert\pm\rangle$	$(c_1\vert 0\rangle \pm c_0\vert 1\rangle)_8(a_0\vert 00\rangle - a_1\vert 11\rangle)_{45}(b_0\vert 11\rangle + b_1\vert 00\rangle)_{67}$
$\vert\psi^{+}\rangle\vert\eta^{2}\rangle\vert\eta^{2}\rangle\vert\pm\rangle$	$(c_1\vert 0\rangle \pm c_0\vert 1\rangle)_8(a_0\vert 00\rangle - a_1\vert 11\rangle)_{45}(b_0\vert 11\rangle - b_1\vert 00\rangle)_{67}$
$\vert\psi^{+}\rangle\vert\eta^{2}\rangle\vert\eta^{3}\rangle\vert\pm\rangle$	$(c_1\vert 0\rangle \pm c_0\vert 1\rangle)_8(a_0\vert 00\rangle - a_1\vert 11\rangle)_{45}(b_0\vert 00\rangle + b_1\vert 11\rangle)_{67}$
$\vert\psi^{+}\rangle\vert\eta^{2}\rangle\vert\eta^{4}\rangle\vert\pm\rangle$	$(c_1\vert 0\rangle \pm c_0\vert 1\rangle)_8(a_0\vert 00\rangle - a_1\vert 11\rangle)_{45}(b_0\vert 00\rangle - b_1\vert 11\rangle)_{67}$
$\vert\psi^{+}\rangle\vert\eta^{3}\rangle\vert\eta^{1}\rangle\vert\pm\rangle$	$(c_1\vert 0\rangle \pm c_0\vert 1\rangle)_8(a_0\vert 11\rangle + a_1\vert 00\rangle)_{45}(b_0\vert 11\rangle + b_1\vert 00\rangle)_{67}$

(continued)

Table 5.2 (continued)

$\vert\psi^+\rangle\vert\eta^3\rangle\vert\eta^2\rangle\vert\pm\rangle$	$(c_1\vert 0\rangle \pm c_0\vert 1\rangle)_8(a_0\vert 11\rangle + a_1\vert 00\rangle)_{45}(b_0\vert 11\rangle - b_1\vert 00\rangle)_{67}$
$\vert\psi^+\rangle\vert\eta^3\rangle\vert\eta^3\rangle\vert\pm\rangle$	$(c_1\vert 0\rangle \pm c_0\vert 1\rangle)_8(a_0\vert 11\rangle + a_1\vert 00\rangle)_{45}(b_0\vert 00\rangle + b_1\vert 11\rangle)_{67}$
$\vert\psi^+\rangle\vert\eta^3\rangle\vert\eta^4\rangle\vert\pm\rangle$	$(c_1\vert 0\rangle \pm c_0\vert 1\rangle)_8(a_0\vert 11\rangle + a_1\vert 00\rangle)_{45}(b_0\vert 00\rangle - b_1\vert 11\rangle)_{67}$
$\vert\psi^+\rangle\vert\eta^4\rangle\vert\eta^1\rangle\vert\pm\rangle$	$(c_1\vert 0\rangle \pm c_0\vert 1\rangle)_8(a_0\vert 11\rangle - a_1\vert 00\rangle)_{45}(b_0\vert 11\rangle + b_1\vert 00\rangle)_{67}$
$\vert\psi^+\rangle\vert\eta^4\rangle\vert\eta^2\rangle\vert\pm\rangle$	$(c_1\vert 0\rangle \pm c_0\vert 1\rangle)_8(a_0\vert 11\rangle - a_1\vert 00\rangle)_{45}(b_0\vert 11\rangle - b_1\vert 00\rangle)_{67}$
$\vert\psi^+\rangle\vert\eta^4\rangle\vert\eta^3\rangle\vert\pm\rangle$	$(c_1\vert 0\rangle \pm c_0\vert 1\rangle)_8(a_0\vert 11\rangle - a_1\vert 00\rangle)_{45}(b_0\vert 00\rangle + b_1\vert 11\rangle)_{67}$
$\vert\psi^+\rangle\vert\eta^4\rangle\vert\eta^4\rangle\vert\pm\rangle$	$(c_1\vert 0\rangle \pm c_0\vert 1\rangle)_8(a_0\vert 11\rangle - a_1\vert 00\rangle)_{45}(b_0\vert 00\rangle - b_1\vert 11\rangle)_{67}$
$\vert\psi^-\rangle\vert\eta^1\rangle\vert\eta^1\rangle\vert +\rangle$ $\vert\psi^-\rangle\vert\eta^1\rangle\vert\eta^1\rangle\vert -\rangle$	$(c_0\vert 1\rangle - c_1\vert 0\rangle)_8(a_0\vert 00\rangle + a_1\vert 11\rangle)_{45}(b_0\vert 11\rangle + b_1\vert 00\rangle)_{67}$ $-(c_1\vert 0\rangle + c_0\vert 1\rangle)_8(a_0\vert 00\rangle + a_1\vert 11\rangle)_{45}(b_0\vert 11\rangle + b_1\vert 00\rangle)_{67}$
$\vert\psi^-\rangle\vert\eta^1\rangle\vert\eta^2\rangle\vert\pm\rangle$	$(c_1\vert 0\rangle \mp c_0\vert 1\rangle)_8(a_0\vert 00\rangle + a_1\vert 11\rangle)_{45}(b_1\vert 00\rangle - b_0\vert 11\rangle)_{67}$
$\vert\psi^-\rangle\vert\eta^1\rangle\vert\eta^3\rangle\vert +\rangle$ $\vert\psi^-\rangle\vert\eta^1\rangle\vert\eta^3\rangle\vert -\rangle$	$(c_0\vert 1\rangle - c_1\vert 0\rangle)_8(a_0\vert 00\rangle + a_1\vert 11\rangle)_{45}(b_0\vert 00\rangle + b_1\vert 11\rangle)_{67}$ $-(c_1\vert 0\rangle + c_0\vert 1\rangle)_8(a_0\vert 00\rangle + a_1\vert 11\rangle)_{45}(b_0\vert 00\rangle + b_1\vert 11\rangle)_{67}$
$\vert\psi^-\rangle\vert\eta^1\rangle\vert\eta^4\rangle\vert\pm\rangle$	$(c_1\vert 0\rangle \mp c_0\vert 1\rangle)_8(a_0\vert 00\rangle + a_1\vert 11\rangle)_{45}(b_1\vert 11\rangle - b_0\vert 00\rangle)_{67}$
$\vert\psi^-\rangle\vert\eta^2\rangle\vert\eta^1\rangle\vert\pm\rangle$	$(c_1\vert 0\rangle \mp c_0\vert 1\rangle)_8(a_1\vert 11\rangle - a_0\vert 00\rangle)_{45}(b_0\vert 11\rangle + b_1\vert 00\rangle)_{67}$
$\vert\psi^-\rangle\vert\eta^2\rangle\vert\eta^2\rangle\vert\pm\rangle$	$(c_1\vert 0\rangle \mp c_0\vert 1\rangle)_8(a_1\vert 11\rangle - a_0\vert 00\rangle)_{45}(b_0\vert 11\rangle - b_1\vert 00\rangle)_{67}$
$\vert\psi^-\rangle\vert\eta^2\rangle\vert\eta^3\rangle\vert\pm\rangle$	$(c_1\vert 0\rangle \mp c_0\vert 1\rangle)_8(a_1\vert 11\rangle - a_0\vert 00\rangle)_{45}(b_0\vert 00\rangle + b_1\vert 11\rangle)_{67}$
$\vert\psi^-\rangle\vert\eta^2\rangle\vert\eta^4\rangle\vert\pm\rangle$	$(c_1\vert 0\rangle \mp c_0\vert 1\rangle)_8(a_1\vert 11\rangle - a_0\vert 00\rangle)_{45}(b_0\vert 00\rangle - b_1\vert 11\rangle)_{67}$
$\vert\psi^-\rangle\vert\eta^3\rangle\vert\eta^1\rangle\vert +\rangle$ $\vert\psi^-\rangle\vert\eta^3\rangle\vert\eta^1\rangle\vert -\rangle$	$(c_0\vert 1\rangle - c_1\vert 0\rangle)_8(a_0\vert 11\rangle + a_1\vert 00\rangle)_{45}(b_0\vert 11\rangle + b_1\vert 00\rangle)_{67}$ $-(c_0\vert 1\rangle + c_1\vert 0\rangle)_8(a_0\vert 11\rangle + a_1\vert 00\rangle)_{45}(b_0\vert 11\rangle + b_1\vert 00\rangle)_{67}$
$\vert\psi^-\rangle\vert\eta^3\rangle\vert\eta^2\rangle\vert\pm\rangle$	$(c_1\vert 0\rangle \mp c_0\vert 1\rangle)_8(a_0\vert 11\rangle + a_1\vert 00\rangle)_{45}(b_1\vert 00\rangle - b_0\vert 11\rangle)_{67}$
$\vert\psi^-\rangle\vert\eta^3\rangle\vert\eta^3\rangle\vert +\rangle$ $\vert\psi^-\rangle\vert\eta^3\rangle\vert\eta^3\rangle\vert -\rangle$	$(c_0\vert 1\rangle - c_1\vert 0\rangle)_8(a_0\vert 11\rangle + a_1\vert 00\rangle)_{45}(b_0\vert 00\rangle + b_1\vert 11\rangle)_{67}$ $-(c_0\vert 1\rangle + c_1\vert 0\rangle)_8(a_0\vert 11\rangle + a_1\vert 00\rangle)_{45}(b_0\vert 00\rangle + b_1\vert 11\rangle)_{67}$
$\vert\psi^-\rangle\vert\eta^3\rangle\vert\eta^4\rangle\vert\pm\rangle$	$(c_1\vert 0\rangle \mp c_0\vert 1\rangle)_8(a_0\vert 11\rangle + a_1\vert 00\rangle)_{45}(b_1\vert 11\rangle - b_1\vert 00\rangle)_{67}$
$\vert\psi^-\rangle\vert\eta^4\rangle\vert\eta^1\rangle\vert\pm\rangle$	$(c_1\vert 0\rangle \mp c_0\vert 1\rangle)_8(a_1\vert 00\rangle - a_0\vert 11\rangle)_{45}(b_0\vert 11\rangle + b_1\vert 00\rangle)_{67}$
$\vert\psi^-\rangle\vert\eta^4\rangle\vert\eta^2\rangle\vert\pm\rangle$	$(c_1\vert 0\rangle \mp c_0\vert 1\rangle)_8(a_1\vert 00\rangle - a_0\vert 11\rangle)_{45}(b_0\vert 11\rangle - b_1\vert 00\rangle)_{67}$
$\vert\psi^-\rangle\vert\eta^4\rangle\vert\eta^3\rangle\vert\pm\rangle$	$(c_1\vert 0\rangle \mp c_0\vert 1\rangle)_8(a_1\vert 00\rangle - a_0\vert 11\rangle)_{45}(b_0\vert 00\rangle + b_1\vert 11\rangle)_{67}$
$\vert\psi^-\rangle\vert\eta^4\rangle\vert\eta^4\rangle\vert\pm\rangle$	$(c_1\vert 0\rangle \mp c_0\vert 1\rangle)_8(a_1\vert 00\rangle - a_0\vert 11\rangle)_{45}(b_0\vert 00\rangle - b_1\vert 11\rangle)_{67}$

5.3.2 *Comparison*

In the C-ACQT scheme we proposed, the controller David is added, which becomes more secure and controllable through the entire communication process. In this part, we compare the asymmetric cyclic quantum teleportation scheme proposed here with other similar schemes, including quantum resource consumption, necessary operations, classical resource consumption, transmitted qubits, and inherent efficiency of transmission. Our formula for calculating transmission efficiency is defined as [7] $\eta = \frac{c}{q}$, where c is the number of quantum states transmitted in the scheme and q is the number of total qubits. The specific comparative data are given in Table 5.4.

QRC quantum resource consumption, Q qubit, CS cluster state, ES entangled state, NO necessary operation, CRC classical resource consumption, QIBT quantum information bits transmitted, SM single-qubit measurement, BSM Bell state

Table 5.3 All measurement results and corresponding unitary operations (in the first column, from left to right, are the measurement results of Charlie, Alice, Bob, and David)

Measurement result	Unitary operation on qubits 8, 4, 5, 6, and 7
$\vert\phi^+\rangle\vert\eta^1\rangle\vert\eta^1\rangle\vert\pm\rangle$	$I_8 \otimes I_{45} \otimes (\sigma_x \otimes \sigma_x)_{67}$ $(\sigma_z)_8 \otimes I_{45} \otimes (\sigma_x \otimes \sigma_x)_{67}$
$\vert\phi^+\rangle\vert\eta^1\rangle\vert\eta^2\rangle\vert\pm\rangle$	$I_8 \otimes I_{45} \otimes (\sigma_z\sigma_x \otimes \sigma_x)_{67}$ $(\sigma_z)_8 \otimes I_{45} \otimes (\sigma_z\sigma_x \otimes \sigma_x)_{67}$
$\vert\phi^+\rangle\vert\eta^1\rangle\vert\eta^3\rangle\vert\pm\rangle$	$I_8 \otimes I_{45} \otimes I_{67}$ $(\sigma_z)_8 \otimes I_{45} \otimes I_{67}$
$\vert\phi^+\rangle\vert\eta^1\rangle\vert\eta^4\rangle\vert\pm\rangle$	$I_8 \otimes I_{45} \otimes (\sigma_z \otimes \sigma_0)_{67}$ $(\sigma_z)_8 \otimes I_{45} \otimes (\sigma_z \otimes \sigma_0)_{67}$
$\vert\phi^+\rangle\vert\eta^2\rangle\vert\eta^1\rangle\vert\pm\rangle$	$I_8 \otimes (\sigma_z \otimes \sigma_0)_{45} \otimes (\sigma_x \otimes \sigma_x)_{67}$ $(\sigma_z)_8 \otimes (\sigma_z \otimes \sigma_0)_{45} \otimes (\sigma_x \otimes \sigma_x)_{67}$
$\vert\phi^+\rangle\vert\eta^2\rangle\vert\eta^2\rangle\vert\pm\rangle$	$I_8 \otimes (-(\sigma_z \otimes \sigma_0))_{45} \otimes (\sigma_x\sigma_z \otimes \sigma_x)_{67}$ $(\sigma_z)_8 \otimes (-(\sigma_z \otimes \sigma_0))_{45} \otimes (\sigma_x\sigma_z \otimes \sigma_x)_{67}$
$\vert\phi^+\rangle\vert\eta^2\rangle\vert\eta^3\rangle\vert\pm\rangle$	$I_8 \otimes (\sigma_z \otimes \sigma_0)_{45} \otimes I_{67}$ $(\sigma_z)_8 \otimes (\sigma_z \otimes \sigma_0)_{45} \otimes I_{67}$
$\vert\phi^+\rangle\vert\eta^2\rangle\vert\eta^4\rangle\vert\pm\rangle$	$I_8 \otimes (\sigma_z \otimes \sigma_0)_{45} \otimes (\sigma_z \otimes \sigma_0)_{67}$ $(\sigma_z)_8 \otimes (\sigma_z \otimes \sigma_0)_{45} \otimes (\sigma_z \otimes \sigma_0)_{67}$
$\vert\phi^+\rangle\vert\eta^3\rangle\vert\eta^1\rangle\vert\pm\rangle$	$I_8 \otimes (\sigma_x \otimes \sigma_x)_{45} \otimes (\sigma_x \otimes \sigma_x)_{67}$ $(\sigma_z)_8 \otimes (\sigma_x \otimes \sigma_x)_{45} \otimes (\sigma_x \otimes \sigma_x)_{67}$
$\vert\phi^+\rangle\vert\eta^3\rangle\vert\eta^2\rangle\vert\pm\rangle$	$I_8 \otimes (\sigma_x \otimes \sigma_x)_{45} \otimes (\sigma_z\sigma_x \otimes \sigma_x)_{67}$ $(\sigma_z)_8 \otimes (\sigma_x \otimes \sigma_x)_{45} \otimes (\sigma_z\sigma_x \otimes \sigma_x)_{67}$
$\vert\phi^+\rangle\vert\eta^3\rangle\vert\eta^3\rangle\vert\pm\rangle$	$I_8 \otimes (\sigma_x \otimes \sigma_x)_{45} \otimes I_{67}$ $(\sigma_z)_8 \otimes (\sigma_x \otimes \sigma_x)_{45} \otimes I_{67}$
$\vert\phi^+\rangle\vert\eta^3\rangle\vert\eta^4\rangle\vert\pm\rangle$	$I_8 \otimes (\sigma_x \otimes \sigma_x)_{45} \otimes (\sigma_z \otimes \sigma_0)_{67}$ $(\sigma_z)_8 \otimes (\sigma_x \otimes \sigma_x)_{45} \otimes (\sigma_z \otimes \sigma_0)_{67}$
$\vert\phi^+\rangle\vert\eta^4\rangle\vert\eta^1\rangle\vert\pm\rangle$	$I_8 \otimes (\sigma_z\sigma_x \otimes \sigma_x)_{45} \otimes (\sigma_x \otimes \sigma_x)_{67}$ $(\sigma_z)_8 \otimes (\sigma_z\sigma_x \otimes \sigma_x)_{45} \otimes (\sigma_x \otimes \sigma_x)_{67}$
$\vert\phi^+\rangle\vert\eta^4\rangle\vert\eta^2\rangle\vert\pm\rangle$	$I_8 \otimes (\sigma_z\sigma_x \otimes \sigma_x)_{45} \otimes (\sigma_z\sigma_x \otimes \sigma_x)_{67}$ $(\sigma_z)_8 \otimes (\sigma_z\sigma_x \otimes \sigma_x)_{45} \otimes (\sigma_z\sigma_x \otimes \sigma_x)_{67}$
$\vert\phi^+\rangle\vert\eta^4\rangle\vert\eta^3\rangle\vert\pm\rangle$	$I_8 \otimes (\sigma_z\sigma_x \otimes \sigma_x)_{45} \otimes I_{67}$ $(\sigma_z)_8 \otimes (\sigma_z\sigma_x \otimes \sigma_x)_{45} \otimes I_{67}$
$\vert\phi^+\rangle\vert\eta^4\rangle\vert\eta^4\rangle\vert\pm\rangle$	$I_8 \otimes (\sigma_z\sigma_x \otimes \sigma_x)_{45} \otimes (\sigma_z \otimes \sigma_0)_{67}$ $(\sigma_z)_8 \otimes (\sigma_z\sigma_x \otimes \sigma_x)_{45} \otimes (\sigma_z \otimes \sigma_0)_{67}$
$\vert\phi^-\rangle\vert\eta^1\rangle\vert\eta^1\rangle\vert\pm\rangle$	$(\sigma_z)_8 \otimes I_{45} \otimes (\sigma_x \otimes \sigma_x)_{67}$ $I_8 \otimes I_{45} \otimes (\sigma_x \otimes \sigma_x)_{67}$
$\vert\phi^-\rangle\vert\eta^1\rangle\vert\eta^2\rangle\vert\pm\rangle$	$(\sigma_z)_8 \otimes I_{45} \otimes (\sigma_z\sigma_x \otimes \sigma_x)_{67}$ $I_8 \otimes I_{45} \otimes (\sigma_z\sigma_x \otimes \sigma_x)_{67}$
$\vert\phi^-\rangle\vert\eta^1\rangle\vert\eta^3\rangle\vert\pm\rangle$	$(\sigma_z)_8 \otimes I_{45} \otimes I_{67}$ $I_8 \otimes I_{45} \otimes I_{67}$
$\vert\phi^-\rangle\vert\eta^1\rangle\vert\eta^4\rangle\vert\pm\rangle$	$(\sigma_z)_8 \otimes I_{45} \otimes (\sigma_z \otimes \sigma_0)_{67}$ $I_8 \otimes I_{45} \otimes (\sigma_z \otimes \sigma_0)_{67}$
$\vert\phi^-\rangle\vert\eta^2\rangle\vert\eta^1\rangle\vert\pm\rangle$	$(\sigma_z)_8 \otimes (\sigma_z \otimes \sigma_0)_{45} \otimes (\sigma_x \otimes \sigma_x)_{67}$ $I_8 \otimes (\sigma_z \otimes \sigma_0)_{45} \otimes (\sigma_x \otimes \sigma_x)_{67}$
$\vert\phi^-\rangle\vert\eta^2\rangle\vert\eta^2\rangle\vert\pm\rangle$	$(\sigma_z)_8 \otimes (\sigma_z \otimes \sigma_0)_{45} \otimes (\sigma_z\sigma_x \otimes \sigma_x)_{67}$ $I_8 \otimes (\sigma_z \otimes \sigma_0)_{45} \otimes (\sigma_z\sigma_x \otimes \sigma_x)_{67}$

(continued)

Table 5.3 (continued)

$\|\phi^-\rangle\|\eta^2\rangle\|\eta^3\rangle\|\pm\rangle$	$(\sigma_z)_8 \otimes (\sigma_z \otimes \sigma_0)_{45} \otimes I_{67}$ $I_8 \otimes (\sigma_z \otimes \sigma_0)_{45} \otimes I_{67}$
$\|\phi^-\rangle\|\eta^2\rangle\|\eta^4\rangle\|\pm\rangle$	$(\sigma_z)_8 \otimes (\sigma_z \otimes \sigma_0)_{45} \otimes (\sigma_z \otimes \sigma_0)_{67}$ $I_8 \otimes (\sigma_z \otimes \sigma_0)_{45} \otimes (\sigma_z \otimes \sigma_0)_{67}$
$\|\phi^-\rangle\|\eta^3\rangle\|\eta^1\rangle\|\pm\rangle$	$(\sigma_z)_8 \otimes (\sigma_x \otimes \sigma_x)_{45} \otimes (\sigma_x \otimes \sigma_x)_{67}$ $I_8 \otimes (\sigma_x \otimes \sigma_x)_{45} \otimes (\sigma_x \otimes \sigma_x)_{67}$
$\|\phi^-\rangle\|\eta^3\rangle\|\eta^2\rangle\|\pm\rangle$	$(\sigma_z)_8 \otimes (\sigma_x \otimes \sigma_x)_{45} \otimes (\sigma_z\sigma_x \otimes \sigma_x)_{67}$ $I_8 \otimes (\sigma_x \otimes \sigma_x)_{45} \otimes (\sigma_z\sigma_x \otimes \sigma_x)_{67}$
$\|\phi^-\rangle\|\eta^3\rangle\|\eta^3\rangle\|\pm\rangle$	$(\sigma_z)_8 \otimes (\sigma_x \otimes \sigma_x)_{45} \otimes I_{67}$ $I_8 \otimes (\sigma_x \otimes \sigma_x)_{45} \otimes I_{67}$
$\|\phi^-\rangle\|\eta^3\rangle\|\eta^4\rangle\|\pm\rangle$	$(\sigma_z)_8 \otimes (\sigma_x \otimes \sigma_x)_{45} \otimes (\sigma_z \otimes \sigma_0)_{67}$ $I_8 \otimes (\sigma_x \otimes \sigma_x)_{45} \otimes (\sigma_z \otimes \sigma_0)_{67}$
$\|\phi^-\rangle\|\eta^4\rangle\|\eta^1\rangle\|\pm\rangle$	$(\sigma_z)_8 \otimes (\sigma_z\sigma_x \otimes \sigma_x)_{45} \otimes (\sigma_x \otimes \sigma_x)_{67}$ $I_8 \otimes (\sigma_z\sigma_x \otimes \sigma_x)_{45} \otimes (\sigma_x \otimes \sigma_x)_{67}$
$\|\phi^-\rangle\|\eta^4\rangle\|\eta^2\rangle\|\pm\rangle$	$(\sigma_z)_8 \otimes (\sigma_z\sigma_x \otimes \sigma_x)_{45} \otimes (\sigma_z\sigma_x \otimes \sigma_x)_{67}$ $I_8 \otimes (\sigma_z\sigma_x \otimes \sigma_x)_{45} \otimes (\sigma_z\sigma_x \otimes \sigma_x)_{67}$
$\|\phi^-\rangle\|\eta^4\rangle\|\eta^3\rangle\|\pm\rangle$	$(\sigma_z)_8 \otimes (\sigma_z\sigma_x \otimes \sigma_x)_{45} \otimes I_{67}$ $I_8 \otimes (\sigma_z\sigma_x \otimes \sigma_x)_{45} \otimes I_{67}$
$\|\phi^-\rangle\|\eta^4\rangle\|\eta^4\rangle\|\pm\rangle$	$(\sigma_z)_8 \otimes (\sigma_z\sigma_x \otimes \sigma_x)_{45} \otimes (\sigma_z \otimes \sigma_0)_{67}$ $I_8 \otimes (\sigma_z\sigma_x \otimes \sigma_x)_{45} \otimes (\sigma_z \otimes \sigma_0)_{67}$
$\|\psi^+\rangle\|\eta^1\rangle\|\eta^1\rangle\|\pm\rangle$	$(\sigma_x)_8 \otimes I_{45} \otimes (\sigma_x \otimes \sigma_x)_{67}$ $-(\sigma_z\sigma_x)_8 \otimes I_{45} \otimes (\sigma_x \otimes \sigma_x)_{67}$
$\|\psi^+\rangle\|\eta^1\rangle\|\eta^2\rangle\|\pm\rangle$	$(\sigma_x)_8 \otimes I_{45} \otimes (\sigma_z\sigma_x \otimes \sigma_x)_{67}$ $-(\sigma_z\sigma_x)_8 \otimes I_{45} \otimes (\sigma_z\sigma_x \otimes \sigma_x)_{67}$
$\|\psi^+\rangle\|\eta^1\rangle\|\eta^3\rangle\|\pm\rangle$	$(\sigma_x)_8 \otimes I_{45} \otimes I_{67}$ $-(\sigma_z\sigma_x)_8 \otimes I_{45} \otimes I_{67}$
$\|\psi^+\rangle\|\eta^1\rangle\|\eta^4\rangle\|\pm\rangle$	$(\sigma_x)_8 \otimes I_{45} \otimes (\sigma_z \otimes \sigma_0)_{67}$ $-(\sigma_z\sigma_x)_8 \otimes I_{45} \otimes (\sigma_z \otimes \sigma_0)_{67}$
$\|\psi^+\rangle\|\eta^2\rangle\|\eta^1\rangle\|\pm\rangle$	$(\sigma_x)_8 \otimes (\sigma_z \otimes \sigma_0)_{45} \otimes (\sigma_x \otimes \sigma_x)_{67}$ $-(\sigma_z\sigma_x)_8 \otimes (\sigma_z \otimes \sigma_0)_{45} \otimes (\sigma_x \otimes \sigma_x)_{67}$
$\|\psi^+\rangle\|\eta^2\rangle\|\eta^2\rangle\|\pm\rangle$	$(\sigma_x)_8 \otimes (\sigma_z \otimes \sigma_0)_{45} \otimes (\sigma_z\sigma_x \otimes \sigma_x)_{67}$ $-(\sigma_z\sigma_x)_8 \otimes (\sigma_z \otimes \sigma_0)_{45} \otimes (\sigma_z\sigma_x \otimes \sigma_x)_{67}$
$\|\psi^+\rangle\|\eta^2\rangle\|\eta^3\rangle\|\pm\rangle$	$(\sigma_x)_8 \otimes (\sigma_z \otimes \sigma_0)_{45} \otimes I_{67}$ $-(\sigma_z\sigma_x)_8 \otimes (\sigma_z \otimes \sigma_0)_{45} \otimes I_{67}$
$\|\psi^+\rangle\|\eta^2\rangle\|\eta^4\rangle\|\pm\rangle$	$(\sigma_x)_8 \otimes (\sigma_z \otimes \sigma_0)_{45} \otimes (\sigma_z \otimes \sigma_0)_{67}$ $-(\sigma_z\sigma_x)_8 \otimes (\sigma_z \otimes \sigma_0)_{45} \otimes (\sigma_z \otimes \sigma_0)_{67}$
$\|\psi^+\rangle\|\eta^3\rangle\|\eta^1\rangle\|\pm\rangle$	$(\sigma_x)_8 \otimes (\sigma_x \otimes \sigma_x)_{45} \otimes (\sigma_x \otimes \sigma_x)_{67}$ $-(\sigma_z\sigma_x)_8 \otimes (\sigma_x \otimes \sigma_x)_{45} \otimes (\sigma_x \otimes \sigma_x)_{67}$
$\|\psi^+\rangle\|\eta^3\rangle\|\eta^2\rangle\|\pm\rangle$	$(\sigma_x)_8 \otimes (\sigma_x \otimes \sigma_x)_{45} \otimes (\sigma_z\sigma_x \otimes \sigma_x)_{67}$ $-(\sigma_z\sigma_x)_8 \otimes (\sigma_x \otimes \sigma_x)_{45} \otimes (\sigma_z\sigma_x \otimes \sigma_x)_{67}$
$\|\psi^+\rangle\|\eta^3\rangle\|\eta^3\rangle\|\pm\rangle$	$(\sigma_x)_8 \otimes (\sigma_x \otimes \sigma_x)_{45} \otimes I_{67}$ $-(\sigma_z\sigma_x)_8 \otimes (\sigma_x \otimes \sigma_x)_{45} \otimes I_{67}$
$\|\psi^+\rangle\|\eta^3\rangle\|\eta^4\rangle\|\pm\rangle$	$(\sigma_x)_8 \otimes (\sigma_x \otimes \sigma_x)_{45} \otimes (\sigma_z \otimes \sigma_0)_{67}$ $-(\sigma_z\sigma_x)_8 \otimes (\sigma_x \otimes \sigma_x)_{45} \otimes (\sigma_z \otimes \sigma_0)_{67}$
$\|\psi^+\rangle\|\eta^4\rangle\|\eta^1\rangle\|\pm\rangle$	$(\sigma_x)_8 \otimes (\sigma_z\sigma_x \otimes \sigma_x)_{45} \otimes (\sigma_x \otimes \sigma_x)_{67}$ $-(\sigma_z\sigma_x)_8 \otimes (\sigma_z\sigma_x \otimes \sigma_x)_{45} \otimes (\sigma_x \otimes \sigma_x)_{67}$

(continued)

Table 5.3 (continued)

$\vert\psi^{+}\rangle\vert\eta^{4}\rangle\vert\eta^{2}\rangle\vert\pm\rangle$	$(\sigma_x)_8 \otimes (\sigma_z\sigma_x \otimes \sigma_x)_{45} \otimes (\sigma_z\sigma_x \otimes \sigma_x)_{67}$ $-(\sigma_z\sigma_x)_8 \otimes (\sigma_z\sigma_x \otimes \sigma_x)_{45} \otimes (\sigma_z\sigma_x \otimes \sigma_x)_{67}$
$\vert\psi^{+}\rangle\vert\eta^{4}\rangle\vert\eta^{3}\rangle\vert\pm\rangle$	$(\sigma_x)_8 \otimes (\sigma_z\sigma_x \otimes \sigma_x)_{45} \otimes I_{67}$ $-(\sigma_z\sigma_x)_8 \otimes (\sigma_z\sigma_x \otimes \sigma_x)_{45} \otimes I_{67}$
$\vert\psi^{+}\rangle\vert\eta^{4}\rangle\vert\eta^{4}\rangle\vert\pm\rangle$	$(\sigma_x)_8 \otimes (\sigma_z\sigma_x \otimes \sigma_x)_{45} \otimes (\sigma_z \otimes \sigma_0)_{67}$ $-(\sigma_z\sigma_x)_8 \otimes (\sigma_z\sigma_x \otimes \sigma_x)_{45} \otimes (\sigma_z \otimes \sigma_0)_{67}$
$\vert\psi^{-}\rangle\vert\eta^{1}\rangle\vert\eta^{1}\rangle\vert\pm\rangle$	$(\sigma_z\sigma_x)_8 \otimes I_{45} \otimes (\sigma_x \otimes \sigma_x)_{67}$ $-(\sigma_x)_8 \otimes I_{45} \otimes (\sigma_x \otimes \sigma_x)_{67}$
$\vert\psi^{-}\rangle\vert\eta^{1}\rangle\vert\eta^{2}\rangle\vert\pm\rangle$	$-(\sigma_z\sigma_x)_8 \otimes I_{45} \otimes (\sigma_x\sigma_z \otimes \sigma_x)_{67}$ $(\sigma_x)_8 \otimes I_{45} \otimes (\sigma_x\sigma_z \otimes \sigma_x)_{67}$
$\vert\psi^{-}\rangle\vert\eta^{1}\rangle\vert\eta^{3}\rangle\vert\pm\rangle$	$(\sigma_z\sigma_x)_8 \otimes I_{45} \otimes I_{67}$ $-(\sigma_x)_8 \otimes I_{45} \otimes I_{67}$
$\vert\psi^{-}\rangle\vert\eta^{1}\rangle\vert\eta^{4}\rangle\vert\pm\rangle$	$-(\sigma_z\sigma_x)_8 \otimes I_{45} \otimes (-(\sigma_z \otimes \sigma_0))_{67}$ $(\sigma_x)_8 \otimes I_{45} \otimes (-(\sigma_z \otimes \sigma_0))_{67}$
$\vert\psi^{-}\rangle\vert\eta^{2}\rangle\vert\eta^{1}\rangle\vert\pm\rangle$	$-(\sigma_z\sigma_x)_8 \otimes (-(\sigma_z \otimes \sigma_0))_{45} \otimes (\sigma_x \otimes \sigma_x)_{67}$ $(\sigma_x)_8 \otimes (-(\sigma_z \otimes \sigma_0))_{45} \otimes (\sigma_x \otimes \sigma_x)_{67}$
$\vert\psi^{-}\rangle\vert\eta^{2}\rangle\vert\eta^{2}\rangle\vert\pm\rangle$	$-(\sigma_z\sigma_x)_8 \otimes (-(\sigma_z \otimes \sigma_0))_{45} \otimes (\sigma_z\sigma_x \otimes \sigma_x)_{67}$ $(\sigma_x)_8 \otimes (-(\sigma_z \otimes \sigma_0))_{45} \otimes (\sigma_z\sigma_x \otimes \sigma_x)_{67}$
$\vert\psi^{-}\rangle\vert\eta^{2}\rangle\vert\eta^{3}\rangle\vert\pm\rangle$	$-(\sigma_z\sigma_x)_8 \otimes (-(\sigma_z \otimes \sigma_0))_{45} \otimes I_{67}$ $(\sigma_x)_8 \otimes (-(\sigma_z \otimes \sigma_0))_{45} \otimes I_{67}$
$\vert\psi^{-}\rangle\vert\eta^{2}\rangle\vert\eta^{4}\rangle\vert\pm\rangle$	$-(\sigma_z\sigma_x)_8 \otimes (-(\sigma_z \otimes \sigma_0))_{45} \otimes (\sigma_z \otimes \sigma_0)_{67}$ $(\sigma_x)_8 \otimes (-(\sigma_z \otimes \sigma_0))_{45} \otimes (\sigma_z \otimes \sigma_0)_{67}$
$\vert\psi^{-}\rangle\vert\eta^{3}\rangle\vert\eta^{1}\rangle\vert\pm\rangle$	$(\sigma_z\sigma_x)_8 \otimes (\sigma_x \otimes \sigma_x)_{45} \otimes (\sigma_x \otimes \sigma_x)_{67}$ $-(\sigma_x)_8 \otimes (\sigma_x \otimes \sigma_x)_{45} \otimes (\sigma_x \otimes \sigma_x)_{67}$
$\vert\psi^{-}\rangle\vert\eta^{3}\rangle\vert\eta^{2}\rangle\vert\pm\rangle$	$-(\sigma_z\sigma_x)_8 \otimes (\sigma_x \otimes \sigma_x)_{45} \otimes (\sigma_x\sigma_z \otimes \sigma_x)_{67}$ $(\sigma_x)_8 \otimes (\sigma_x \otimes \sigma_x)_{45} \otimes (\sigma_x\sigma_z \otimes \sigma_x)_{67}$
$\vert\psi^{-}\rangle\vert\eta^{3}\rangle\vert\eta^{3}\rangle\vert\pm\rangle$	$(\sigma_z\sigma_x)_8 \otimes (\sigma_x \otimes \sigma_x)_{45} \otimes I_{67}$ $-(\sigma_x)_8 \otimes (\sigma_x \otimes \sigma_x)_{45} \otimes I_{67}$
$\vert\psi^{-}\rangle\vert\eta^{3}\rangle\vert\eta^{4}\rangle\vert\pm\rangle$	$-(\sigma_z\sigma_x)_8 \otimes (\sigma_x \otimes \sigma_x)_{45} \otimes (-(\sigma_z \otimes \sigma_0))_{67}$ $(\sigma_x)_8 \otimes (\sigma_x \otimes \sigma_x)_{45} \otimes (-(\sigma_z \otimes \sigma_0))_{67}$
$\vert\psi^{-}\rangle\vert\eta^{4}\rangle\vert\eta^{1}\rangle\vert\pm\rangle$	$-(\sigma_z\sigma_x)_8 \otimes (\sigma_x\sigma_z \otimes \sigma_x)_{45} \otimes (\sigma_x \otimes \sigma_x)_{67}$ $(\sigma_x)_8 \otimes (\sigma_x\sigma_z \otimes \sigma_x)_{45} \otimes (\sigma_x \otimes \sigma_x)_{67}$
$\vert\psi^{-}\rangle\vert\eta^{4}\rangle\vert\eta^{2}\rangle\vert\pm\rangle$	$-(\sigma_z\sigma_x)_8 \otimes (\sigma_x\sigma_z \otimes \sigma_x)_{45} \otimes (\sigma_z\sigma_x \otimes \sigma_x)_{67}$ $(\sigma_x)_8 \otimes (\sigma_x\sigma_z \otimes \sigma_x)_{45} \otimes (\sigma_z\sigma_x \otimes \sigma_x)_{67}$
$\vert\psi^{-}\rangle\vert\eta^{4}\rangle\vert\eta^{3}\rangle\vert\pm\rangle$	$-(\sigma_z\sigma_x)_8 \otimes (\sigma_x\sigma_z \otimes \sigma_x)_{45} \otimes I_{67}$ $(\sigma_x)_8 \otimes (\sigma_x\sigma_z \otimes \sigma_x)_{45} \otimes I_{67}$
$\vert\psi^{-}\rangle\vert\eta^{4}\rangle\vert\eta^{4}\rangle\vert\pm\rangle$	$-(\sigma_z\sigma_x)_8 \otimes (\sigma_x\sigma_z \otimes \sigma_x)_{45} \otimes (\sigma_x\sigma_z \otimes \sigma_x)_{67}$ $(\sigma_x)_8 \otimes (\sigma_x\sigma_z \otimes \sigma_x)_{45} \otimes (\sigma_x\sigma_z \otimes \sigma_x)_{67}$

measurement, FQVNM four-qubit von Neumann measurement, GHZSM GHZ state measurement.

There is another way to calculate the inherent efficiency of transmission, which is defined as [14] $\eta = \frac{q_s}{q_u + b_t}$, where q_s is the number of qubits that are exchanged in the protocol, q_u is the number of qubits that are used to form the quantum channel, and $_{nn}$ is the number of classical bits used to transmit measurement information. Under this calculation method, the efficiency of C-ACQT scheme is also considerable.

Table 5.4 Comparison between the proposed C-ACQT scheme and previous works

Scheme	QRC	NO	CRC	QIBT	η
Ref. [9]	Seven-QES	BSM, FQVNM, SSM	7	3	3/10
Ref. [10]	Seven-QCS	2BSM, SSM	5	3	3/10
Ref. [11]	Six-QES	SSM, 2BSM	5	3	1/3
Ref. [12]	Seven-QES	3BSM, SSM	7	3	3/10
Ref. [13]	Seven-QES	3BSM, SSM	7	3	3/10
C-ACQT	Nine-QES	BSM, 2GHZSM, SSM	7	5	5/14

5.3.3 *Summary*

We propose an ideal method for quantum teleportation of asymmetric cyclic transmission. In this C-ACQT scheme, we use the nine-qubit entangled state as the quantum channel. Alice, Bob, and Charlie are the sender of the message and the receiver of the message. Also, we have added the controller David to the protocol to make the communication process more secure and controllable. In the communication process, Charlie first measures the Bell state and tells the measurement results to other participants through the classical channel, and then Alice and Bob conduct a GHZ state measurement, respectively. At the same time, if there is no information about the measurement result of the controller David, Alice, Bob, and Charlie cannot complete the information transmission. In the scheme, we have selected Bell state measurement, GHZ state measurement, and single-event measurement. These measurement methods are experimentally feasible. At the same time, we compared the proposed communication scheme with other similar quantum teleportation schemes. Under the two internal efficiency calculation methods, the C-ACQT scheme has good performance.

References

1. Chen YX, Du J, Liu SY et al (2017) Cyclic quantum teleportation [J]. Quantum Inf Process 16(8)
2. Zhi-wen S (2018) Cyclic controlled teleportation by using a seven-qubit entangled state [J]. Int J Theor Phys 57:3835
3. Sisodia M, Shukla A, Thapliyal K, Pathak A (2017) Design and experimental realization of an optimal scheme for teleportation of an n-qubit quantum state. Quantum Inf Process 16(12):1–19
4. Sabir DJM (2018) Efficient schemes for the quantum teleportation of a sub-class of tripartite entangled states. Quantum Inf Process 17(7):1–11
5. Zhou R-G, Qian C, Ian H (2019) Cyclic and bidirectional quantum teleportation via pseudo multi-qubit states. IEEE Access 7(1):42445–42449
6. Ri-Gui Z, Ling C (2021) Asymmetric cyclic controlled quantum teleportation by using nine-qubit entangled state. Int J Theor Phys 60(9):3435–3459
7. Kao SH, Chia-Wei T, Hwang T (2011) Enhanced multiparty controlled QSDC using GHZ state [J]. Commun Theor Phys 55(6):1007

8. Banerjee A, Pathak A (2012) Maximally efficient protocols for direct secure quantum communication. Phys Lett A 376(45):2944–2950
9. Zha XW, Zou ZC, Qi JX et al (2013) Bidirectional quantum controlled teleportation via five-qubit cluster state [J]. Int J Theor Phys 52(6):1740–1744
10. Duan YJ, Zha XW (2014) Bidirectional quantum controlled teleportation via a six-qubit entangled state [J]. Int J Theor Phys 53(11):3780–3786
11. Zhang D, Zha XW, Duan YJ (2015) Bidirectional and asymmetric quantum controlled teleportation [J]. Int J Theor Phys 54(5):1711–1719
12. Yang YQ, Zha XW, Yu Y (2016) Asymmetric bidirectional controlled teleportation via seven-qubit cluster state [J]. Int J Theor Phys 55(10):4197–4204
13. Choudhury BS, Samanta S (2017) Asymmetric bidirectional 3⇔ 2 qubit teleportation protocol between Alice and bob via 9-qubit cluster state [J]. Int J Theor Phys 56(10):3285–3296
14. Yuan H, Liu YM, Zhang W et al (2008) Optimizing resource consumption, operation complexity and efficiency in quantum-state sharing [J]. J Phys B Atomic Mol Phys 41(14):145506

Chapter 6
Quantum Teleportation in Noisy Environment

Abstract Due to the fragility of quantum entanglement, quantum teleportation is highly susceptible to noise, which can originate from both the internal and external environments of the quantum system, leading to a significant reduction in the fidelity of the transmitted state. Quantum teleportation in noisy environments holds broad application prospects in the fields of quantum communication, quantum computation, and quantum networks. However, realizing the widespread application of this technology necessitates overcoming a series of technical and practical challenges, including enhancing the efficiency and reliability of quantum error correction codes, designing more effective decoherence-free subspaces, and optimizing hybrid entanglement techniques. With the continuous development and refinement of quantum technologies, it is believed that quantum teleportation in noisy environments will play a crucial role in the future.

6.1 Introduction

The noise existing in the quantum channel is one of the main issues for reliable teleportation of the information. The preshared entangled state that interacted with the external environment might evolve into a mixed state during transmission, leading to the loss of quantum information recovered by the output system. The majority of research on quantum noise is predicated on the assumption of memoryless channel, where the noisy transformations are performed on the sequences of quantum carriers independently and identically [1–3]. However, the assumption of vanishing correlation or memory in the system-environment interaction is not always justified. With increasing transmission rates in communication channel, for example, successive transmissions happen so rapidly that the environment may retain a memory between individual transmissions. The experimental counterparts of memory effects in quantum communication schemes are limited to the direct transmission of each carrier, such as the photons traveling through optical fiber that can be affected by the birefringence fluctuations [4, 5] whose characteristic time scales are much longer than the temporal separation between consecutive light pulses; the information transmission in a spin chain channel where if the chain is not

R.-G. Zhou et al., *Design of Quantum Teleportation Schemes*,
https://doi.org/10.1007/978-3-031-82725-9_6

reset after each use of the channel, the first transmission of information state could influence the second [6]; and the bosonic channel considering the loss of energy en route where the much higher use rate of the channel would induce the unwanted overlaps between consecutive input pulses or interfere with the finite relaxation time of the local environment, leading to the memory effects of attenuation [7, 8]. Interestingly, several models of memory channels have shown that the quantum or classical capacity can be enhanced compared to the results of memoryless channels. Therefore, it is worth considering how to construct a model of quantum memory channel during the entanglement distribution for improving the communication quality of teleportation.

In this chapter, three quantum teleportation schemes in noisy environment are proposed.

In Sect. 6.2, a novel scheme of bidirectional controlled quantum teleportation (C-BQT) via a seven-qubit state is proposed [9], where Alice and Bob, two legitimate users, can teleport two-qubit states to each other. In the whole process, users achieve the initial state based on preprocessing of quantum channel, Bell state measurement (BSM), single-qubit measurement (SM), unitary operations, and so on. The main superiority of the proposed protocol is more efficient compared with previous work. In addition, the proposed C-BQT scheme is considered in noisy channel, and it shows that the fidelities under amplitude-damping (AD) and phase-damping (PD) noise only rest with the amplitude parameter of the initial state and the decoherence noisy rate.

In Sect. 6.3, a novel scheme for asymmetric bidirectional controlled quantum teleportation (C-ABQT) by using a nine-qubit entangled state as a quantum channel is proposed [10]. Alice transmits an arbitrary three-qubit entangled state to Bob, and at the same time, Bob transmits an arbitrary single-qubit entangled state to Alice under the permission of a third party as a controller. This scheme is based on the Hadamard (H) gates, controlled-NOT (CNOT) operations, appropriate unitary operations, and single-qubit von Neumann measurements, which are more efficient than multi-qubit joint measurements (JM). This scheme explicitly shows how the channel of a nine-qubit entangled state prepares, and it is discussed in six types of noisy scenarios (amplitude-damping, phase-damping, bit-flip, phase-flip, bit-phase-flip, and depolarizing noisy scenarios). Then, we analytically derive the fidelities of the NEC-ABQT process and show that they only depend on the amplitude parameter of the initial state and the decoherence noisy rate. Finally, compared with previous schemes, this scheme also shows remarkable advantages.

In Sect. 6.4, a model of quantum memory channel during the entanglement distribution is constructed for better understanding the action of memory effects of channel on qutrit teleportation [11]. Then based on the model, a common representation of the dynamical evolution of a maximally two-qutrit entangled state through different kinds of noise is derived. Here, Pauli errors and energy dissipation in the channel are paid much attention. Once the operators describing various noises are given, the evolution of a maximally qutrit entangled state can be numerically analyzed. In particular, the Kraus operators of describing correlated amplitude-damping channel cannot be directly represented like that of correlated Pauli channel

[12]. It is creatively derived by solving the Lindblad equation representing the dynamical evolution of the compound system of two three-level V-type atoms. Finally, the communication quality of teleportation is evaluated by average fidelity, which is the function of noise intensity and the degree of memory. It is discovered that qutrit teleportation with Pauli channel (except for phase-flip) or amplitude-damping channel has some robustness to noise when memory effects of quantum channel are considered.

6.2 NEC-BQT Scheme of Two-Qubit States via a Seven-Qubit Entangled State

A both novel and efficient scheme for NEC-BQT is proposed, where Alice and Bob can teleport an arbitrary two-qubit state and a pure two-qubit state to each other based on sharing a seven-qubit entangled state simultaneously. They can obtain 27 results with the same possibility after BSM and SM; they share the respective measurement results via classical channel and do corresponding unitary operation to reconstruct the initial states under the control of Charlie (based on Charlie's measurement results). Then, the proposed scheme is analyzed in typical noisy channels, including amplitude-damping noise and phase-damping noise. A conclusion that the fidelity of NEC-BQT process can only be affected by amplitude parameter of initial state and the decoherence rate is shown clearly by analysis and figures in Sect. 6.2.3.

6.2.1 Generation of Quantum Channel

The quantum channel (a seven-qubit entangled state) for realizing the proposed protocol can be achieved experimentally, whose circuit is clear as shown in Fig. 6.1. The seven-qubit state can be specifically projected by means of two

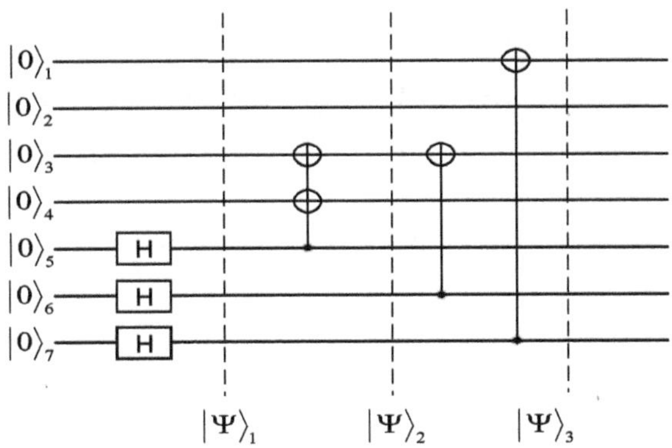

Fig. 6.1 Illustration of quantum circuit of the proposed seven-qubit entangled state

Hadamard gates and five controlled-NOT (CNOT) gates, which is designed at great length as the following:

Step 1. Preparing seven zero states as initial state like (6.1)

$$|\psi_0\rangle = |0\rangle_1 \otimes |0\rangle_2 \otimes |0\rangle_3 \otimes |0\rangle_4 \otimes |0\rangle_5 \otimes |0\rangle_6 \otimes |0\rangle_7 \tag{6.1}$$

Step 2. Three Hadamard gates are implemented to qubits 5, 6, and 7; the $|\psi\rangle_0$ is transformed into

$$\begin{aligned} |\psi_0\rangle &= |0\rangle_1 \otimes |0\rangle_2 \otimes |0\rangle_3 \otimes |0\rangle_4 \otimes \frac{(|0\rangle + |1\rangle)}{\sqrt{2}}_5 \otimes \frac{(|0\rangle + |1\rangle)}{\sqrt{2}}_6 \otimes \frac{(|0\rangle + |1\rangle)}{\sqrt{2}}_7 \\ &= \frac{1}{2\sqrt{2}}(|0000000\rangle + |0000001\rangle + |0000010\rangle + |0000011\rangle \\ &\quad +|0000100\rangle + |0000101\rangle + |0000110\rangle + |1111111\rangle)_{1234567} \end{aligned} \tag{6.2}$$

Step 3. In this step, CNOT operations are implemented where qubit 5 serves as controlled qubit and qubits 2 and 4 are employed as target qubits. The particular process is shown as the following and the state is transformed into $|\psi\rangle_2$

$$\begin{aligned} |\psi_2\rangle &= \mathrm{CNOT}[5,(2,4)]|0\rangle_1 \otimes |0\rangle_2 \otimes |0\rangle_3 \otimes |0\rangle_4 \otimes \frac{(|0\rangle + |1\rangle)}{\sqrt{2}}_5 \otimes \frac{(|0\rangle + |1\rangle)}{\sqrt{2}}_6 \otimes \frac{(|0\rangle + |1\rangle)}{\sqrt{2}}_7 \\ &= \frac{1}{2\sqrt{2}}(|0000000\rangle + |0000001\rangle + |0000010\rangle + |0000011\rangle \\ &\quad +|0101100\rangle + |0101101\rangle + |0101110\rangle + |0101111\rangle)_{1234567} \end{aligned} \tag{6.3}$$

Step 4. Controller implements other two CNOT operations on qubits 1 and 3, where qubits 7 and 6 act as controlled qubits. Then the state will convert into $|\psi\rangle_3$ shown as follows:

$$\begin{aligned} |\psi_3\rangle &= \frac{1}{2\sqrt{2}}(|0000000\rangle + |1000001\rangle + |0010010\rangle + |1010011\rangle \\ &\quad +|0101100\rangle + |1101101\rangle + |0111110\rangle + |1111111\rangle)_{1234567} \end{aligned} \tag{6.4}$$

The state $|\psi\rangle_3$ above is prepared and utilized as quantum channel in this scheme.

6.2.2 C-BQT Scheme

In this subsection, Alice and Bob, the legitimate users, want to teleport two-qubit states to each other by utilizing seven-qubit entangled states mentioned in Sect. 6.2.1. The two-qubit states (including a pure two-qubit state and an arbitrary two-qubit state) that they would like to teleport to each other simultaneously can be presented as

$$\begin{aligned} |\Phi\rangle_{A_1A_2} &= \alpha_0|00\rangle + \alpha_1|11\rangle \\ |\Phi\rangle_{B_1B_2} &= \beta_0|00\rangle + \beta_1|01\rangle + \beta_2|10\rangle + \beta_3|11\rangle \end{aligned} \tag{6.5}$$

where $|\alpha_0|^2 + |\alpha_1|^2 = 1$ and $|\beta_0|^2 + |\beta_1|^2 + |\beta_2|^2 + |\beta_3|^2 = 1$. Alice wants to teleport her qubits A_1 and A_2 to Bob; meanwhile, Bob teleports his qubits B_1 and B_2 to Alice. In the process, it is necessary for them to do some appropriate CNOT operations, BSM, SM, and unitary operations to make the scheme come true. The schematic demonstration is illustrated in Fig. 6.2.

The C-BQT scheme can be divided into the following steps in detail:

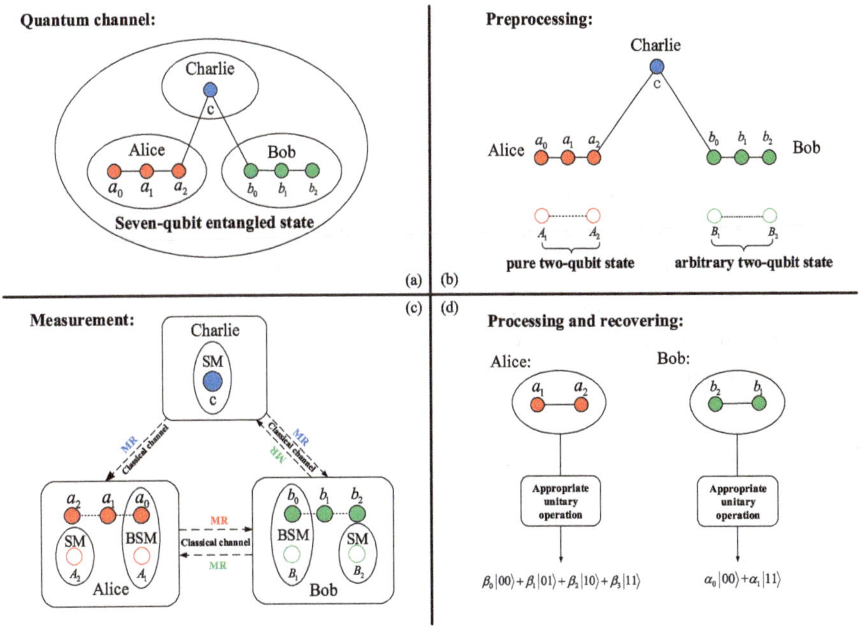

Fig. 6.2 The schematic demonstration of the whole system: the quantum channel seven-qubit entangled state and its preprocessing, the C-BQT scheme of measurements, and the process of reconstructing the initial state are shown in (**a**), (**b**), (**c**), and (**d**), respectively. The MR represents measurement results

Step 1. Suppose that Alice, Bob, and supervisor Charlie share the state $|\psi\rangle_3$ as quantum channel where Alice holds qubits 2, 6, and 7; Bob possesses qubits 3, 4, and 5; and qubit 1 belongs to Charlie. And the state can be rewritten as (6.6)

$$|\Psi\rangle_{ca_0b_0b_1b_2a_1a_2} = \frac{1}{2\sqrt{2}}(|0000000\rangle + |1000001\rangle + |0010010\rangle+|1010011\rangle+ \\ |0101100\rangle + |1101101\rangle+|0111110\rangle + |1111111\rangle)_{ca_0b_0b_1b_2a_1a_2}. \tag{6.6}$$

The whole system can be expressed by

$$|\Phi\rangle_{ca_0b_0b_1b_2a_1a_2A_1A_2B_1B_2} = |\Psi\rangle_{ca_0b_0b_1b_2a_1a_2} \otimes |\Phi\rangle_{A_1A_2} \otimes |\Phi\rangle_{B_1B_2}. \tag{6.7}$$

Step 2. In this step, Alice performs a Bell state measurement on qubits A_2 and a_0. The remaining unmeasured particles will collapse into one of the following possible states that are given in Table 6.1. Meanwhile, the Bell state measurement basis is shown as follows (6.8):

$$\begin{aligned} |\psi^\pm\rangle &= \frac{\sqrt{2}}{2}(|00\rangle \pm |11\rangle) \\ |\phi^\pm\rangle &= \frac{\sqrt{2}}{2}(|01\rangle \pm |10\rangle) \end{aligned}. \tag{6.8}$$

Step 3. According to Table 6.1, if Alice's measurement result is $|\psi^+\rangle_{A_2a_0}$, she will announce it to Bob via classical channel. Then Bob measures his qubits B_1, b_0 based on the same basic Bell states, and the measurements are given in Table 6.2.

Similarly, Bob replies his results to Alice via classical channel. If Bob's measurement result is $|\psi^+\rangle_{B_1b_0}$, the state will collapse to the state as (6.9) under the mentioned operations.

$$\begin{aligned} |\Phi''\rangle_{a_1a_2cb_1b_2a_1a_2A_2B_2} = &[\alpha_0\beta_0(|0000000\rangle + |1000100\rangle) + \alpha_0\beta_1(|0000001\rangle + |1000101\rangle) \\ &+\alpha_0\beta_2(|0001000\rangle + |1001100\rangle) + \alpha_0\beta_3(|0001000\rangle + |1001100\rangle) \\ &+\alpha_1\beta_0(|0110010\rangle + |1110110\rangle) + \alpha_1\beta_1(|0110011\rangle + |1110111\rangle) \\ &+\alpha_1\beta_2(|0111010\rangle + |1111110\rangle) + \alpha_1\beta_3(|0111011\rangle + |1111111\rangle) \end{aligned}. \tag{6.9}$$

Step 4. In this step, Alice and Bob perform X-basis single-qubit measurements on qubits A_2 and B_2. Then Alice and Bob announce their measurement results to each other via classical channel. As shown in Table 6.3, the other qubits will collapse to one of the four states with the same possibility.

Table 6.1 Alice's measurement results and the collapsed states of the remaining particles

Alice's measurement results	The collapsed state of particles $c, b_0, b_1, b_2, a_1, a_2, A_2, B_1, B_2$
$\|\psi^+\rangle_{A_1a_0}$	$\alpha_0\beta_0(\|000000000\rangle + \|100001000\rangle + \|010010000\rangle + \|110011000\rangle)$ $+\alpha_0\beta_1(\|000000001\rangle + \|100001001\rangle + \|010010001\rangle + \|110011001\rangle)$ $+\alpha_0\beta_2(\|000000010\rangle + \|100001010\rangle + \|010010010\rangle + \|110011010\rangle)$ $+\alpha_0\beta_3(\|000000011\rangle + \|100001011\rangle + \|010010011\rangle + \|110011011\rangle)$ $+\alpha_1\beta_0(\|001100100\rangle + \|101101100\rangle + \|011110100\rangle + \|111111100\rangle)$ $+\alpha_1\beta_1(\|001100101\rangle + \|101101101\rangle + \|011110101\rangle + \|111111101\rangle)$ $+\alpha_1\beta_2(\|001100110\rangle + \|101101110\rangle + \|011110110\rangle + \|111111110\rangle)$ $+\alpha_1\beta_3(\|001100111\rangle + \|101101111\rangle + \|011110111\rangle + \|111111111\rangle)$
$\|\psi^-\rangle_{A_1a_0}$	$\alpha_0\beta_0(\|000000000\rangle + \|100001000\rangle + \|010010000\rangle + \|110011000\rangle)$ $+\alpha_0\beta_1(\|000000001\rangle + \|100001001\rangle + \|010010001\rangle + \|110011001\rangle)$ $+\alpha_0\beta_2(\|000000010\rangle + \|100001010\rangle + \|010010010\rangle + \|110011010\rangle)$ $+\alpha_0\beta_3(\|000000011\rangle + \|100001011\rangle + \|010010011\rangle + \|110011011\rangle)$ $-\alpha_1\beta_0(\|001100100\rangle + \|101101100\rangle + \|011110100\rangle + \|111111100\rangle)$ $-\alpha_1\beta_1(\|001100101\rangle + \|101101101\rangle + \|011110101\rangle + \|111111101\rangle)$ $-\alpha_1\beta_2(\|001100110\rangle + \|101101110\rangle + \|011110110\rangle + \|111111110\rangle)$ $-\alpha_1\beta_3(\|001100111\rangle + \|101101111\rangle + \|011110111\rangle + \|111111111\rangle)$
$\|\phi^+\rangle_{A_1a_0}$	$\alpha_0\beta_0(\|001100000\rangle + \|101101000\rangle + \|011110000\rangle + \|111111000\rangle)$ $+\alpha_0\beta_1(\|001100001\rangle + \|101101001\rangle + \|011110001\rangle + \|111111001\rangle)$ $+\alpha_0\beta_2(\|001100010\rangle + \|101101010\rangle + \|011110010\rangle + \|111111010\rangle)$ $+\alpha_0\beta_3(\|001100011\rangle + \|101101011\rangle + \|011110011\rangle + \|111111011\rangle)$ $+\alpha_1\beta_0(\|000000100\rangle + \|100001100\rangle + \|010010100\rangle + \|110011100\rangle)$ $+\alpha_1\beta_1(\|000000101\rangle + \|100001101\rangle + \|010010101\rangle + \|110011101\rangle)$ $+\alpha_1\beta_2(\|000000110\rangle + \|100001110\rangle + \|010010110\rangle + \|110011110\rangle)$ $+\alpha_1\beta_3(\|000000111\rangle + \|100001111\rangle + \|010010111\rangle + \|110011111\rangle)$
$\|\phi^-\rangle_{A_1a_0}$	$\alpha_0\beta_0(\|001100000\rangle + \|101101000\rangle + \|011110000\rangle + \|111111000\rangle)$ $+\alpha_0\beta_1(\|001100001\rangle + \|101101001\rangle + \|011110001\rangle + \|111111001\rangle)$ $+\alpha_0\beta_2(\|001100010\rangle + \|101101010\rangle + \|011110010\rangle + \|111111010\rangle)$ $+\alpha_0\beta_3(\|001100011\rangle + \|101101011\rangle + \|011110011\rangle + \|111111011\rangle)$ $-\alpha_1\beta_0(\|000000100\rangle + \|100001100\rangle + \|010010100\rangle + \|110011100\rangle)$ $-\alpha_1\beta_1(\|000000101\rangle + \|100001101\rangle + \|010010101\rangle + \|110011101\rangle)$ $-\alpha_1\beta_2(\|000000110\rangle + \|100001110\rangle + \|010010110\rangle + \|110011110\rangle)$ $-\alpha_1\beta_3(\|000000111\rangle + \|100001111\rangle + \|010010111\rangle + \|110011111\rangle)$

Step 5. Here, it is necessary for Bob to do Z-basis single-qubit measurement on his qubit B_2 and notify results to supervisor Charlie. Charlie firstly performs X-Pauli gate on his qubit c if Bob's measurement results are $|1\rangle$. Then he performs Z-basis single-qubit measurement on qubit c. If Alice's and Bob's results are both $|+\rangle$, the state can be transformed into (6.10), and Charlie's measurement results are shown in Table 6.4.

Table 6.2 Bob's measurement results and the collapsed states of the remaining particles $c,b_1,b_2,a_1,a_2,A_2,B_2$

Bell measurement results	The collapsed state of qubits $c,b_1,b_2,a_1,a_2,A_2,B_2$
$\|\psi^+\rangle_{B_1b_0}$	$\alpha_0\beta_0(\|0000000\rangle+\|1000100\rangle)+\alpha_0\beta_1(\|0000001\rangle+\|1000101\rangle)$ $+\alpha_0\beta_2(\|0001000\rangle+\|1001100\rangle)+\alpha_0\beta_3(\|0001001\rangle+\|1001101\rangle)$ $+\alpha_1\beta_0(\|0110010\rangle+\|1110110\rangle)+\alpha_1\beta_1(\|0110011\rangle+\|1110111\rangle)$ $+\alpha_1\beta_2(\|0111010\rangle+\|1111110\rangle)+\alpha_1\beta_3(\|0111011\rangle+\|1111111\rangle)$
$\|\psi^-\rangle_{B_1b_0}$	$\alpha_0\beta_0(\|0000000\rangle+\|1000100\rangle)+\alpha_0\beta_1(\|0000001\rangle+\|1000101\rangle)$ $-\alpha_0\beta_2(\|0001000\rangle+\|1001100\rangle)-\alpha_0\beta_3(\|0001001\rangle+\|1001101\rangle)$ $+\alpha_1\beta_0(\|0110010\rangle+\|1110110\rangle)+\alpha_1\beta_1(\|0110011\rangle+\|1110111\rangle)$ $-\alpha_1\beta_2(\|0111010\rangle+\|1111110\rangle)-\alpha_1\beta_3(\|0111011\rangle+\|1111111\rangle)$
$\|\phi^+\rangle_{B_1b_0}$	$\alpha_0\beta_0(\|0001000\rangle+\|1001100\rangle)+\alpha_0\beta_1(\|0001001\rangle+\|1001101\rangle)$ $+\alpha_0\beta_2(\|0000000\rangle+\|1000100\rangle)+\alpha_0\beta_3(\|0000001\rangle+\|1000101\rangle)$ $+\alpha_1\beta_0(\|0111010\rangle+\|1111110\rangle)+\alpha_1\beta_1(\|0111011\rangle+\|1111111\rangle)$ $+\alpha_1\beta_2(\|0110010\rangle+\|1110110\rangle)+\alpha_1\beta_3(\|0110011\rangle+\|1110111\rangle)$
$\|\phi^-\rangle_{B_1b_0}$	$\alpha_0\beta_0(\|0001000\rangle+\|1001100\rangle)+\alpha_0\beta_1(\|0001001\rangle+\|1001101\rangle)$ $-\alpha_0\beta_2(\|0000000\rangle+\|1000100\rangle)-\alpha_0\beta_3(\|0000001\rangle+\|1000101\rangle)$ $+\alpha_1\beta_0(\|0111010\rangle+\|1111110\rangle)+\alpha_1\beta_1(\|0111011\rangle+\|1111111\rangle)$ $-\alpha_1\beta_2(\|0110010\rangle+\|1110110\rangle)-\alpha_1\beta_3(\|0110011\rangle+\|1110111\rangle)$

Table 6.3 The X-basis single-qubit measurement results and the collapsed state of the remaining qubits

Alice's results	Bob's results	The collapsed state of qubits c,b_1,b_2,a_1,a_2
+	+	$\alpha_0\beta_0(\|00000\rangle+\|10001\rangle)+\alpha_0\beta_1(\|00000\rangle+\|10001\rangle)$ $+\alpha_0\beta_2(\|00010\rangle+\|10011\rangle)+\alpha_0\beta_3(\|00010\rangle+\|10011\rangle)$ $+\alpha_1\beta_0(\|01100\rangle+\|11101\rangle)+\alpha_1\beta_1(\|01100\rangle+\|11101\rangle)$ $+\alpha_1\beta_2(\|01110\rangle+\|11111\rangle)+\alpha_1\beta_3(\|01110\rangle+\|11111\rangle)$
+	−	$\alpha_0\beta_0(\|00000\rangle+\|10001\rangle)-\alpha_0\beta_1(\|00000\rangle+\|10001\rangle)$ $+\alpha_0\beta_2(\|00010\rangle+\|10011\rangle)-\alpha_0\beta_3(\|00010\rangle+\|10011\rangle)$ $+\alpha_1\beta_0(\|01100\rangle+\|11101\rangle)-\alpha_1\beta_1(\|01100\rangle+\|11101\rangle)$ $+\alpha_1\beta_2(\|01110\rangle+\|11111\rangle)-\alpha_1\beta_3(\|01110\rangle+\|11111\rangle)$
−	+	$\alpha_0\beta_0(\|00000\rangle+\|10001\rangle)+\alpha_0\beta_1(\|00000\rangle+\|10001\rangle)$ $+\alpha_0\beta_2(\|00010\rangle+\|10011\rangle)+\alpha_0\beta_3(\|00010\rangle+\|10011\rangle)$ $-\alpha_1\beta_0(\|01100\rangle+\|11101\rangle)-\alpha_1\beta_1(\|01100\rangle+\|11101\rangle)$ $-\alpha_1\beta_2(\|01110\rangle+\|11111\rangle)-\alpha_1\beta_3(\|01110\rangle+\|11111\rangle)$
−	−	$\alpha_0\beta_0(\|00000\rangle+\|10001\rangle)-\alpha_0\beta_1(\|00000\rangle+\|10001\rangle)$ $+\alpha_0\beta_2(\|00010\rangle+\|10011\rangle)-\alpha_0\beta_3(\|00010\rangle+\|10011\rangle)$ $-\alpha_1\beta_0(\|01100\rangle+\|11101\rangle)+\alpha_1\beta_1(\|01100\rangle+\|11101\rangle)$ $-\alpha_1\beta_2(\|01110\rangle+\|11111\rangle)+\alpha_1\beta_3(\|01110\rangle+\|11111\rangle)$

Table 6.4 Charlie's measurement results and the collapsed state of the remaining qubits

Charlie's measurement results	The collapsed state of remaining qubits b_1,b_2,a_1,a_2
0	$\alpha_0\beta_0\|0000\rangle + \alpha_0\beta_1\|0001\rangle + \alpha_0\beta_2\|0010\rangle + \alpha_0\beta_3\|0011\rangle$ $+\alpha_1\beta_0\|1100\rangle + \alpha_1\beta_1\|1101\rangle + \alpha_1\beta_2\|1110\rangle + \alpha_1\beta_3\|1111\rangle$
1	$\alpha_0\beta_0\|0001\rangle + \alpha_0\beta_1\|0000\rangle + \alpha_0\beta_2\|0011\rangle + \alpha_0\beta_3\|0010\rangle$ $+\alpha_1\beta_0\|1101\rangle + \alpha_1\beta_1\|1100\rangle + \alpha_1\beta_2\|1111\rangle + \alpha_1\beta_3\|1110\rangle$

Table 6.5 The corresponding relationship between unitary operation and three users' measurement results

Alice's results	Bob's results	Charlie's results	Unitary operation on a_1,a_2,b_1,b_2
+	+	0	$I \otimes I \otimes I \otimes I$
+	+	1	$I \otimes I \otimes I \otimes X$
+	−	0	$I \otimes I \otimes I \otimes Z$
+	−	1	$I \otimes I \otimes I \otimes iY$
−	+	0	$Z \otimes I \otimes I \otimes I$
−	+	1	$Z \otimes I \otimes I \otimes X$
−	−	0	$Z \otimes I \otimes I \otimes Z$
−	−	1	$Z \otimes I \otimes I \otimes iY$

$$\begin{aligned} \left|\Phi''\right\rangle_{cb_1b_2a_1a_2} = & [\alpha_0\beta_0(|00000\rangle + |10001\rangle) + \alpha_0\beta_1(|10000\rangle + |00001\rangle) \\ & +\alpha_0\beta_2(|00010\rangle + |10011\rangle) + \alpha_0\beta_3(|10010\rangle + |00011\rangle) \\ & +\alpha_1\beta_0(|01100\rangle + |11101\rangle) + \alpha_1\beta_1(|11100\rangle + |01101\rangle) \\ & +\alpha_1\beta_2(|01110\rangle + |11111\rangle) + \alpha_1\beta_3(|11110\rangle + |01111\rangle) \end{aligned} . \tag{6.10}$$

Step 6. Charlie announces his measurement results to both Alice and Bob, in this step. Upon receiving the results, Alice and Bob implement the appropriate unitary operation for reconstructing the initial state. As shown in Table 6.5, Alice and Bob perform the unitary operation according to the mentioned results above.

According to Charlie's results, Alice and Bob implement the X, Z, and *i*Y unitary operations on the relevant particles. Finally, Alice and Bob can reconstruct the transmitted state as (6.11). There are 27 possible results in this scheme that are all confirmed both feasibly and legitimately. By now the C-BQT is successfully completed.

$$\begin{aligned} |\Phi\rangle_{A_1A_2} &= \beta_0|00\rangle + \beta_1|01\rangle + \beta_2|10\rangle + \beta_3|11\rangle \\ |\Phi\rangle_{B_1B_2} &= \alpha_0|00\rangle + \alpha_1|11\rangle \end{aligned} . \tag{6.11}$$

6.2.3 Effects of NEC-BQT Scheme

In this subsection, the effects of amplitude-damping (AD) noise and phase-damping (PD) noise are taken into consideration. The two noise models can be characterized by Kraus operations as follows [13]:

$$E_0^A = \begin{bmatrix} 1 & 0 \\ 0 & \sqrt{1-\eta_A} \end{bmatrix}, E_1^A = \sqrt{\eta_A}\begin{bmatrix} 0 & 1 \\ 0 & 0 \end{bmatrix} \tag{6.12}$$

$$E_0^P = \sqrt{1-\eta_P} I, E_1^P = \sqrt{\eta_P}\begin{bmatrix} 1 & 0 \\ 0 & 0 \end{bmatrix}, E_2^P = \sqrt{\eta_P}\begin{bmatrix} 0 & 0 \\ 0 & 1 \end{bmatrix}. \tag{6.13}$$

where $\eta_A (0 \leq \eta_A \leq 1)$ and $\eta_P (0 \leq \eta_P \leq 1)$ are decoherence rates of AD noise and PD noise, respectively, which describe the possibility of occurring error in quantum state due to travel qubit.

The quantum channel mentioned above is a pure seven-qubit entangled state $|\psi\rangle$. It is well known that the corresponding density matrix is defined as $\rho = |\psi\rangle\langle\psi|$. Here, after being teleported in noisy channel, the ρ can be rewritten as (6.14)

$$\begin{aligned} \zeta^r(\rho) = \sum_m \left(E_m^{ra_0}\right)\left(E_m^{rb_0}\right)\left(E_m^{rb_1}\right)\left(E_m^{rb_2}\right)\left(E_m^{ra_1}\right)\left(E_m^{ra_2}\right) \\ \rho\left(E_m^{ra_0}\right)^\dagger\left(E_m^{rb_0}\right)^\dagger\left(E_m^{rb_1}\right)^\dagger\left(E_m^{rb_2}\right)^\dagger\left(E_m^{ra_1}\right)^\dagger\left(E_m^{ra_2}\right)^\dagger \end{aligned} \tag{6.14}$$

where $r \in \{A, P\}$, when $r = A$, *i. e.* , $m \in \{0, 1\}$ and when $r = P$, *i. e.* , $m \in \{0, 1, 2\}$. The superscripts a_i and b_i represent which qubits are performed on operator E, and ζdenotes a quantum operation, which maps ρ to $\zeta(\rho)$ due to the noise. In (6.14), it is assumed that qubits $(a_0, a_1, a_2, b_0, b_1, b_2)$ teleported in noisy environment between Alice and Bob are affected by the same Kraus operator. Meanwhile, it is clear that Charlie's qubit c is not affected by noise, because controller Charlie creates the channel and his qubit does not need to be teleported in this channel. As a result of the noisy environment, the shared state would become a mixed state after distributing these qubits, which are given by

$$\begin{aligned}\zeta^A(\rho)=&\frac{1}{8}\Big\{\Big[|0000000\rangle+\sqrt{1-\eta_A}|1000001\rangle+(1-\eta_A)|0010010\rangle+\\&\sqrt{(1-\eta_A)^3}|0010011\rangle+\sqrt{(1-\eta_A)^3}|1101100\rangle+(1-\eta_A)^2|1101101\rangle\\&+\sqrt{(1-\eta_A)^5}|0111110\rangle+(1-\eta_A)^6|1111111\rangle\Big]\times\\&[\langle 0000000|+(1-\eta_A)\langle 1000001|+(1-\eta_A)\langle 0010010|+\sqrt{(1-\eta_A)^3}\langle 0010011|\\&+\sqrt{(1-\eta_A)^3}\langle 1101100|+(1-\eta_A)^2\langle 1101101|+\sqrt{(1-\eta_A)^5}\langle 0111110|\\&+(1-\eta_A)^6\langle 1111111|\Big]+\eta_A^6|1000000\rangle\langle 0000001|\Big\}\end{aligned} \tag{6.15}$$

and

$$\begin{aligned}\zeta^P(\rho)=&\frac{1}{8}\Big\{(1-\eta_P)^6[|0000000\rangle+|1000001\rangle+|0010010\rangle+|0010011\rangle\\&+|1101100\rangle+|1101101\rangle+|0111110\rangle+|1111111\rangle]\times\\&[\langle 0000000|+\langle 1000001|+\langle 0010010|+\langle 0010011|+\\&\langle 1101100|+\langle 1101101|+\langle 0111110|+\langle 1111111|]+\\&\eta_P^6(|0000000\rangle\langle 0000000|+|1111111\rangle\langle 1111111|)\Big\}\end{aligned} \tag{6.16}$$

For recovering the initial state, Alice and Bob perform appropriate unitary operations on their own qubits according to measurement results shared before. Afterward, the density matrix of the final output state can be expressed as

$$\rho_{out}^r=Tr_{A_1A_2B_1B_2ca_0b_0}\left\{U\left[\rho_{A_1A_2}\otimes\rho_{B_1B_2}\otimes\zeta^r(\rho)\right]U^\dagger\right\} \tag{6.17}$$

where $Tr_{A_1A_2B_1B_2ca_0b_0}$ is the partial trace over qubits $(A_1,A_2,B_1,B_2,c,a_0,b_0)$ and U is a unitary operation to describe the C-BQT process, which can be expressed as (6.18)

$$\begin{aligned}U=&\left\{I_{A_1A_2a_0}\otimes I_{B_1B_2b_0}\otimes I_c\otimes\sigma_{b_1b_2a_1a_2}^{ljynzm}\right\}\\&\left\{I_{A_1A_2a_0}\otimes I_{B_1b_0}\otimes|\phi\rangle_{B_2}^m\langle\phi|_{B_2}^m\otimes|\phi\rangle_c^z\langle\phi|_c^z\otimes I_{b_1b_2a_1a_2}\right\}\\&\left\{I_{A_1a_0}\otimes I_{B_1b_0}\otimes|\phi\rangle_{A_2}^n\langle\phi|_{A_2}^n\otimes|\phi\rangle_{B_2}^y\langle\phi|_{B_2}^y\otimes I_c\otimes I_{b_1b_2a_1a_2}\right\}\\&\left\{|\phi\rangle_{A_1a_0}^j\langle\phi|_{A_1a_0}^j\otimes|\phi\rangle_{B_1b_0}^l\langle\phi|_{B_1b_0}^l\otimes I_{A_2}\otimes I_{B_2}\otimes I_c\otimes I_{b_1b_2a_1a_2}\right\}\end{aligned} \tag{6.18}$$

where $l,j\in\{1,2\}$, $y,n,z,m\in\{1,2,3,4\}$, $|\phi\rangle_{A_1a_0}^j\langle\phi|_{A_1a_0}^j$, and $|\phi\rangle_{B_1b_0}^l\langle\phi|_{B_1b_0}^l$ denote the Bell state measurement results and $|\phi\rangle_{A_2}^n\langle\phi|_{A_2}^n$, $|\phi\rangle_{B_2}^y\langle\phi|_{B_2}^y$, $|\phi\rangle_{B_2}^m\langle\phi|_{B_2}^m$ and

$|\phi\rangle_c^z\langle\phi|_c^z$ denote the single-qubit measurement results. $\sigma_{b_1b_2a_1a_2}^{ljynzm}$ represents Alice and Bob's recovering unitary operation, which is performed after reciprocally notifying measurement results.

It is easy to obtain the final quantum state ρ_{out}^r, which is the product of quantum states received by Alice and Bob, under the effects of AD and PD noises. The ideal final states without noises were known as (6.11). Alice holds qubit a_1, a_2 in the form of $(\beta_0|00\rangle+\beta_1|01\rangle+\beta_2|10\rangle+\beta_3|11\rangle)_{a_1a_2}$ and Bob holds qubit b_1, b_2 in the state of $(\alpha_0|00\rangle+\alpha_1|11\rangle)_{b_1b_2}$.

The fidelity is an important metric to indicate the distance between the initial state and actual state. Namely, it can visualize the effects of noises by comparing the state ρ_{out}^r with state $|\psi\rangle$ in this section, which is given by

$$F=\langle\Psi|\rho_{out}^r|\Psi\rangle. \tag{6.19}$$

The fidelity ranges from 0 to 1, where 0 indicates two orthogonal states and 1 indicates two equal states before and after teleporting (the perfect teleportation). The smaller the F, the more information lost in noisy environment.

It is easy to obtain the state ρ_{out}^r according to (6.17) and (6.18), and it is unaffected by Alice, Bob, and Charlie's measurement results after detailed calculation. Here, ρ_{out}^r is shown as (6.20) and (6.21).

$$\begin{aligned}\rho_{out}^A=&\{[\alpha_0\beta_0|0000\rangle+\left(\sqrt{1-\eta_A}\right)\alpha_0\beta_1|0001\rangle+(1-\eta_A)\alpha_0\beta_2|0010\rangle+\\&\sqrt{(1-\eta_A)^3}\alpha_0\beta_3|0011\rangle+\sqrt{(1-\eta_A)^3}\alpha_1\beta_0|1100\rangle+(1-\eta_A)^2\alpha_1\beta_1|1101\rangle\\&+\sqrt{(1-\eta_A)^5}\alpha_1\beta_2|1110\rangle+(1-\eta_A)^6\alpha_1\beta_3|1111\rangle]\times[\alpha_0\beta_0\langle 0000|+\\&\left(\sqrt{1-\eta_A}\right)\alpha_0\beta_1\langle 0001|+(1-\eta_A)\alpha_0\beta_2\langle 0010|+\sqrt{(1-\eta_A)^3}\alpha_0\beta_3\langle 0011|\\&+\sqrt{(1-\eta_A)^3}\alpha_1\beta_0\langle 1100|+(1-\eta_A)^2\alpha_1\beta_1\langle 1101|+\sqrt{(1-\eta_A)^5}\alpha_1\beta_2\langle 1110|\\&+(1-\eta_A)^6\alpha_1\beta_3\langle 1111|]+\eta_A^6\alpha_1^2\beta_3^2|0000\rangle\langle 0000|\}\end{aligned} \tag{6.20}$$

and

$$\begin{aligned}\rho_{out}^P=&\Big\{(1-\eta_P)^6[\alpha_0\beta_0|0000\rangle+\alpha_0\beta_1|0001\rangle+\alpha_0\beta_2|0010\rangle+\alpha_0\beta_3|0011\rangle\\&+\alpha_1\beta_0|1100\rangle+\alpha_1\beta_1|1101\rangle+\alpha_1\beta_2|1110\rangle+\alpha_1\beta_3|1111\rangle]\\&\times[\alpha_0\beta_0\langle 0000|+\alpha_0\beta_1\langle 0001|+\alpha_0\beta_2\langle 0010|+\alpha_0\beta_3\langle 0011|\\&+\alpha_1\beta_0\langle 1100|+\alpha_1\beta_1\langle 1101|+\alpha_1\beta_2\langle 1110|+\alpha_1\beta_3\langle 1111|]\\&+\eta_P^6\alpha_0^2\beta_0^2|0000\rangle\langle 0000|+\eta_P^6\alpha_1^2\beta_3^2|1111\rangle\langle 1111|\}\end{aligned} \tag{6.21}$$

According to (6.19) and (6.20), the fidelity of the quantum states teleported in AD noisy channel can be easily obtained as

$$\begin{aligned} F^A &= \{ [\alpha_0{}^2\beta_0{}^2 + \sqrt{1-\eta_A}\alpha_0{}^2\beta_1{}^2 + (1-\eta_A)\alpha_0{}^2\beta_2{}^2 \\ &+\sqrt{(1-\eta_A)^3}\alpha_0{}^2\beta_3{}^2 + \sqrt{(1-\eta_A)^3}\alpha_1{}^2\beta_0{}^2 + (1-\eta_A)^2\alpha_1{}^2\beta_1{}^2 . \\ &+\sqrt{(1-\eta_A)^5}\alpha_1{}^2\beta_2{}^2 + (1-\eta_A)^6\alpha_1{}^2\beta_3{}^2]^2 + \eta_A\alpha_0{}^2\alpha_1^2\beta_0^2\beta_3{}^2 \} \end{aligned} \tag{6.22}$$

Similarly, from (6.19) and (6.21), the fidelity of the quantum states teleported in PD noisy channel can be obtained as

$$F^P = (1-\eta_P)^6 + \eta_P{}^6\alpha_0{}^4\beta_0^4 + \eta_P{}^6\alpha_1^4\beta_3{}^4 . \tag{6.23}$$

It is clear that the amplitude parameter of the initial state and the decoherence rate, serving as factors, can affect the fidelities of both cases. Specifically, Figure 6.3a–f clearly illustrates the effect of AD (PD) noise on the fidelity $F^A(F^P)$ and variation of the fidelity with amplitude parameter of the initial state and decoherence rate η_r. It is obvious that the fidelities decrease with decoherence rates, respectively (Figure 6.3a, b, d, e).

Similar nature is also observed in Fig. 6.4, where the AD noise and PD noise are compared by assuming $\eta_A = \eta_P = \eta, \alpha_0 = \alpha_1 = \frac{\sqrt{2}}{2}, \beta_0 = \beta_1 = \beta_2 = \beta_3 = \frac{1}{2}$ $(\alpha_0, \alpha_1, \beta_0, \beta_1, \beta_2, \beta_3 \in R)$. As shown in Fig. 6.4, the fidelity of AD channel (solid line in Fig. 6.4a) is always more than PD channel (dashed line in Fig. 6.4a) for the same decoherence rate η. Thus, in the case of particular amplitude parameter of the initial state, the loss of information is less when the travel qubits are teleported in AD channel compared with PD channel. However, as shown in Fig. 6.4b, when $\eta > 0.7018$ for $\alpha_0 = \beta_0 = \frac{1}{2}, \alpha_1 = \beta_3 = \frac{\sqrt{3}}{2}, \beta_1 = \beta_2 = 0$, the fidelity of PD channel is more than AD channel, and the effect of AD noise is more.

Thus, in a real physical situation, fidelity decreases with increased decoherence rate (more noises). However, even in a noisy environment, C-BQT may be implemented with unit fidelity, i.e., as perfect, if states with particular α_i and β_j are teleported. This fact can be visualized through the peak of Figure 6.3c–f.

6.2.4 Efficiency Analysis and Comparison

Arbitrary and pure two-qubit states serving as transmitted state in this scheme are the superiority compared with previous schemes [14–17]. In this section, the number of qubits for sending, qubits of quantum channel, Bell state measurements, single-qubit measurements, and unitary operation (quantum gates) are utilized as elements for comparing efficiency. Here, efficiency [18] is defined as

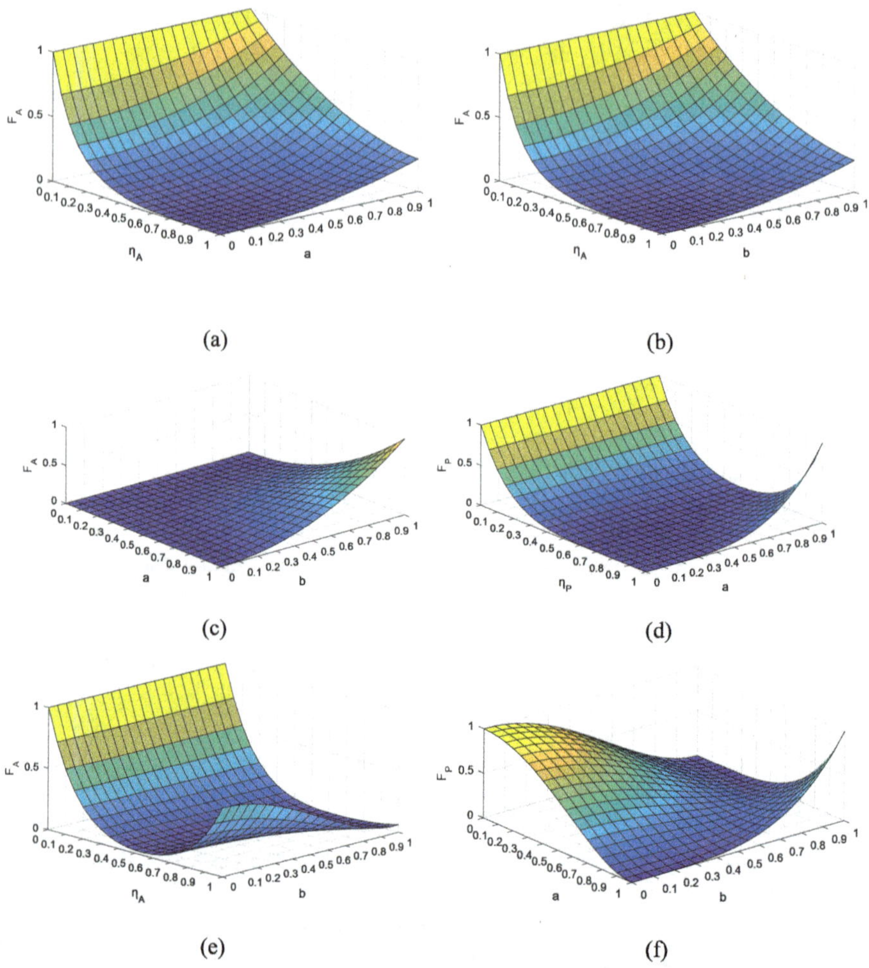

Fig. 6.3 The effect of NEC-BQT scheme visualized through variation of fidelity F^A (for AD noise model) and F^P (for PD noise model) with respect to amplitude information of the states to be teleported and decoherence rates for various situations: (**a**) AD noise with $a=\alpha_0, \beta_0=\beta_3=\frac{\sqrt{2}}{2}, \beta_1=\beta_2=0$; (**b**) AD noise with $b=\beta_0, \alpha_0=\alpha_1=\frac{\sqrt{2}}{2}, \beta_1=\beta_2=0$; (**c**) AD noise with $b=\beta_0$, $\eta_A=1$, $a=\alpha_0$, $\beta_1=\beta_2=0$; (**d**) PD noise with $a=\alpha_0$, $\beta_0=1$, $\beta_1=\beta_2=0$; (**e**) PD noise with $b=\beta_0, \alpha_0=\frac{1}{2}, \beta_1=\beta_2=0, \alpha_1=\frac{\sqrt{3}}{2}$; and (**f**) PD noise with $a=\alpha_0$, $b=\beta_0$, $\beta_1=\beta_2=0$, $\eta_P=1$

$$\eta=\frac{c}{q}. \tag{6.24}$$

Here, the transmitted efficiency η depends on the ratio of c (the number of qubits for sending) and q (qubit number of quantum channel).

As shown in Table 6.6, the proposed scheme realizes the teleportation with forms of $2 \leftrightarrow 2$ via a seven-qubit entangled state, which is different from previous work.

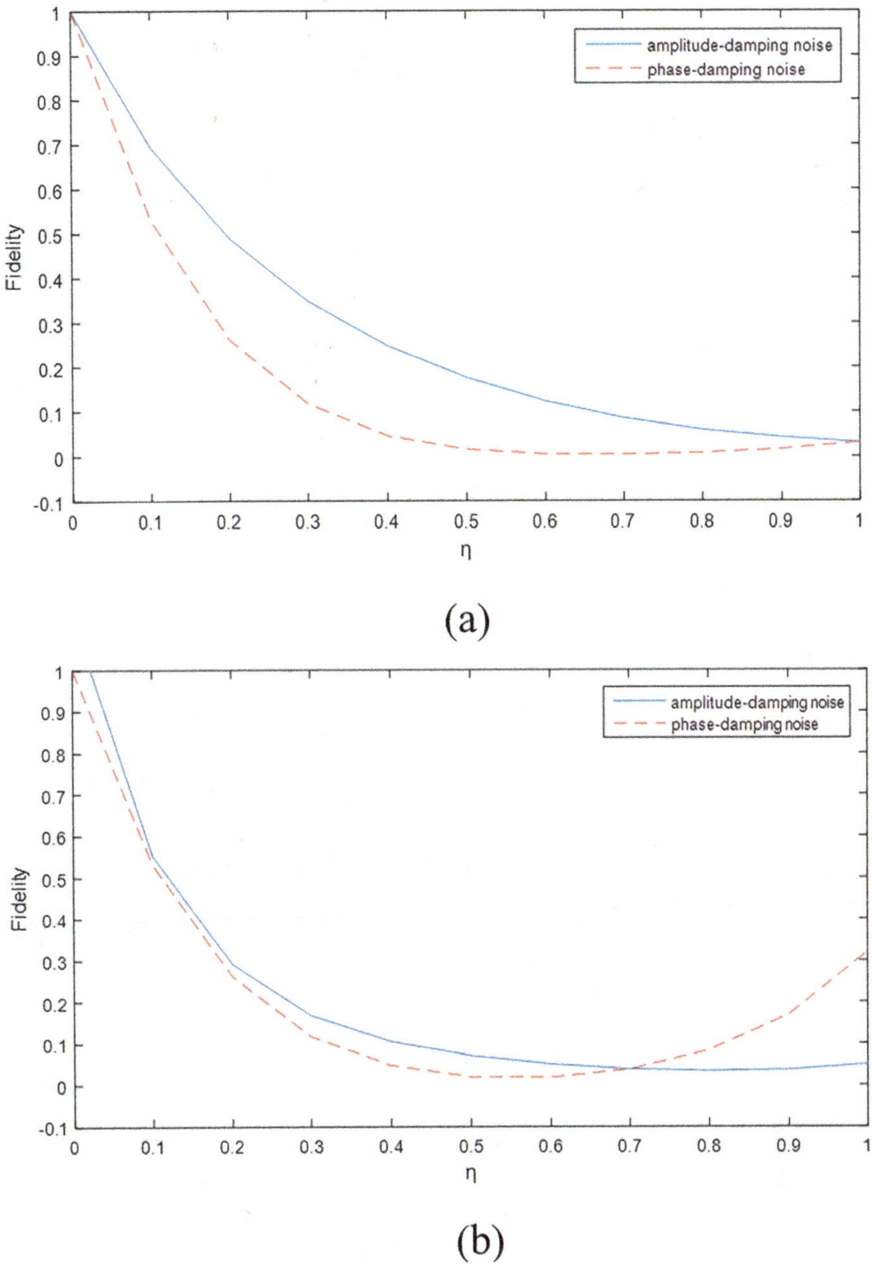

Fig. 6.4 The fidelities of amplitude-damping noise and phase-damping noise. The blue solid line represents the amplitude-damping noise, and the red dashed line represents the phase-damping noise

Table 6.6 Comparison of previous C-BQT schemes ((A) represents arbitrary state and (P) represents pure state)

References	Type of scheme	The number of transmitted qubits	The number of quantum channel	Efficiency	BSMs	SMs	Noisy analysis or not
[15]	C-BQT	1 and 1	7	2/7	2	3	No
[16]	C-BQT	1 and (P)2	7	3/7	3	1	No
[17]	BQT	(A)2 and (A)2	8	1/2	0	8	No
NEC-BQT	C-BQT	(P)2 and (A)2	7	4/7	2	4	Yes

Twice Bell state measurements and third single-qubit states are employed in this scheme. Also, some CNOT gates and a Hadamard gate are employed to generate and preprocess the quantum channel to make the teleportation favorable.

It is clear that the proposed scheme is more efficient according to the definition in (6.24). Meanwhile, it needs fewer classical bits while announcing measurement results via classical channel.

6.2.5 *Summary*

A novel scheme is proposed for C-BQT of a pure two-qubit state and an arbitrary two-qubit state simultaneously by utilizing seven-qubit entangled states as quantum channel. It is clear that it realizes teleportation of more qubits by implementing the quantum channel with the same length. To make the scheme proceed smoothly, Alice and Bob need to make CNOT operations on sharing quantum channel and do corresponding unitary operations after BSM and SM. Then, the proposed scheme is analyzed in typical noisy environment, and it concluded that the fidelities of the C-BQT scheme only depend on two factors, amplitude parameter of initial state and decoherence rate. In addition, it is a safer scheme because of the existence of supervisor Charlie. In a word, the scheme is both efficient and dependable, and at the same time, it is also feasible in experiment.

6.3 NEC-ABQT Scheme by Using a Nine-Qubit Entangled State

A scheme is proposed for implementing NEC-ABQT by using a nine-qubit entangled state as quantum channel. Alice transmits an arbitrary three-qubit entangled state A_0, A_1, and A_2 to Bob, and Bob simultaneously transmits an arbitrary single-qubit entangled state B to Alice via the control of the supervisor. It's worth noting that our nine-qubit entangled state channel can be prepared from nine single-qubit state product states by using H gates and CNOT operations. In this scheme, controlled-NOT operation and appropriate unitary operations are necessary. To save resources of classical channel, this scheme only uses single-qubit measurement basis. In an actual communication, noise is a necessary factor to be considered. However, only few QT scheme was considered in noisy environment. Hence, we review the proposed NEC-ABQT scheme in six types of noisy environments including the amplitude-damping, phase-damping, bit-flip, phase-flip, bit-phase-flip, and depolarizing noisy environment. A key ingredient in C-QT or C-ABQT is a quantum channel that is supposed to be the entangled pure state and connect three parties (senders, receivers, and controller). However, qubits that make up the entangled channel will inevitably suffer from quantum noise during the distribution among three parties. Consequently, the pure state transforms into mixed state, and it affects the efficiency (the fidelity of output state) of NEC-ABQT scheme seriously. This scheme only considers that noise only affects the travel qubits of the quantum channel. Moreover, we analytically derive the output states and fidelities of the NEC-ABQT process under the noisy environments. Finally, it is found that the fidelities in these cases only related to the amplitude parameter of the initial state and the decoherence noisy rate.

6.3.1 The Construction of the Quantum Channel

In NEC-ABQT scheme, an arbitrary single-qubit entangled state and a three-qubit entangled state can be transmitted by using a nine-qubit entangled state as a quantum channel. In this subsection, the process of construction of the quantum channel [19] is given. The proposed quantum channel (nine-qubit entangled state) is not only theoretically existent but also practically feasible. Quantum circuit of quantum channel can be created by using four H gates and five CNOT gates.

The nine-qubit product state $|\psi_0\rangle$ as an input state, where

$$|\psi_0\rangle = |0\rangle_{a_0} \otimes |0\rangle_{a_1} \otimes |0\rangle_{a_2} \otimes |0\rangle_{b_0} \otimes |0\rangle_{b_1} \otimes |0\rangle_{b_2} \otimes |0\rangle_{a_3} \otimes |0\rangle_c \otimes |0\rangle_{b_3}. \quad (6.25)$$

The construction of the quantum channel is as follows:

1. Executing H operation on qubits a_0, a_1, a_2, and b_3. $|\psi_0\rangle$ is transformed into $|\psi_1\rangle$.

$$\begin{aligned} |\psi_1\rangle &= \frac{1}{4}(|0\rangle + |1\rangle)_{a_0} \otimes (|0\rangle + |1\rangle)_{a_1} \otimes (|0\rangle + |1\rangle)_{a_2} \\ &\otimes |0\rangle_{b_0} \otimes |0\rangle_{b_1} \otimes |0\rangle_{b_2} \otimes |0\rangle_{a_3} \otimes |0\rangle_c \otimes (|0\rangle + |1\rangle)_{b_3} \end{aligned} \quad (6.26)$$

2. Performing CNOT operations on qubit pairs (a_0, b_0), (a_1, b_1), (a_2, b_2), (b_3, a_3), (b_3, c), respectively, where the first qubit serves as controlled qubit and the second qubit as target qubit. After that, the quantum channel can be constructed as (6.27)

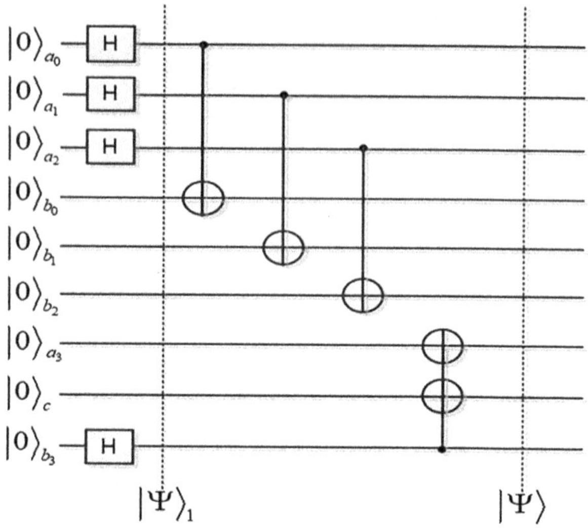

Fig. 6.5 The construction of quantum circuit for generating quantum channel

$$
\begin{aligned}
|\psi\rangle_{a_0a_1a_2b_0b_1b_2a_3cb_3} = & \frac{1}{4}(|000000000\rangle+|000000111\rangle+|001001000\rangle+|001001111\rangle \\
& +|010010000\rangle+|010010111\rangle+|011011000\rangle+|011011111\rangle \\
& +|100100000\rangle+|100100111\rangle+|101101000\rangle+|101101111\rangle \\
& +|110110000\rangle+|110110111\rangle+|111111000\rangle+|111111111\rangle)_{a_0a_1a_2b_0b_1b_2a_3cb_3}
\end{aligned}
\tag{6.27}
$$

The construction circuit of the quantum channel is shown in Fig. 6.5. Although the generation and maintenance of the multi-qubit entangled state is a difficult task, some experiments [34–37] have been carried out to study the multi-qubit entangled state. Therefore, we believe that advanced technology can make this task no longer difficult in the future.

6.3.2 NEC-ABQT Scheme

In this subsection, the details of NEC-ABQT scheme will be described. Alice wants to transmit an arbitrary three-qubit entangled state $|\Phi\rangle_{A_0A_1A_2}$ to Bob, and Bob simultaneously wants to transmit an arbitrary single-qubit entangled state $|\Phi\rangle_B$ to Alice under the permission of the controller Charlie. The entangled state to be transmitted is described by (6.28).

$$|\Phi\rangle_{A_0A_1A_2} = \begin{pmatrix} a_0|000\rangle + a_1|001\rangle + a_2|010\rangle + a_3|011\rangle \\ +a_4|100\rangle + a_5|101\rangle + a_6|110\rangle + a_7|111\rangle \end{pmatrix}_{A_0A_1A_2}$$
$$|\Phi\rangle_B = (b_0|0\rangle + b_1|1\rangle)_B \tag{6.28}$$

where the coefficients of the original state satisfy $|a_0^2 + a_1^2 + a_2^2 + a_3^2 + a_4^2 + a_5^2 + a_6^2 + a_7^2| = 1$ and $|b_0^2 + b_1^2| = 1$. This scheme consists of the following steps:

Step 1. Assume Alice, Bob, and Charlie share a nine-qubit entangled state as a quantum channel, as shown in (6.27). The qubits a_0, a_1, a_2 and a_3 belong to Alice, the qubits b_0, b_1, b_2 and b_3 belong to Bob, and the qubit c belongs to Charlie. Therefore, the state of the whole system can be expressed by (6.29).

$$|\Omega\rangle_{a_0a_1a_2b_0b_1b_2a_3cb_3A_0A_1A_2B} = |\psi\rangle_{a_0a_1a_2b_0b_1b_2a_3cb_3} \otimes |\Phi\rangle_{A_0A_1A_2} \otimes |\Phi\rangle_B \tag{6.29}$$

Step 2. To implement the NEC-ABQT scheme, Alice and Bob make CONT operations with qubits A_0, A_1, A_2 and B as controlled qubits and qubits a_0, a_1, a_2 and b as target qubits, respectively. The state will be transformed in the form of (6.30).

$$\begin{aligned} |\Omega'\rangle_{a_0a_1a_2b_0b_1b_2a_3cb_3A_0A_1A_2B} = \frac{1}{4}[& a_0b_0|\eta_0\rangle\otimes|0000\rangle + a_0b_1|\eta_1\rangle\otimes|0001\rangle + a_1b_0|\eta_2\rangle\otimes|0010\rangle \\ & + a_1b_1|\eta_3\rangle\otimes|0011\rangle + a_2b_0|\eta_4\rangle\otimes|0100\rangle + a_2b_1|\eta_5\rangle\otimes|0101\rangle \\ & + a_3b_0|\eta_6\rangle\otimes|0110\rangle + a_3b_1|\eta_7\rangle\otimes|0111\rangle + a_4b_0|\eta_8\rangle\otimes|1000\rangle \\ & + a_4b_1|\eta_9\rangle\otimes|1001\rangle + a_5b_0|\eta_{10}\rangle\otimes|1010\rangle + a_5b_1|\eta_{11}\rangle\otimes|1011\rangle \\ & + a_6b_0|\eta_{12}\rangle\otimes|1100\rangle + a_6b_1|\eta_{13}\rangle\otimes|1101\rangle \\ & + a_7b_0|\eta_{14}\rangle\otimes|1110\rangle + a_7b_1|\eta_{15}\rangle\otimes|1111\rangle] \end{aligned} \tag{6.30}$$

where $|\eta_i\rangle\{i = 0, 1, 2, \cdots, 16\}$ can be given in (6.3.6 Appendix A Formula) in Appendix.

Step 3. Alice and Bob perform a single-qubit measurement in Z-basis $\{|0\rangle, |1\rangle\}$ on qubits a_0, a_1, a_2 and b_3. The unmeasured qubits b_0, b_1, b_2, a_3, c, A_0, A_1, A_2 and B may collapse into one of the 16 possible states with the same probability, as shown in Table A.6.1 in Appendix.

Step 4. Alice (Bob) tells the measurement results to Bob (Alice) and Charlie via a classical channel. Then Alice and Bob apply appropriate unitary operation on qubits b_0, b_1, b_2, a_3, c, A_0, A_1, A_2 and B, as shown in Table A.6.2 in Appendix. The state of the qubits b_0, b_1, b_2 and a_3 will be converted to the same form

$$\begin{aligned}&a_0b_0|000000000\rangle + a_0b_1|000110001\rangle + a_1b_0|001000010\rangle + a_1b_1|001110011\rangle\\&+a_2b_0|010000100\rangle + a_2b_1|010110101\rangle + a_3b_0|011000110\rangle + a_3b_1|011110111\rangle\\&+a_4b_0|100001000\rangle + a_4b_1|100111001\rangle + a_5b_0|101001010\rangle + a_5b_1|101111011\rangle\\&+a_6b_0|110001100\rangle + a_6b_1|110111101\rangle + a_7b_0|111001110\rangle + a_7b_1|111111111\rangle\end{aligned} \tag{6.31}$$

Step 5. Alice and Bob perform a single-qubit measurement in X-basis $\{|+\rangle, |-\rangle\}$ on qubits A_0, A_1, A_2 and B. The other qubits may collapse into one of the 16 possible states with the same probability, as shown in Table A.6.3 in Appendix.

Step 6. Alice (Bob) tells the measurement results to Bob (Alice) and Charlie via a classical channel. Then they apply appropriate unitary operation on qubits b_0, b_1, b_2, a_3 and a_3 as shown in Table A.6.4 in Appendix. The state of the other qubits b_0, b_1, b_2 and c will be converted to the same form

$$\begin{aligned}&a_0b_0|00000\rangle + a_0b_1|00011\rangle + a_1b_0|00100\rangle + a_1b_1|00111\rangle\\&+a_2b_0|01000\rangle + a_2b_1|01011\rangle + a_3b_0|01100\rangle + a_3b_1|01111\rangle\\&+a_4b_0|10000\rangle + a_4b_1|10011\rangle + a_5b_0|10100\rangle + a_5b_1|10111\rangle\\&+a_6b_0|11000\rangle + a_6b_1|11011\rangle + a_7b_0|11100\rangle + a_7b_1|11111\rangle\end{aligned}. \tag{6.32}$$

Step 7. Whether NEC-ABQT scheme can be achieved depends on the controller Charlie. If Charlie allows for Alice and Bob to reconstruct the initial unknown state, he needs to carry out the single-qubit measurement in X-basis on qubit c and tells his measured result to the receivers Alice and Bob via a classical channel. By combining information from Alice, Bob, and Charlie, Alice and Bob can realize the exchange of their secret quantum information if they make the appropriate unitary transformations on the qubits at hand. These operations are shown in Table A.6.5 in Appendix. Bob receives quantum information of particles b_0, b_1 and b_2, and Alice receives quantum information of particle a_3. As shown in (6.33),

$$\begin{aligned}|\Phi\rangle_{b_0b_1b_2} &= \begin{pmatrix} a_0|000\rangle + a_1|001\rangle + a_2|010\rangle + a_3|011\rangle + \\ a_4|100\rangle + a_5|101\rangle + a_6|110\rangle + a_7|111\rangle \end{pmatrix}_{b_0b_1b_2}.\\ |\Phi\rangle_{a_3} &= (b_0|0\rangle + b_1|1\rangle)_{a_3}\end{aligned} \tag{6.33}$$

The scheme of asymmetric bidirectional controlled quantum teleportation is successfully realized. The theoretical quantum circuit of this scheme is shown in Fig. 6.6.

Up to now, single-qubit measurements and the single-qubit unitary operations have already been realized in various quantum systems, such as the cavity QED system [20], ion-trap system [21], and optical system [22]. So, we hope that our

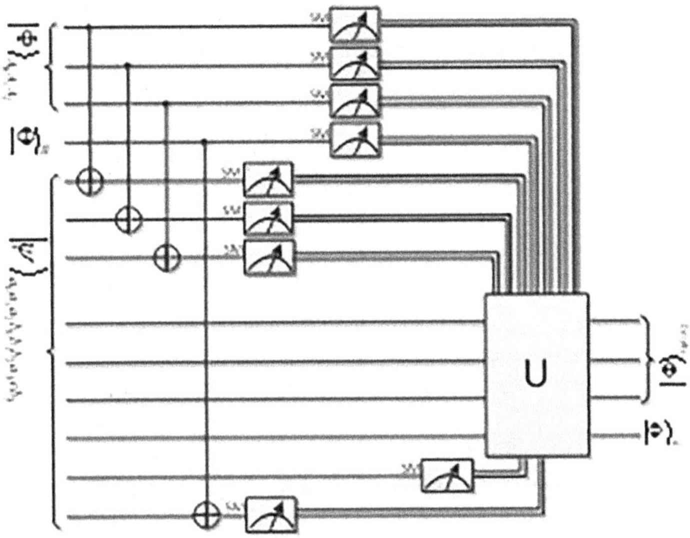

Fig. 6.6 The quantum circuit diagram for NEC-ABQT by using four CNOT operations and nine single-qubit measurements. "SM" corresponds to the single measurement. The line "——" carries classical bits, which corresponds to classical channel of transmitting the measurement outcomes from Alice, Bob, and Charlie, which enables Alice and Bob to apply unitary operation U. The line "" carries quantum bits. Alice receives $|\Phi\rangle_{a_3}$, and Bob receives $|\Phi\rangle_{b_0b_1b_2}$

scheme will be implemented in the future with the advances in quantum information technology.

6.3.3 *NEC-ABQT Scheme in Six Noisy Environments*

In an actual quantum communication, it is impossible to have a noiseless environment. In this section, considering the effect of noise on the NEC-ABQT process, here, we review the proposed scheme in six types of noisy environments including the amplitude-damping, phase-damping, bit-flip, phase-flip, bit-phase-flip, and depolarizing noisy environment. By providing the output states in the noisy environment, and calculating and analyzing the fidelities of the output states, we can evaluate the performance of the system.

In this scheme, qubit c belongs to Charlie in quantum channel. Charlie as a controller, this work has considered that the qubit c is not transmitted through the noisy environment, and it is not affected by the noise. These the six environment noise can be characterized by Kraus operators [23], so we discuss the situation that the qubits a_0, a_1, a_2, a_3, b_0, b_1, b_2 and b_3 are affected by the same Kraus operator.

In quantum noisy environment, a pure input state will be transformed into a mixed state, which is more effective to be represented by density operator rather than vector state. The proposed nine-qubit entangled state $|\Psi\rangle_{a_0a_1a_2b_0b_1b_2a_3b_3}$ is a pure state, and it is straightforward to obtain the corresponding density matrix $\rho = |\Psi\rangle_{a_0a_1a_2b_0b_1b_2a_3b_3}\langle\Psi|_{a_0a_1a_2b_0b_1b_2a_3b_3}$. The effect of the noise on the density matrix ρ can be stated as

$$\begin{aligned}\xi^r(\rho) = \sum_m \left(E_m^{ra_0}\right)\left(E_m^{ra_1}\right)\left(E_m^{ra_2}\right)\left(E_m^{rb_0}\right)\left(E_m^{rb_1}\right)\left(E_m^{rb_2}\right)\left(E_m^{ra_3}\right)\left(E_m^{ra_3}\right)\rho \\ \left(E_m^{ra_0}\right)^\dagger\left(E_m^{ra_1}\right)^\dagger\left(E_m^{ra_2}\right)^\dagger\left(E_m^{rb_0}\right)^\dagger\left(E_m^{rb_1}\right)^\dagger\left(E_m^{rb_2}\right)^\dagger\left(E_m^{ra_3}\right)^\dagger\left(E_m^{ra_3}\right)^\dagger\end{aligned} \tag{6.34}$$

where r $\in \{A, P, B, W, F, D\}$, for $r = A$, i.e., for amplitude-damping noise $m \in \{0, 1\}$; for $r = P$, i.e., for phase-damping noise $m \in \{0, 1, 2\}$; for $r = B$, i.e., for bit-flip noise $m \in \{0, 1\}$; for $r = W$, i.e., for phase-flip noise $m \in \{0, 1\}$; for $r = F$, i.e., for bit-phase-flip noise $m \in \{0, 1\}$; and for $r = D$, i.e., for depolarizing noise $m \in \{0, 1, 2, 3\}$. The superscripts $(a_0, a_1, a_2, b_0, b_1, b_2, a_3, b_3)$ represent the operator E act on which qubit, ξ denotes a quantum operation, which maps from ρ to $\xi^{\mathrm{r}}(\rho)$ due to the noise. As a consequence of the noisy environment, the shared state would become a mixed state after the distribution of the qubits.

For recovering the original state, Alice and Bob can make appropriate operation on own qubits according to Alice's and Bob's and Charlie's measured results. Then, the final state ρ_{out}^r can be expressed as density matrix as shown in (6.34). The density matrix that corresponding to quantum state $|\Phi\rangle_{A_0A_1A_2}$ is $\rho_{A_0A_1A_2} = |\Phi\rangle_{A_0A_1A_2}\langle\Phi|_{A_0A_1A_2}$, and corresponding to quantum state $|\Phi\rangle_B$ is $\rho_B = |\Phi\rangle_B\langle\Phi|_B$.

$$\rho_{out}^r = Tr_{A_0A_1A_2Ba_0a_1a_2cb_3}\left\{U\left[\rho_{A_0A_1A_2} \otimes \rho_B \otimes \xi^r(\rho)\right]U^\dagger\right\} \tag{6.35}$$

where is the partial trace $Tr_{A_0A_1A_2Ba_0a_1a_2cb_3}$ over qubits $(A_0, A_1, A_2, B, a_0, a_1, a_2, c, b_3)$ and U is a unitary operation to describe the NEC-ABQT process, which is given by (6.36).

$$\begin{aligned}U = &\left\{I_{A_0A_1A_2a_0a_1a_2} \otimes I_{Bb_3} \otimes I_c \otimes \sigma_{b_0b_1b_2a_3}^{klmnopqst}\right\}\left\{I_{A_0A_1A_2a_0a_1a_2} \otimes I_{Bb_3} \otimes |\varphi\rangle_c^t\langle\varphi|_c^t \otimes I_{b_0b_1b_2a_3}\right\} \\ &\left\{|\varphi\rangle_{A_0}^s\langle\varphi|_{A_0}^s \otimes |\varphi\rangle_{A_1}^q\langle\varphi|_{A_1}^q \otimes |\varphi\rangle_{A_2}^p\langle\varphi|_{A_2}^p \otimes |\varphi\rangle_B^o\langle\varphi|_B^o \otimes I_{a_0a_1a_2} \otimes I_{b_3} \otimes I_c \otimes I_{b_0b_1b_2a_3}\right\} \\ &\left\{I_{A_0A_1A_2} \otimes I_B \otimes |\varphi\rangle_{a_0}^n\langle\varphi|_{a_0}^n \otimes |\varphi\rangle_{a_1}^m\langle\varphi|_{a_1}^m \otimes |\varphi\rangle_{a_2}^l\langle\varphi|_{a_2}^l \otimes |\varphi\rangle_{b_3}^k\langle\varphi|_{b_3}^k \otimes I_{a_0a_1a_2} \otimes I_c \otimes I_{b_0b_1b_2a_3}\right\} \\ &\left\{U_{A_0a_0} \otimes U_{A_1a_1} \otimes U_{A_2a_2} \otimes I_{Bb_3} \otimes I_c \otimes I_{b_0b_1b_2a_3}\right\}\end{aligned} \tag{6.36}$$

where $k, l, m, n, o, p, q, s, t \in \{1, 2\}$, with U_{ij} stands for CNOT operations on the qubits A_0, A_1, A_2 and B as control qubits and qubits a_0, a_1, a_2 and b_3 as target qubits. $|\varphi\rangle_{A_0}^s\langle\varphi|_{A_0}^s$, $|\varphi\rangle_{A_1}^q\langle\varphi|_{A_1}^q$, $|\varphi\rangle_{A_2}^p\langle\varphi|_{A_2}^p$, $|\varphi\rangle_{a_0}^n\langle\varphi|_{a_0}^n$, $|\varphi\rangle_{a_1}^m\langle\varphi|_{a_1}^m$ and $|\varphi\rangle_{a_2}^l\langle\varphi|_{a_2}^l$ denote Alice's single-qubit measurement results, $|\varphi\rangle_B^o\langle\varphi|_B^o$ and $|\varphi\rangle_{b_3}^k\langle\varphi|_{b_3}^k$ denote Bob's single-qubit measurement results, and $|\varphi\rangle_c^t\langle\varphi|_c^t$ represents Charlie's single-qubit measurement results. $\sigma_{b_0b_1b_2a_3}^{klmnopqst}$ is Alice's and Bob's recovery operation after knowing the measurement results.

Depending upon the choice, the final output state ρ_{out}^{r} can be obtained in noise. This NEC-ABQT scheme has been discussed in ideal situation in Sect. 6.3.2; the final output state is $|\Theta\rangle = (b_0|0\rangle + b_1|1\rangle)_{a_2} \otimes (a_0|000\rangle + a_1|001\rangle + a_2|010\rangle + a_3|011\rangle + a_4|100\rangle + a_5|101\rangle + a_6|110\rangle + a_7|111\rangle)_{b_0b_1b_2}$. The effect of noise can be visualized by comparing the quantum state in the noisy environment with the state $|\Phi\rangle$ by using fidelity ρ_{out}^{r}. Fidelity is a distance-based measure corresponding to the closeness of two quantum states, which can be described as

$$F = \langle\Theta|\rho_{out}^{r}|\Theta\rangle. \tag{6.37}$$

When passing through the noisy environment, the information of the quantum state will be lost, so by calculating the fidelity is an indirect description of how much information is lost. $F = 1$ means perfect communication. If F becomes smaller and smaller, it indicates that we have lost more and more information in this NEC-ABQT process.

Next, the effect of six noisy environments (amplitude-damping, phase-damping, bit-flip, phase-flip, bit-phase-flip, and depolarizing noisy environment) will be discussed on the NEC-ABQT scheme.

6.3.3.1 Amplitude-Damping Noisy Environment

An important application of quantum computing is to describe energy dissipation (an effect caused by the loss of energy from a quantum system); because of the interaction with amplitude-damping noisy environment, a system undergoes energy dissipation. The Kraus operators of amplitude-damping noise are (6.38)

$$E_0^A = \begin{bmatrix} 1 & 0 \\ 0 & \sqrt{1-\eta_A} \end{bmatrix} \quad E_1^A = \sqrt{\eta_A}\begin{bmatrix} 0 & 1 \\ 0 & 0 \end{bmatrix} \tag{6.38}$$

where $\eta_A (0 \leq \eta_A \leq 1)$ describes the probability of missing a photon and η_A also is the decoherence rate. After the qubits $a_0, a_1, a_2, a_3 (b_0, b_1, b_2, b_3)$ are transmitted through the amplitude-damping noisy environment, the quantum channel would become a mixed state, and the density matrix ρ is rewritten as (6.39)

$$\begin{aligned}\xi^A(\rho) = &\frac{1}{16}\{[|000000000\rangle + (1-\eta_A)|000000111\rangle + (1-\eta_A)|001001000\rangle + \\ &(1-\eta_A)^2|001001111\rangle + (1-\eta_A)|010010000\rangle + (1-\eta_A)^2|010010111\rangle + \\ &(1-\eta_A)^2|011011000\rangle + (1-\eta_A)^3|011011111\rangle + (1-\eta_A)|100100000\rangle + \\ &(1-\eta_A)^2|100100111\rangle + (1-\eta_A)^2|101101000\rangle + (1-\eta_A)^3|101101111\rangle + \\ &(1-\eta_A)^2|110110000\rangle + (1-\eta_A)^3|110110111\rangle + (1-\eta_A)^3|111111000\rangle \\ &+(1-\eta_A)^4|111111111\rangle\Big] \times [\langle 000000000| + (1-\eta_A)\langle 000000111| + \\ &(1-\eta_A)\langle 001001000| + (1-\eta_A)^2\langle 001001111| + (1-\eta_A)\langle 010010000| \\ &+(1-\eta_A)^2\langle 010010111| + (1-\eta_A)^2\langle 011011000| + (1-\eta_A)^3\langle 011011111| \\ &+(1-\eta_A)\langle 100100000| + (1-\eta_A)^2\langle 100100111| + (1-\eta_A)^2\langle 101101000| + \\ &(1-\eta_A)^3\langle 101101111| + (1-\eta_A)^2\langle 110110000| + (1-\eta_A)^3\langle 110110111| + \\ &(1-\eta_A)^3\langle 111111000| + (1-\eta_A)^4\langle 111111111|\Big] + \eta_A^8|000000010\rangle\langle 000000010|\}\end{aligned} \tag{6.39}$$

In amplitude-damping noise, the output state $(\rho_{out}^A)_{b_0b_1b_2a_3}$ can be calculated as

$$\begin{aligned}(\rho_{out}^A)_{b_0b_1b_2a_3} = &\{[a_0b_0|0000\rangle + (1-\eta_A)a_0b_1|0001\rangle + (1-\eta_A)a_1b_0|0010\rangle + \\ &(1-\eta_A)^2a_1b_1|0011\rangle + (1-\eta_A)a_2b_0|0100\rangle + (1-\eta_A)^2a_2b_1|0101\rangle \\ &+(1-\eta_A)^2a_3b_0|0110\rangle + (1-\eta_A)^3a_3b_1|0111\rangle + (1-\eta_A)a_4b_0|1000\rangle \\ &+(1-\eta_A)^2a_4b_1|1001\rangle + (1-\eta_A)^2a_5b_0|1010\rangle + (1-\eta_A)^3a_5b_1|1011\rangle \\ &+(1-\eta_A)^2a_6b_0|1100\rangle + (1-\eta_A)^3a_6b_1|1101\rangle + (1-\eta_A)^3a_7b_0|1110\rangle \\ &+(1-\eta_A)^4a_7b_1|1111\rangle] \times [a_0b_0\langle 0000| + (1-\eta_A)a_0b_1\langle 0001| + \\ &(1-\eta_A)a_1b_0\langle 0010| + (1-\eta_A)^2a_1b_1\langle 0011| + (1-\eta_A)a_2b_0\langle 0100| + \\ &(1-\eta_A)^2a_2b_1\langle 0101| + (1-\eta_A)^2a_3b_0\langle 0110| + (1-\eta_A)^3a_3b_1\langle 0111| \\ &+(1-\eta_A)a_4b_0\langle 1000| + (1-\eta_A)^2a_4b_1\langle 1001| + (1-\eta_A)^2a_5b_0\langle 1010| \\ &+(1-\eta_A)^3a_5b_1\langle 1011| + (1-\eta_A)^2a_6b_0\langle 1100| + (1-\eta_A)^3a_6b_1\langle 1101| \\ &+(1-\eta_A)^3a_7b_0\langle 1110| + (1-\eta_A)^4a_7b_1\langle 1111|] + \eta_A^8a_7^2b_1^2|0000\rangle\langle 0000|\}\end{aligned} \tag{6.40}$$

In amplitude-damping noise, the fidelity F^A of the output state can be calculated according to (6.37) and (6.40).

$$
\begin{aligned}
F^A &= \langle \Theta | \rho_{out}^A | \Theta \rangle \\
&= \left[{a_0}^2{b_0}^2 + (1-\eta_A){a_0}^2{b_1}^2 + (1-\eta_A){a_1}^2{b_0}^2 \right. \\
&\quad +(1-\eta_A)^2 a_1 b_1 |0011\rangle + (1-\eta_A){a_2}^2{b_0}^2 + \\
&\quad (1-\eta_A)^2{a_2}^2{b_1}^2 + (1-\eta_A)^2{a_3}^2{b_0}^2|0110\rangle + \\
&\quad (1-\eta_A)^3{a_3}^2{b_1}^2 + (1-\eta_A){a_4}^2{b_0}^2 + (1-\eta_A)^2{a_4}^2{b_1}^2 \\
&\quad +(1-\eta_A)^2{a_5}^2{b_0}^2|1010\rangle + (1-\eta_A)^3{a_5}^2{b_1}^2 + \\
&\quad (1-\eta_A)^2{a_6}^2{b_0}^2 + (1-\eta_A)^3{a_6}^2{b_1}^2 + (1-\eta_A)^3{a_7}^2{b_0}^2|1110\rangle \\
&\quad \left. +(1-\eta_A)^4{a_7}^2{b_1}^2 \right] + \eta_A^8 {a_0}^2 a_7^2 b_0^2 {b_1}^2
\end{aligned} \quad . \tag{6.41}
$$

(6.41) shows that the fidelity depends on the amplitude parameter of the initial state and the decoherence rate in amplitude-damping noise. Figure 6.7a–c clearly illustrates the effect of amplitude-damping noise on the fidelity F^A, and it also illustrates the variation of the fidelity with amplitude parameter of the initial state and the decoherence rate η_A. In Fig. 6.7, it can be easily observed that fidelity F^A always decreases with the decoherence η_A and the fidelity slowly increases with the amplitude parameters a and b.

6.3.3.2 Phase-Damping Noisy Environment

Phase damping is a noise process with purely quantum mechanical properties. It describes the loss of relative phase information in a quantum state without energy dissipation. The Kraus operators of phase-damping noise are (6.42)

$$
E_0^P = \sqrt{1-\eta_P} I \; E_1^P = \sqrt{\eta_P}\begin{bmatrix} 1 & 0 \\ 0 & 0 \end{bmatrix} \; E_2^P = \sqrt{\eta_P}\begin{bmatrix} 0 & 0 \\ 0 & 1 \end{bmatrix} \tag{6.42}
$$

where $\eta_P (0 \le \eta_P \le 1)$ is the possibility of loss of relative phase information. After the qubits a_0, a_1, a_2, $a_3 (b_0, b_1, b_2, b_3)$ are transmitted through the phase-damping noisy environment, the density matrix ρ is rewritten as (6.43)

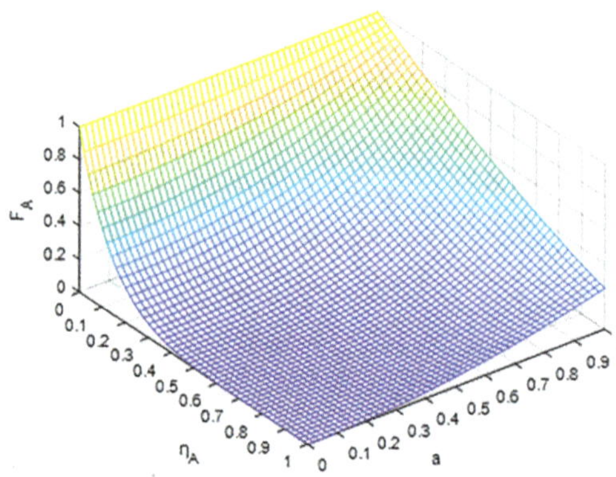

(a)

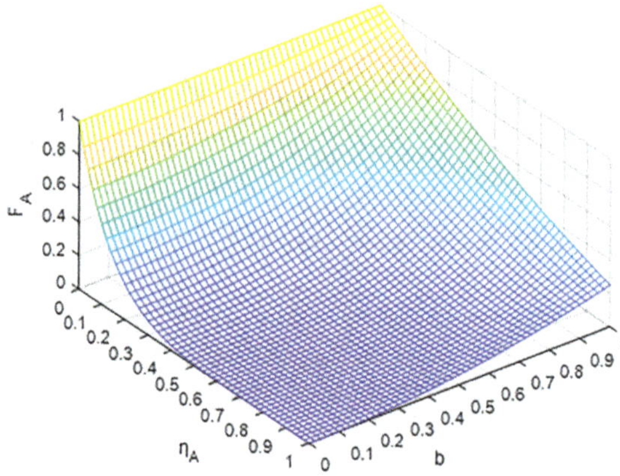

(b)

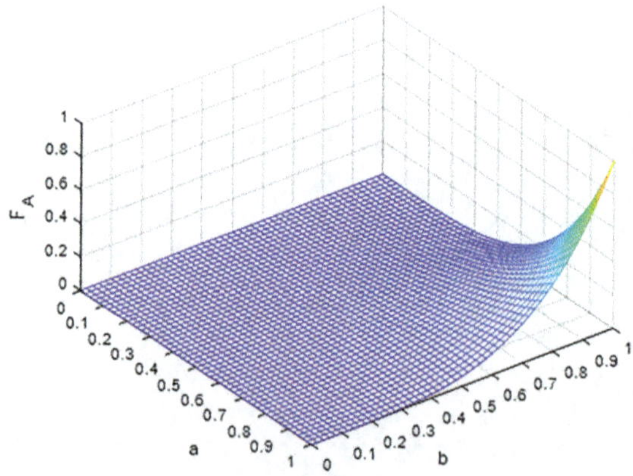

(c)

Fig. 6.7 The effect of amplitude-damping noise on NEC-ABQT scheme visualized through variation of fidelity F^A, which is related to the amplitude information of

$$\begin{aligned}\xi^P(\rho) = \frac{1}{16}\Big\{&(1-\eta_P)^8[|000000000\rangle + |000000111\rangle + |001001000\rangle + |001001111\rangle\\
&+|010010000\rangle + |010010111\rangle + |011011000\rangle + |011011111\rangle\\
&+|100100000\rangle + |100100111\rangle + |101101000\rangle + |101101111\rangle\\
&+|110110000\rangle + |110110111\rangle + |111111000\rangle + |111111111\rangle]\\
&\times[\langle 000000000| + \langle 000000111| + \langle 001001000| + \langle 001001111|\\
&+\langle 010010000| + \langle 010010111| + \langle 011011000| + \langle 011011111|\\
&+\langle 100100000| + \langle 100100111| + \langle 101101000| + \langle 101101111|\\
&+\langle 110110000| + \langle 110110111| + \langle 111111000| + \langle 111111111|]\\
&+\eta_P^8|000000000\rangle\langle 000000000| + \eta_P^8|111111111\rangle\langle 111111111|\Big\}\end{aligned} \tag{6.43}$$

In phase-damping noise, the output state can be calculated as

$$\begin{aligned}\left(\rho^P_{out}\right)_{b_0b_1b_2a_3} = \Big\{&(1-\eta_P)^8[a_0b_0|0000\rangle + a_0b_1|0001\rangle + a_1b_0|0010\rangle\\
&+a_1b_1|0011\rangle + a_2b_0|0100\rangle + a_2b_1|0101\rangle + a_3b_0|0110\rangle\\
&+a_3b_1|0111\rangle + a_4b_0|1000\rangle + a_4b_1|1001\rangle + a_5b_0|1010\rangle\\
&+a_5b_1|1011\rangle + a_6b_0|1100\rangle + a_6b_1|1101\rangle + a_7b_0|1110\rangle\\
&+a_7b_1|1111\rangle] \times [a_0b_0\langle 0000| + a_0b_1\langle 0001| + a_1b_0\langle 0010|\\
&+a_1b_1\langle 0011| + a_2b_0\langle 0100| + a_2b_1\langle 0101| + a_3b_0\langle 0110|\\
&+a_3b_1\langle 0111| + a_4b_0\langle 1000| + a_4b_1\langle 1001| + a_5b_0\langle 1010|\\
&+a_5b_1\langle 1011| + a_6b_0\langle 1100| + a_6b_1\langle 1101| + a_7b_0\langle 1110|\\
&+a_7b_1\langle 1111|] + {\eta_P}^8{a_0}^2{b_0}^2|0000\rangle\langle 0000| + {\eta_P}^8{a_7}^2{b_1}^2|1111\rangle\langle 1111|\Big\}\end{aligned} \tag{6.44}$$

In phase-damping noise, the fidelity $F^P$of the output state can be calculated according to (6.37) and (6.44).

$$F^P = \langle\Theta|\rho^P_{out}|\Theta\rangle = (1-\eta_P)^8 + {\eta_P}^8{a_0}^4{b_0}^4 + {\eta_P}^8{a_7}^4{b_1}^4. \tag{6.45}$$

(6.45) shows that the fidelity depends on the amplitude parameter of the initial state and the decoherence rate in bit-flip noise. Figure 6.8a–c clearly illustrates the effect of phase-damping noise on the fidelity F^P; it also illustrates the variation of the fidelity with amplitude parameter of the initial state and the decoherence rate η_P. In

Fig. 6.7 (continued) the initial states and decoherence rate for various situations: (**a**) with $a = a_0, b_0 = b_1 = \sqrt{2}/2,\ a_1 = a_2 = a_3 = a_4 = a_5 = a_6 = 0$; (**b**) with $b = b_0, a_0 = a_7 = \sqrt{2}/2, a_1 = a_2 = a_3 = a_4 = a_5 = a_6 = 0$; (**c**) with $a = a_0$, $b = b_0$, $\eta_A = 1$, $a_1 = a_2 = a_3 = a_4 = a_5 = a_6 = 0$

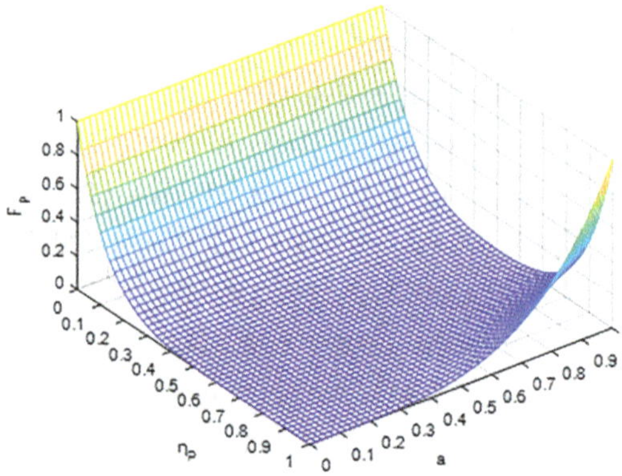

(a)

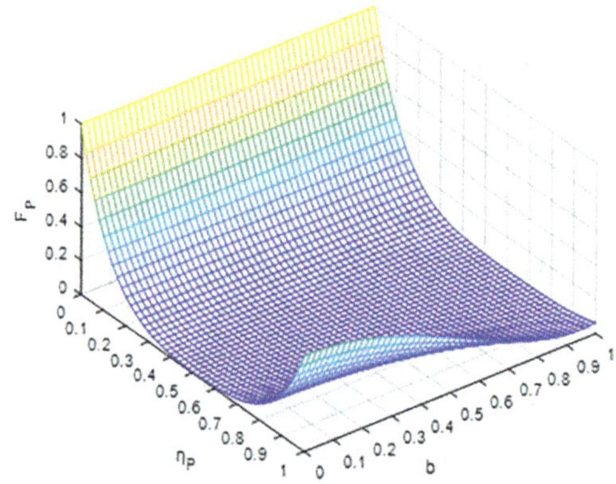

(b)

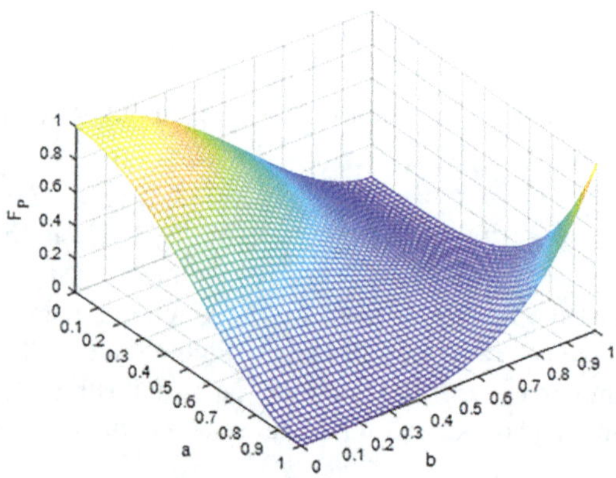

(c)

Fig. 6.8 The effect of phase-damping noise on NEC-ABQT scheme visualized through variation of fidelity F^P, which is related to amplitude information of the initial states and decoherence rate for various situations: (**a**) with $a = a_0$, $b_0 = 1$, $b_1 = a_1 = a_2 = a_3 = a_4 = a_5 = a_6 = 0$; (**b**) with

Fig. 6.8, it can be easily observed that the fidelity F^P always decreases with the decoherence η_P, and when the decoherence rate and the amplitude parameters of the original state are in a specific combination, it also shows an upward trend to $F^P = 1$. The fidelity takes on a variety of changes with the change of amplitude parameters a and b.

6.3.3.3 C. Bit-Flip Noisy Environment

In bit-flip noisy environment, the bit-flip noise generates a bit-flip error with the probability $\eta_B(0 \leq \eta_B \leq 1)$ and is left invariant with the probability $1 - \eta_B$. The Kraus operators of bit-flip noise are (6.46)

$$\begin{aligned} E_0^B &= \sqrt{1-\eta_B} I \\ E_1^B &= \sqrt{\eta_B} X = \sqrt{\eta_B} \begin{bmatrix} 0 & 1 \\ 1 & 0 \end{bmatrix} \end{aligned} . \tag{6.46}$$

After the qubits a_0, a_1, a_2, $a_3(b_0, b_1, b_2, b_3)$ are transmitted through the bit-flip noisy environment, the density matrix ρ is rewritten as (6.47)

Fig. 6.8 (continued) $b = b_0, a_0 = 1/2, a_7 = \sqrt{3}/2, a_1 = a_2 = a_3 = a_4 = a_5 = a_6 = 0$; **(c)** with $a = a_0$, $b = b_0$, $\eta_P = 1$, $a_1 = a_2 = a_3 = a_4 = a_5 = a_6 = 0$

$$
\begin{aligned}
\xi^B(\rho) = \frac{1}{16}\Big\{&(1-\eta_B)^8[|000000000\rangle + |000000111\rangle + |001001000\rangle + |001001111\rangle \\
&+|010010000\rangle + |010010111\rangle + |011011000\rangle + |011011111\rangle \\
&+|100100000\rangle + |100100111\rangle + |101101000\rangle + |101101111\rangle \\
&+|110110000\rangle + |110110111\rangle + |111111000\rangle + |111111111\rangle] \\
&\times [\langle 000000000| + \langle 000000111| + \langle 001001000| + \langle 001001111| \\
&+\langle 010010000| + \langle 010010111| + \langle 011011000| + \langle 011011111| \\
&+\langle 100100000| + \langle 100100111| + \langle 101101000| + \langle 101101111| \\
&+\langle 110110000| + \langle 110110111| + \langle 111111000| + \langle 111111111|] \\
&+\eta_B^8[|111111101\rangle + |111111010\rangle + |110110101\rangle + |110110010\rangle \\
&+|101101101\rangle + |100100010\rangle + |100100101\rangle + |101101010\rangle \\
&+|011011101\rangle + |010010010\rangle + |010010101\rangle + |011011010\rangle \\
&+|001001101\rangle + |001001010\rangle + |000000101\rangle + |000000010\rangle] \\
&\times [\langle 111111101| + \langle 111111010| + \langle 110110010| + \langle 110110101| \\
&+\langle 101101101| + \langle 100100010| + \langle 100100101| + \langle 101101010| \\
&+\langle 011011101| + \langle 010010010| + \langle 010010101| + \langle 011011010| \\
&+\langle 001001101| + \langle 000000101| + \langle 001001010| + \langle 000000010|]\Big\}
\end{aligned} \tag{6.47}
$$

In bit-flip noise, the output state $\left(\rho^B_{out}\right)_{b_0b_1b_2a_3}$ can be calculated as

$$
\begin{aligned}
\left(\rho^B_{out}\right)_{b_0b_1b_2a_3} = \Big\{&(1-\eta_P)^8[a_0b_0|0000\rangle + a_0b_1|0001\rangle + a_1b_0|0010\rangle + \\
&a_1b_1|0011\rangle + a_2b_0|0100\rangle + a_2b_1|0101\rangle + a_3b_0|0110\rangle + \\
&a_3b_1|0111\rangle + a_4b_0|1000\rangle + a_4b_1|1001\rangle + a_5b_0|1010\rangle + \\
&a_5b_1|1011\rangle + a_6b_0|1100\rangle + a_6b_1|1101\rangle + a_7b_0|1110\rangle + \\
&a_7b_1|1111\rangle] \times [a_0b_0\langle 0000| + a_0b_1\langle 0001| + a_1b_0\langle 0010| + \\
&a_1b_1\langle 0011| + a_2b_0\langle 0100| + a_2b_1\langle 0101| + a_3b_0\langle 0110| + \\
&a_3b_1\langle 0111| + a_4b_0\langle 1000| + a_4b_1\langle 1001| + a_5b_0\langle 1010| + \\
&a_5b_1\langle 1011| + a_6b_0\langle 1100| + a_6b_1\langle 1101| + a_7b_0\langle 1110| + \\
&a_7b_1\langle 1111|]{\eta_P}^8[a_0b_0|1111\rangle + a_0b_1|1110\rangle + a_1b_0|1101\rangle \\
&+a_1b_1|1100\rangle + a_2b_0|1011\rangle + a_2b_1|1010\rangle + a_3b_0|1001\rangle + \\
&a_3b_1|1000\rangle + a_4b_0|0111\rangle + a_4b_1|0110\rangle + a_5b_0|0101\rangle + \\
&a_5b_1|0100\rangle + a_6b_0|0011\rangle + a_6b_1|0010\rangle + a_7b_0|0111\rangle + \\
&a_7b_1|0000\rangle] \times [a_0b_0\langle 1111| + a_0b_1\langle 1000| + a_1b_0\langle 1101| + \\
&a_1b_1\langle 1100| + a_2b_0\langle 1011| + a_2b_1\langle 1010| + a_3b_0\langle 1001| + \\
&a_3b_1\langle 1000| + a_4b_0\langle 0111| + a_4b_1\langle 0110| + a_5b_0\langle 0101| + \\
&a_5b_1\langle 0100| + a_6b_0\langle 0011| + a_6b_1\langle 0010| + a_7b_0\langle 0111| + a_7b_1\langle 0000|]\Big\}
\end{aligned} \tag{6.48}
$$

In bit-flip noise, the fidelity F^B of the output state can be calculated according to (6.37) and (6.48).

$$\begin{aligned} F^B &= \langle\Theta|\rho^B_{out}|\Theta\rangle \\ &= (1-\eta_B)^8 + 16{\eta_B}^8{b_0}^2{b_1}^2(a_0a_7+a_1a_6+a_2a_5+a_3a_4)^2 \end{aligned}. \tag{6.49}$$

(6.49) shows that the fidelity depends on the amplitude parameter of the initial state and the decoherence rate in bit-flip noise. Figure 6.9a–c clearly illustrates the effect of bit-flip noise on the fidelity F^B, and it also illustrates the variation of the fidelity with amplitude parameter of the initial state and the decoherence rate η_B. In Fig. 6.9, it can be easily observed that the fidelity F^B always decreases with the decoherence η_B and when the decoherence rate is large, it also shows an upward trend. The fidelity F^B peaks at point $(a,b)=\left(\sqrt{2}/2,\sqrt{2}/2\right)$ with the change of amplitude parameters a and b.

6.3.3.4 Phase-Flip Noisy Environment

In this noisy environment, the relative phase will flip with the probability $\eta_W(0\leq\eta_W\leq1)$ and is left invariant with the probability $1-\eta_W$. The Kraus operators of phase-flip noise are (6.50)

$$E^W_0=\sqrt{1-\eta_W}I \quad E^W_1=\sqrt{\eta_W}Z=\sqrt{\eta_W}\begin{bmatrix}1 & 0\\ 0 & -1\end{bmatrix}. \tag{6.50}$$

After the qubits $a_0, a_1, a_2, a_3(b_0, b_1, b_2, b_3)$ are transmitted through the phase-flip noisy environment, the density matrix ρ can be rewritten as (6.51)

$$\begin{aligned} \xi^W(\rho)=&\frac{1}{16}\Big\{\left[(1-\eta_W)^8+{\eta_W}^8\right][|000000000\rangle+|000000111\rangle+|001001000\rangle \\ &+|001001111\rangle+|010010000\rangle+|010010111\rangle+|011011000\rangle+ \\ &|011011111\rangle+|100100000\rangle+|100100111\rangle+|101101000\rangle+ \\ &|101101111\rangle+|110110000\rangle+|110110111\rangle+|111111000\rangle \\ &+|111111111\rangle]\times[\langle000000000|+\langle000000111|+\langle001001000| \\ &+\langle001001111|+\langle010010000|+\langle010010111|+\langle011011000|+ \\ &\langle011011111|+\langle100100000|+\langle100100111|+\langle101101000|+ \\ &\langle101101111|+\langle110110000|+\langle110110111|+\langle111111000|+\langle111111111|]\Big\} \end{aligned} \tag{6.51}$$

In phase-flip noise, the output state $\left(\rho^W_{out}\right)_{b_0b_1b_2a_3}$ can be calculated as

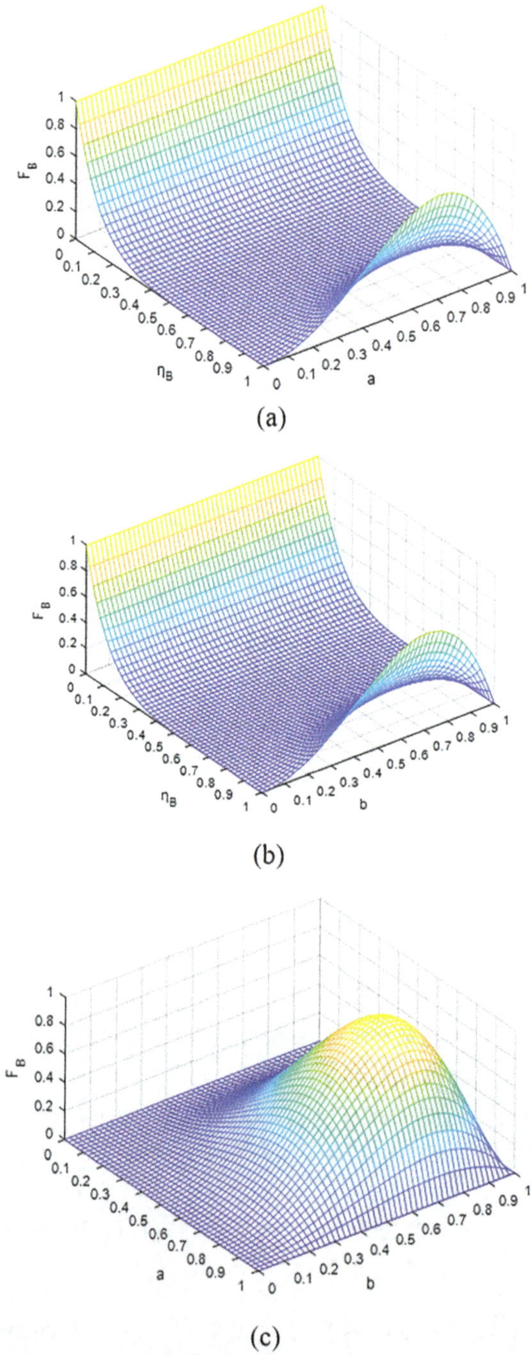

Fig. 6.9 The effect of bit-flip noise on NEC-ABQT scheme visualized through variation of fidelity F^B, which is related to amplitude information of the initial states and decoherence rate for various situations: (**a**) with $a = a_0, b_0 = 1/2, b_1 = \sqrt{3}/2, a_1 = a_2 = a_3 = a_4 = a_5 = a_6 = 0$; (**b**) with

$$\begin{aligned}\rho_{out}^{W} = \Big\{ \Big[(1-\eta_W)^8 + {\eta_W}^8 \Big] & [a_0b_0|0000\rangle + a_0b_1|0001\rangle + a_1b_0|0010\rangle \\ & + a_1b_1|0011\rangle + a_2b_0|0100\rangle + a_2b_1|0101\rangle + a_3b_0|0110\rangle + \\ & a_3b_1|0111\rangle + a_4b_0|1000\rangle + a_4b_1|1001\rangle + a_5b_0|1010\rangle + \\ & a_5b_1|1011\rangle + a_6b_0|1100\rangle + a_6b_1|1101\rangle + a_7b_0|1110\rangle + a_7b_1|1111\rangle] \\ & \times [a_0b_0\langle 0000| + a_0b_1\langle 0001| + a_1b_0\langle 0010| + a_1b_1\langle 0011| + \\ & a_2b_0\langle 0100| + a_2b_1\langle 0101| + a_3b_0\langle 0110| + a_3b_1\langle 0111| + \\ & a_4b_0\langle 1000| + a_4b_1\langle 1001| + a_5b_0\langle 1010| + a_5b_1\langle 1011| + \\ & a_6b_0\langle 1100| + a_6b_1\langle 1101| + a_7b_0\langle 1110| + a_7b_1\langle 1111|] + \\ & {\eta_P}^8{a_0}^2{b_0}^2|0000\rangle\langle 0000| + {\eta_P}^8{a_7}^2{b_1}^2|1111\rangle\langle 1111|\Big\}\end{aligned} \quad (6.52)$$

In phase-flip noise, the fidelity F^W of the output state can be calculated according to (6.37) and (6.52).

$$F^W = \langle\Theta|\rho_{out}^W|\Theta\rangle = (1-\eta_W)^8 + {\eta_W}^8 \quad (6.53)$$

(6.53) shows that the fidelity only depends on the decoherence rate in phase-flip noise. Figure 6.10 clearly illustrates the effect of bit-flip noise on the fidelity F^W and variation of the fidelity with the decoherence rate η_W. In Fig. 6.10, it can be easily observed that the fidelity F^W shows a trend of decreasing first and then rising with the increase of decoherence η_W; when $\eta_W = 1/2$, F^W reaches the minimum.

6.3.3.5 Bit-Phase-Flip Noisy Environment

The Kraus operators of bit-phase-flip noise are (6.54)

$$\begin{aligned} E_0^F &= \sqrt{1-\eta_F} I \\ E_1^F &= \sqrt{\eta_F} Y = \sqrt{\eta_F}\begin{bmatrix} 0 & -i \\ i & 0 \end{bmatrix} \end{aligned} . \quad (6.54)$$

In this noisy environment, the bit-phase will flip with the probability $\eta_F (0 \leq \eta_F \leq 1)$ and is left invariant with the probability 1 - η_F. After the qubits a_0, a_1, a_2, $a_3(b_0, b_1, b_2, b_3)$ are transmitted through the bit-phase-flip noisy environment, the density matrix ρ is rewritten as (6.55)

Fig. 6.9 (continued) $b = b_0, a_0 = \sqrt{3}/2, a_7 = 1/2, a_1 = a_2 = a_3 = a_4 = a_5 = a_6 = 0$; **(c)** with $a = a_0$, $b = b_0$, $\eta_B = 1$, $a_1 = a_2 = a_3 = a_4 = a_5 = a_6 = 0$

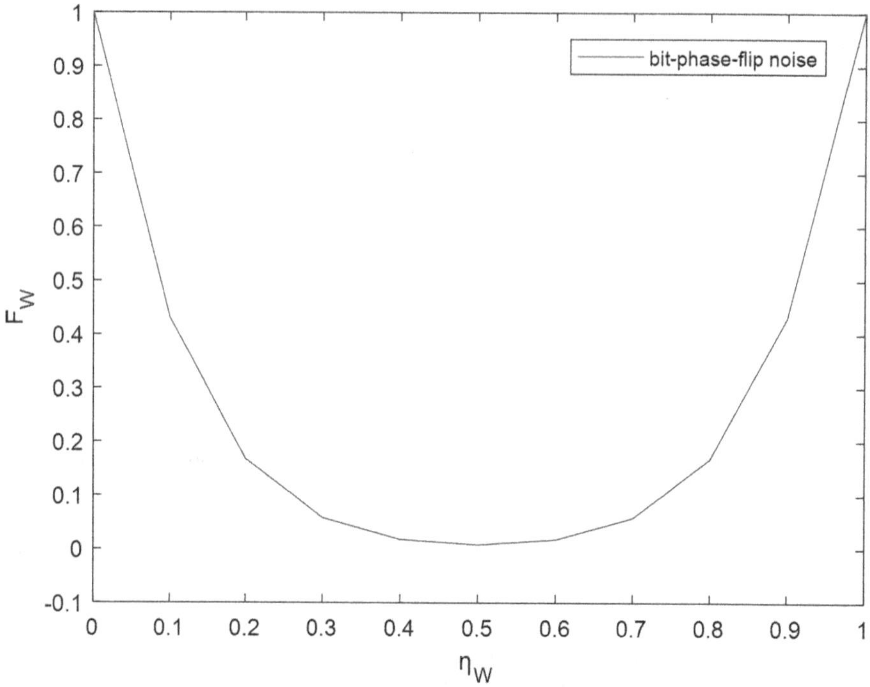

Fig. 6.10 The effect of phase-flip noise on NEC-ABQT scheme visualized through variation of fidelity F^W

$$
\begin{aligned}
\xi^F(\rho)=\frac{1}{16}\Big\{&(1-\eta_F)^8[|000000000\rangle+|000000111\rangle+|001001000\rangle+|001001111\rangle\\
&+|010010000\rangle+|010010111\rangle+|011011000\rangle+|011011111\rangle\\
&+|100100000\rangle+|100100111\rangle+|101101000\rangle+|101101111\rangle\\
&+|110110000\rangle+|110110111\rangle+|111111000\rangle+|111111111\rangle]\\
&\times[\langle000000000|+\langle000000111|+\langle001001000|+\langle001001111|\\
&+\langle010010000|+\langle010010111|+\langle011011000|+\langle011011111|\\
&+\langle100100000|+\langle100100111|+\langle101101000|+\langle101101111|\\
&+\langle110110000|+\langle110110111|+\langle111111000|+\langle111111111|]\\
&-\eta_F^8[|111111101\rangle+|111111010\rangle+|110110101\rangle+|110110010\rangle\\
&+|101101101\rangle+|100100010\rangle+|100100101\rangle+|101101010\rangle\\
&+|011011101\rangle+|010010010\rangle+|010010101\rangle+|011011010\rangle\\
&+|001001101\rangle+|001001010\rangle+|000000101\rangle+|000000010\rangle]\\
&\times[\langle111111101|+\langle111111010|+\langle110110010|+\langle110110101|\\
&+\langle101101101|+\langle100100010|+\langle100100101|+\langle101101010|\\
&+\langle011011101|+\langle010010010|+\langle010010101|+\langle011011010|\\
&+\langle001001101|+\langle000000101|+\langle001001010|+\langle000000010|]\Big\}.
\end{aligned}
\tag{6.55}
$$

In bit-phase-flip noise, the output state $\left(\rho_{out}^{F}\right)_{b_0 b_1 b_2 a_3}$ can be calculated as

$$\begin{aligned}
\rho_{out}^{F} = \Big\{ & (1-\eta_F)^8 [a_0 b_0 |0000\rangle + a_0 b_1 |0001\rangle + a_1 b_0 |0010\rangle + \\
& a_1 b_1 |0011\rangle + a_2 b_0 |0100\rangle + a_2 b_1 |0101\rangle + a_3 b_0 |0110\rangle + \\
& a_3 b_1 |0111\rangle + a_4 b_0 |1000\rangle + a_4 b_1 |1001\rangle + a_5 b_0 |1010\rangle + \\
& a_5 b_1 |1011\rangle + a_6 b_0 |1100\rangle + a_6 b_1 |1101\rangle + a_7 b_0 |1110\rangle + \\
& a_7 b_1 |1111\rangle] \times [a_0 b_0 \langle 0000| + a_0 b_1 \langle 0001| + a_1 b_0 \langle 0010| + \\
& a_1 b_1 \langle 0011| + a_2 b_0 \langle 0100| + a_2 b_1 \langle 0101| + a_3 b_0 \langle 0110| + \\
& a_3 b_1 \langle 0111| + a_4 b_0 \langle 1000| + a_4 b_1 \langle 1001| + a_5 b_0 \langle 1010| + \\
& a_5 b_1 \langle 1011| + a_6 b_0 \langle 1100| + a_6 b_1 \langle 1101| + a_7 b_0 \langle 1110| + \\
& a_7 b_1 \langle 1111|] - {\eta_F}^8 [a_0 b_0 |1111\rangle + a_0 b_1 |1110\rangle + a_1 b_0 |1101\rangle \\
& + a_1 b_1 |1100\rangle + a_2 b_0 |1011\rangle + a_2 b_1 |1010\rangle + a_3 b_0 |1001\rangle + \\
& a_3 b_1 |1000\rangle + a_4 b_0 |0111\rangle + a_4 b_1 |0110\rangle + a_5 b_0 |0101\rangle + \\
& a_5 b_1 |0100\rangle + a_6 b_0 |0011\rangle + a_6 b_1 |0010\rangle + a_7 b_0 |0111\rangle + \\
& a_7 b_1 |0000\rangle] \times [a_0 b_0 \langle 1111| + a_0 b_1 \langle 1000| + a_1 b_0 \langle 1101| + \\
& a_1 b_1 \langle 1100| + a_2 b_0 \langle 1011| + a_2 b_1 \langle 1010| + a_3 b_0 \langle 1001| + \\
& a_3 b_1 \langle 1000| + a_4 b_0 \langle 0111| + a_4 b_1 \langle 0110| + a_5 b_0 \langle 0101| + \\
& a_5 b_1 \langle 0100| + a_6 b_0 \langle 0011| + a_6 b_1 \langle 0010| + a_7 b_0 \langle 0111| + a_7 b_1 \langle 0000|] \Big\}
\end{aligned} \tag{6.56}$$

In bit-phase-flip noise, the fidelity F^F of the output state can be calculated according to (6.37) and (6.56).

$$F^F = \langle \Theta | \rho_{out}^{F} | \Theta \rangle = (1-\eta_F)^8 + 16{\eta_F}^8 {b_0}^2 {b_1}^2 (a_0 a_7 + a_1 a_6 + a_2 a_5 + a_3 a_4)^2. \tag{6.57}$$

(6.57) shows that the fidelity depends on the amplitude parameter of the initial state and the decoherence rate in bit-phase-flip noise. Figure 6.11(a)–(c) clearly illustrates the effect of bit-phase-flip noise on the fidelity F^F, and it also illustrates the variation of the fidelity with amplitude parameter of the initial state and the decoherence rate η_F. Similar to the bit-noise case, Fig. 6.11 and Fig. 6.9 coincide exactly can be easily observe that fidelity F^F always decrease and when the decoherence rate is large, it also shows an upward trend. The graph (c) shows a convex transformation about the change of amplitude parameters a and b.

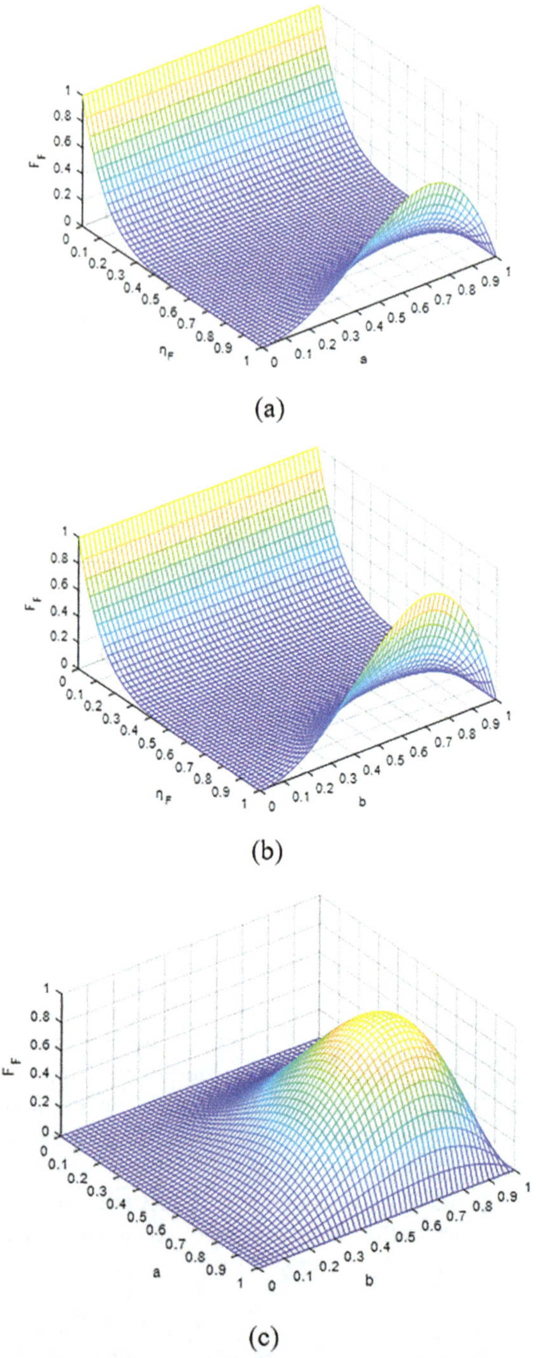

Fig. 6.11 The effect of bit-phase-flip noise on NEC-ABQT scheme visualized through variation of fidelity F^F, which is related to amplitude information of the initial states and decoherence rate for various situations: (**a**) with$a = a_0, b_0 = \sqrt{3}/2, b_1 = 1/2, a_1 = a_2 = a_3 = a_4 = a_5 = a_6 = 0$; (**b**) with

6.3.3.6 Depolarizing Noisy Environment

Depolarizing noise is an important kind of quantum noise; in depolarizing noisy environment, the qubits are depolarized with probability η_D, and the system is left invariant with the probability $1 - \eta_D$, while the operators X, Y, and Z act on the system with the probability $\eta_D/3$. The Kraus operators of depolarizing noise are (6.58)

$$\begin{aligned} E_0^D &= \sqrt{1-\eta_D} I \\ E_1^D &= \sqrt{\frac{\eta_D}{3}} X = \sqrt{\frac{\eta_D}{3}} \begin{bmatrix} 0 & 1 \\ 1 & 0 \end{bmatrix} \\ E_2^D &= \sqrt{\frac{\eta_D}{3}} Y = \sqrt{\frac{\eta_D}{3}} \begin{bmatrix} 0 & -i \\ i & 0 \end{bmatrix} \\ E_3^D &= \sqrt{\frac{\eta_D}{3}} Z = \sqrt{\frac{\eta_D}{3}} \begin{bmatrix} 1 & 0 \\ 0 & -1 \end{bmatrix} \end{aligned} . \tag{6.58}$$

After the qubits a_0, a_1, a_2, $a_3(b_0, b_1, b_2, b_3)$ are transmitted through the depolarizing noisy environment, the density matrix ρ can be rewritten as (6.59)

$$\begin{aligned} \xi^D(\rho) = \frac{1}{16}\Bigg\{ & \left[(1-\eta_D)^8 + \left(\frac{\sqrt{\eta_D}}{3}\right)^8 \right] [|000000000\rangle + |000000111\rangle + |001001000\rangle \\ & + |001001111\rangle + |010010000\rangle + |010010111\rangle + |011011000\rangle + |011011111\rangle \\ & + |100100000\rangle + |100100111\rangle + |101101000\rangle + |101101111\rangle + |110110000\rangle \\ & + |110110111\rangle + |111111000\rangle + |111111111\rangle] \times [(\langle 000000000| + \langle 000000111| \\ & + \langle 001001000| + + \langle 001001111| \langle 010010000| + \langle 010010111| + \langle 011011000| \\ & + \langle 011011111| + \langle 100100000| + \langle 100100111| + \langle 101101000| + \langle 101101111| \\ & + \langle 110110000| + \langle 110110111| + \langle 111111000| + \langle 111111111| |] \Bigg\} \end{aligned} \tag{6.59}$$

In depolarizing noise, the output state $\left(\rho_{out}^D\right)_{b_0 b_1 b_2 a_3}$ can be calculated as

Fig. 6.11 (continued) $b = b_0, a_0 = a_7 = \sqrt{2}/2, a_1 = a_2 = a_3 = a_4 = a_5 = a_6 = 0$; (**c**) with $a = a_0$, $b = b_0$, $\eta_F = 1$, $a_1 = a_2 = a_3 = a_4 = a_5 = a_6 = 0$

$$\begin{aligned}(\rho_{out}^{D})_{b_0b_1b_2a_3} = & \left\{ \left[(1-\eta_D)^8 + \left(\frac{\sqrt{\eta_D}}{3} \right)^8 \right] [a_0b_0|0000\rangle + a_0b_1|0001\rangle + a_1b_0|0010\rangle + \right. \\ & a_1b_1|0011\rangle + a_2b_0|0100\rangle + a_2b_1|0101\rangle + a_3b_0|0110\rangle + a_3b_1|0111\rangle + \\ & a_4b_0|1000\rangle + a_4b_1|1001\rangle + a_5b_0|1010\rangle + a_5b_1|1011\rangle + a_6b_0|1100\rangle + \\ & a_6b_1|1101\rangle + a_7b_0|1110\rangle + a_7b_1|1111\rangle] \times [a_0b_0\langle 0000| + a_0b_1\langle 0001| + \\ & a_1b_0\langle 0010| + a_1b_1\langle 0011| + a_2b_0\langle 0100| + a_2b_1\langle 0101| + a_3b_0\langle 0110| + \\ & a_3b_1\langle 0111| + a_4b_0\langle 1000| + a_4b_1\langle 1001| + a_5b_0\langle 1010| + a_5b_1\langle 1011| + \\ & a_6b_0\langle 1100| + a_6b_1\langle 1101| + a_7b_0\langle 1110| + a_7b_1\langle 1111|] + \eta_P{}^8 a_0{}^2 b_0{}^2 |0000\rangle\langle 0000| \\ & \left. + \eta_P{}^8 a_7{}^2 b_1{}^2 |1111\rangle\langle 1111| \right\}. \end{aligned} \tag{6.60}$$

In depolarizing noise, the fidelity F^D of the output state can be calculated according to (6.37) and (6.60).

$$\begin{aligned}\xi^D(\rho) = \frac{1}{16} & \left\{ \left[(1-\eta_D)^8 + \left(\frac{\sqrt{\eta_D}}{3} \right)^8 \right] [|000000000\rangle + |000000111\rangle + |001001000\rangle \right. \\ & + |001001111\rangle + |010010000\rangle + |010010111\rangle + |011011000\rangle + |011011111\rangle \\ & + |100100000\rangle + |100100111\rangle + |101101000\rangle + |101101111\rangle + |110110000\rangle \\ & + |110110111\rangle + |111111000\rangle + |111111111\rangle] \times [\langle 000000000| + \langle 000000111| \\ & + \langle 001001000| + + \langle 001001111| \langle 010010000| + \langle 010010111| + \langle 011011000| \\ & + \langle 011011111| + \langle 100100000| + \langle 100100111| + \langle 101101000| + \langle 101101111| \\ & \left. + \langle 110110000| + \langle 110110111| + \langle 111111000| + \langle 111111111|] \right\} \end{aligned} \tag{6.61}$$

(6.61) shows that the fidelity only depends on the decoherence rate in depolarizing noise. Figure 6.12 clearly illustrates the effect of depolarizing noise on the fidelity F^D and variation of the fidelity with the decoherence rate η_D. In Fig. 6.12, it can be easily observed that the fidelity F^D shows a trend of decreasing with the increase of decoherence η_D.

Finally, in the amplitude-damping, phase-damping, and phase-flip noisy environment, it can be seen that the fidelity of the output state depends on the amplitude parameter of the initial state and the decoherence rate. However, in the bit-flip and depolarizing noisy environment, the fidelity of the output state only depends on the decoherence rate only. In order to compare the trend of the fidelity of the output state with the change of the decoherence rate in six types of noisy environment, a complex two-dimensional graph (Fig. 6.13) of different curve colors is drawn.

In Fig. 6.13, it can be observed that the curves exactly coincide in bit-flip, phase-flip, and bit-phase-flip noisy environment and the curves "almost" exactly coincide in phase-damping and depolarizing noise. With the same decoherence rate in the amplitude-damping, phase-damping, and depolarizing noise, the fidelity (F^A, F^P, F^D) of the output state decreases as the decoherence rate increases; (F^A, F^P, F^D) can achieve 1 when $\eta = 0$, and (F^A, F^P, F^D) reaches minimum when $\eta = 1$. Furthermore, in the bit-flip, phase-flip, and bit-phase-flip noisy environment, the fidelity $(F^B, F^W,$

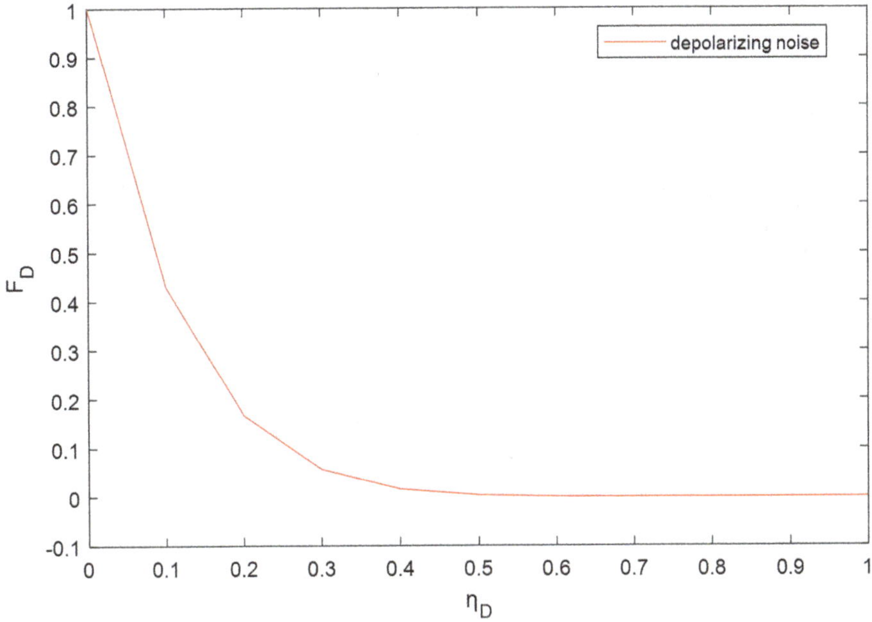

Fig. 6.12 The effect of depolarizing noise on NEC-ABQT scheme visualized through variation of fidelity F^D

F^F) of the output state decreases firstly and then increases as the decoherence rate increases; (F^B, F^W, F^F) can achieve 1 when $\eta = 0$ or $\eta = 1$. (F^B, F^W, F^F) reaches minimum when $\eta = 0.5$. In a word, information of quantum state loss is the most serious when fidelity reaches minimum.

It's worth noting that the above analysis can be applied to real physical scenarios, and perfect NEC-ABQT can be achieved by assigning specific values of amplitude parameter of the initial state; this fact can be seen from the peak of Figs. 6.7, 6.8, 6.9, 6.10, 6.11, and 6.12.

6.3.4 Comparison

In this subsection, the proposed NEC-ABQT scheme is compared with the previous QT schemes in terms of the six aspects including the type of scheme (TOP), the number of qubits transmitted (NQT), the quantum resource consumption (QRC), the classical resource consumption (CRC), the necessary operation (NO), and the intrinsic efficiency as shown in Table 6.7.

From Table 6.7, we can observe the following four points:

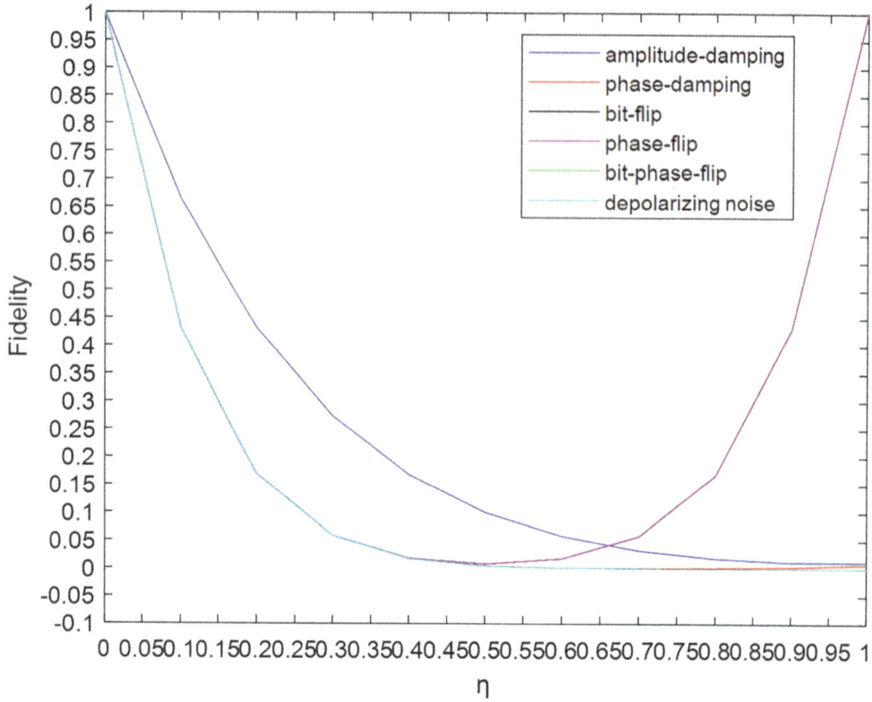

Fig. 6.13 The trend of the fidelity of the output state with the change of the decoherence rate in six types of noisy environments. Here, we assume that $\eta_A = \eta_P = \eta_B = \eta_W = \eta_F = \eta_D = \eta$, $a_0 = a_1 = a_2 = a_3 = a_4 = a_5 = a_6 = a_7 = \sqrt{2}/4$, and $b_0 = b_1 = \sqrt{2}/2$

Table 6.7 Comparison between eight schemes

Ref.	TOP	NOT	QRC	CRC	NO	η
[24]	ABCQT	Arbitrary 1 and 2	7	7	3 BM, 1 SM	3/14
[25]	BQT	Arbitrary 1 and pure 3	6	5	1 GHZ, 1 BM, 1 SM	4/11
[26]	BCQT	Arbitrary 2 and 2	9	9	4 BM, 1 SM	2/9
[27]	ABCQT	Pure 2 and 3	9	9	2 GHZ, 1 BM, 1 SM	5/21
[28]	QT	Pure 3	6	6	3 BM	1/4
[29]	QT	Pure 4	6	4	1 six-qubit JM	2/5
[30]	QT	Pure 5	7	4	5 CNOT, 4 SM	3/14
Ours	NEC-ABQT	Arbitrary 1 and 3	9	9	3 CNOT, 9 SM	2/9

(a) The proposed NEC-ABQT scheme only uses single-qubit measurement basis, which is more efficient than multi-qubit measurements. For example, it is well known that Bell state measurements can be decomposed into an ordering combination of a single-qubit Hadamard operation and a two-qubit CNOT operation as well as two single-qubit measurements.
(b) In NEC-ABQT scheme, Alice can transmit an arbitrary three-qubit state to Bob, and at the same time, Bob can transmit an arbitrary single-qubit state to Alice under the permission of controller Charlie. However, the sender only transmits pure n qubit ($n \geq 3$) to the receiver in the previous scheme.
(c) Although the efficiency of NEC-ABQT scheme has no obvious advantage over other schemes, it is optimal for the transmitted quantum information of arbitrary single-qubit and three-qubit states.
(d) The most remarkable advantage is that this scheme is analyzed in six noisy environments; it can lay the foundation for the practical application in the future.

The efficiency of the schemes can be defined as (6.62)

$$\eta = \frac{q_s}{q_u + b_t} \tag{6.62}$$

where q_s is the number of qubits that consist of the quantum information to be exchanged, q_u is the number of qubits that are used as the quantum channel, and b_t is the classical bits transmitted. Next, a description of the abbreviations is provided: Bell basis measurement (BM), single-qubit measurement (SM), joint measurement (JM), reference (ref.), and efficiency (η).

6.3.5 *Summary*

In view of the above, a summary is made in this subsection. In this subsection, a novel scheme for NEC-ABQT by using a nine-qubit entangled state as a quantum channel is proposed. Alice transmits an arbitrary three-qubit entangled state to Bob, and Bob simultaneously transmits an arbitrary single-qubit entangled state to Alice under the permission of the controller. The construction of the channel of a nine-qubit entangled state is shown explicitly, and it provides reference significance to prepare the nine-qubit entangled channel in the experimental environment in the future. This scheme is discussed in six types of noisy environments including amplitude-damping, phase-damping, bit-flip, phase-flip, bit-phase-flip, and depolarizing noise. We analytically derive the fidelities of the NEC-ABCT process only depend on the amplitude parameter of the initial state and the decoherence noisy rate and then show the variation of fidelity with the amplitude parameter of the original state and the decoherence rate in the form of a stereo image. Finally, compared with previous schemes, this scheme also shows remarkable advantages. We hope that such NEC-ABQT scheme can be realized experimentally and this scheme can be extended to multi-party teleportation [31] in the future.

6.3.6 Appendix

Some formulas and tables are not appropriate for the body of the scheme, so we add them to the Appendix.

$$
\begin{aligned}
|\eta_0\rangle = (&|000000000\rangle + |000000111\rangle + |001001000\rangle + |001001111\rangle \\
&+|010010000\rangle + |010010111\rangle + |011011000\rangle + |011011111\rangle \\
&+|100100000\rangle + |100100111\rangle + |101101000\rangle + |101101111\rangle \\
&+|110110000\rangle + |110110111\rangle + |111111000\rangle + |111111111\rangle)_{a_0a_1a_2b_0b_1b_2a_3cb_3} \\
|\eta_1\rangle = (&|000000001\rangle + |000000110\rangle + |001001001\rangle + |001001110\rangle \\
&+|010010001\rangle + |010010110\rangle + |011011001\rangle + |011011110\rangle \\
&+|100100001\rangle + |100100110\rangle + |101101001\rangle + |101101110\rangle \\
&+|110110001\rangle + |110110110\rangle + |111111001\rangle + |111111110\rangle)_{a_0a_1a_2b_0b_1b_2a_3cb_3} \\
|\eta_2\rangle = (&|001000000\rangle + |001000111\rangle + |000001000\rangle + |000001111\rangle \\
&+|011010000\rangle + |011010111\rangle + |010011000\rangle + |010011111\rangle \\
&+|101100000\rangle + |101100111\rangle + |100101000\rangle + |100101111\rangle \\
&+|111110000\rangle + |111110111\rangle + |110111000\rangle + |110111111\rangle)_{a_0a_1a_2b_0b_1b_2a_3cb_3} \\
|\eta_3\rangle = (&|001000001\rangle + |001000110\rangle + |000001001\rangle + |000001110\rangle \\
&+|011010001\rangle + |011010110\rangle + |010011001\rangle + |010011110\rangle \\
&+|101100001\rangle + |101100110\rangle + |100101001\rangle + |100101110\rangle \\
&+|111110001\rangle + |111110110\rangle + |110111001\rangle + |110111110\rangle)_{a_0a_1a_2b_0b_1b_2a_3cb_3} \\
|\eta_4\rangle = (&|010000000\rangle + |010000111\rangle + |011001000\rangle + |011001111\rangle \\
&+|000010000\rangle + |000010111\rangle + |001011000\rangle + |001011111\rangle \\
&+|110100000\rangle + |110100111\rangle + |111101000\rangle + |111101111\rangle \\
&+|100110000\rangle + |100110111\rangle + |101111000\rangle + |101111111\rangle)_{a_0a_1a_2b_0b_1b_2a_3cb_3} \\
|\eta_5\rangle = (&|010000001\rangle + |010000110\rangle + |011001001\rangle + |011001110\rangle \\
&+|000010001\rangle + |000010110\rangle + |001011001\rangle + |001011110\rangle \\
&+|110100001\rangle + |110100110\rangle + |111101001\rangle + |111101110\rangle \\
&+|100110001\rangle + |100110110\rangle + |101111001\rangle + |101111110\rangle)_{a_0a_1a_2b_0b_1b_2a_3cb_3} \\
|\eta_6\rangle = (&|011000000\rangle + |011000111\rangle + |010001000\rangle + |010001111\rangle \\
&+|001010000\rangle + |001010111\rangle + |000011000\rangle + |000011111\rangle \\
&+|111100000\rangle + |111100111\rangle + |110101000\rangle + |110101111\rangle \\
&+|101110000\rangle + |101110111\rangle + |100111000\rangle + |100111111\rangle)_{a_0a_1a_2b_0b_1b_2a_3cb_3} \\
|\eta_7\rangle = (&|011000001\rangle + |011000110\rangle + |010001001\rangle + |010001110\rangle \\
&+|001010001\rangle + |001010110\rangle + |000011001\rangle + |000011110\rangle \\
&+|111100001\rangle + |111100110\rangle + |110101001\rangle + |110101110\rangle \\
&+|101110001\rangle + |101110110\rangle + |100111001\rangle + |100111110\rangle)_{a_0a_1a_2b_0b_1b_2a_3cb_3}
\end{aligned}
$$

$$
\begin{aligned}
|\eta_8\rangle = (&|100000000\rangle + |100000111\rangle + |101001000\rangle + |101001111\rangle \\
&+|110010000\rangle + |110010111\rangle + |111011000\rangle + |111011111\rangle \\
&+|000100000\rangle + |000100111\rangle + |001101000\rangle + |001101111\rangle \\
&+|010110000\rangle + |010110111\rangle + |011111000\rangle + |011111111\rangle)_{a_0a_1a_2b_0b_1b_2a_3cb_3} \\
|\eta_9\rangle = (&|100000001\rangle + |100000110\rangle + |101001001\rangle + |101001110\rangle \\
&+|110010001\rangle + |110010110\rangle + |111011001\rangle + |111011110\rangle \\
&+|000100001\rangle + |000100110\rangle + |001101001\rangle + |001101110\rangle \\
&+|010110001\rangle + |010110110\rangle + |011111001\rangle + |011111110\rangle)_{a_0a_1a_2b_0b_1b_2a_3cb_3} \\
|\eta_{10}\rangle = (&|101000000\rangle + |101000111\rangle + |100001000\rangle + |100001111\rangle \\
&+|111010000\rangle + |111010111\rangle + |110011000\rangle + |110011111\rangle \\
&+|001100000\rangle + |001100111\rangle + |000101000\rangle + |000101111\rangle \\
&+|011110000\rangle + |011110111\rangle + |010111000\rangle + |010111111\rangle)_{a_0a_1a_2b_0b_1b_2a_3cb_3} \\
|\eta_{11}\rangle = (&|101000001\rangle + |101000110\rangle + |100001001\rangle + |100001110\rangle \\
&+|111010001\rangle + |111010110\rangle + |110011001\rangle + |110011110\rangle \\
&+|001100001\rangle + |001100110\rangle + |000101001\rangle + |000101110\rangle \\
&+|011110001\rangle + |011110110\rangle + |010111001\rangle + |010111110\rangle)_{a_0a_1a_2b_0b_1b_2a_3cb_3} \\
|\eta_{12}\rangle = (&|110000000\rangle + |110000111\rangle + |111001000\rangle + |111001111\rangle \\
&+|100010000\rangle + |100010111\rangle + |101011000\rangle + |101011111\rangle \\
&+|010100000\rangle + |010100111\rangle + |011101000\rangle + |011101111\rangle \\
&+|000110000\rangle + |000110111\rangle + |001111000\rangle + |001111111\rangle)_{a_0a_1a_2b_0b_1b_2a_3cb_3} \\
|\eta_{13}\rangle = (&|110000001\rangle + |110000110\rangle + |111001001\rangle + |111001110\rangle \\
&+|100010001\rangle + |100010110\rangle + |101011001\rangle + |101011110\rangle \\
&+|010100001\rangle + |010100110\rangle + |011101001\rangle + |011101110\rangle \\
&+|000110001\rangle + |000110110\rangle + |001111001\rangle + |001111110\rangle)_{a_0a_1a_2b_0b_1b_2a_3cb_3} \\
|\eta_{14}\rangle = (&|111000000\rangle + |111000111\rangle + |110001000\rangle + |110001111\rangle \\
&+|101010000\rangle + |101010111\rangle + |100011000\rangle + |100011111\rangle \\
&+|011100000\rangle + |011100111\rangle + |010101000\rangle + |010101111\rangle \\
&+|001110000\rangle + |001110111\rangle + |000111000\rangle + |000111111\rangle)_{a_0a_1a_2b_0b_1b_2a_3cb_3} \\
|\eta_{15}\rangle = (&|111000001\rangle + |111000110\rangle + |110001001\rangle + |110001110\rangle \\
&+|101010001\rangle + |101010110\rangle + |100011001\rangle + |100011110\rangle \\
&+|011100001\rangle + |011100110\rangle + |010101001\rangle + |010101110\rangle \\
&+|001110001\rangle + |001110110\rangle + |000111001\rangle + |000111110\rangle)_{a_0a_1a_2b_0b_1b_2a_3cb_3}
\end{aligned}
$$

Table A.6.1 The Z-basis measurement results and the corresponding collapsed state

Alice's results	Bob's results	The collapsed results of qubits $b_0b_1b_2a_3cA_0A_1A_2B$
000	0	$a_0b_1\|000000000\rangle + a_0b_1\|000110001\rangle + a_1b_0\|001000010\rangle + a_1b_1\|001110011\rangle+$ $a_2b_0\|010000100\rangle + a_2b_1\|010110101\rangle + a_3b_0\|011000110\rangle + a_3b_1\|011110111\rangle+$ $a_4b_0\|100001000\rangle + a_4b_1\|100111001\rangle + a_5b_0\|101001010\rangle + a_5b_1\|101111011\rangle+$ $a_6b_0\|110001100\rangle + a_6b_1\|110111101\rangle + a_7b_0\|111001110\rangle + a_7b_1\|111111111\rangle$
000	1	$a_0b_1\|000110000\rangle + a_0b_1\|000000001\rangle + a_1b_0\|001110010\rangle + a_1b_1\|001000011\rangle+$ $a_2b_0\|010110100\rangle + a_2b_1\|010000101\rangle + a_3b_0\|011110110\rangle + a_3b_1\|011000111\rangle+$ $a_4b_0\|100111000\rangle + a_4b_1\|100001001\rangle + a_5b_0\|101111010\rangle + a_5b_1\|101001011\rangle+$ $a_6b_0\|110111100\rangle + a_6b_1\|110001101\rangle + a_7b_0\|111111110\rangle + a_7b_1\|111001111\rangle$
001	0	$a_0b_1\|001000000\rangle + a_0b_1\|001110001\rangle + a_1b_0\|000000010\rangle + a_1b_1\|000110011\rangle+$ $a_2b_0\|011000100\rangle + a_2b_1\|011110101\rangle + a_3b_0\|010000110\rangle + a_3b_1\|010110111\rangle+$ $a_4b_0\|101001000\rangle + a_4b_1\|101111001\rangle + a_5b_0\|100001010\rangle + a_5b_1\|100111011\rangle+$ $a_6b_0\|111001100\rangle + a_6b_1\|111111101\rangle + a_7b_0\|110001110\rangle + a_7b_1\|110111111\rangle$
001	1	$a_0b_1\|001110000\rangle + a_0b_1\|001000001\rangle + a_1b_0\|000110010\rangle + a_1b_1\|000000011\rangle+$ $a_2b_0\|011110100\rangle + a_2b_1\|011000101\rangle + a_3b_0\|010110110\rangle + a_3b_1\|010000111\rangle+$ $a_4b_0\|101111000\rangle + a_4b_1\|101001001\rangle + a_5b_0\|100111010\rangle + a_5b_1\|100001011\rangle+$ $a_6b_0\|111111100\rangle + a_6b_1\|111001101\rangle + a_7b_0\|110111110\rangle + a_7b_1\|110001111\rangle$
010	0	$a_0b_1\|010000000\rangle + a_0b_1\|010110001\rangle + a_1b_0\|011000010\rangle + a_1b_1\|011110011\rangle+$ $a_2b_0\|000000100\rangle + a_2b_1\|000110101\rangle + a_3b_0\|001000110\rangle + a_3b_1\|001110111\rangle+$ $a_4b_0\|110001000\rangle + a_4b_1\|110111001\rangle + a_5b_0\|111001010\rangle + a_5b_1\|111111011\rangle+$ $a_6b_0\|100001100\rangle + a_6b_1\|100111101\rangle + a_7b_0\|101001110\rangle + a_7b_1\|101111111\rangle$
010	1	$a_0b_1\|010110000\rangle + a_0b_1\|010000001\rangle + a_1b_0\|011110010\rangle + a_1b_1\|011000011\rangle+$ $a_2b_0\|000110100\rangle + a_2b_1\|000000101\rangle + a_3b_0\|001110110\rangle + a_3b_1\|001000111\rangle+$ $a_4b_0\|110111000\rangle + a_4b_1\|110001001\rangle + a_5b_0\|111111010\rangle + a_5b_1\|111001011\rangle+$ $a_6b_0\|100111100\rangle + a_6b_1\|100001101\rangle + a_7b_0\|101111110\rangle + a_7b_1\|101001111\rangle$

011	0	$a_0b_1\|011000000\rangle + a_0b_1\|011110001\rangle + a_1b_0\|010000010\rangle + a_1b_1\|010110011\rangle +$ $a_2b_0\|001000100\rangle + a_2b_1\|001110101\rangle + a_3b_0\|000000110\rangle + a_3b_1\|000110111\rangle +$ $a_4b_0\|111001000\rangle + a_4b_1\|111111001\rangle + a_5b_0\|110001010\rangle + a_5b_1\|110111011\rangle +$ $a_6b_0\|101001100\rangle + a_6b_1\|101111101\rangle + a_7b_0\|100001110\rangle + a_7b_1\|100111111\rangle$
011	1	$a_0b_1\|011110000\rangle + a_0b_1\|011000001\rangle + a_1b_0\|010110010\rangle + a_1b_1\|010000011\rangle +$ $a_2b_0\|001110100\rangle + a_2b_1\|001000101\rangle + a_3b_0\|000110110\rangle + a_3b_1\|000000111\rangle +$ $a_4b_0\|111111000\rangle + a_4b_1\|111001001\rangle + a_5b_0\|110111010\rangle + a_5b_1\|110001011\rangle +$ $a_6b_0\|101111100\rangle + a_6b_1\|101001101\rangle + a_7b_0\|100111110\rangle + a_7b_1\|100001111\rangle$
100	0	$a_0b_1\|100000000\rangle + a_0b_1\|100110001\rangle + a_1b_0\|101000010\rangle + a_1b_1\|101110011\rangle +$ $a_2b_0\|110000100\rangle + a_2b_1\|110110101\rangle + a_3b_0\|111000110\rangle + a_3b_1\|111110111\rangle +$ $a_4b_0\|000001000\rangle + a_4b_1\|000111001\rangle + a_5b_0\|001001010\rangle + a_5b_1\|001111011\rangle +$ $a_6b_0\|010001100\rangle + a_6b_1\|010111101\rangle + a_7b_0\|011001110\rangle + a_7b_1\|011111111\rangle$
100	1	$a_0b_1\|100110000\rangle + a_0b_1\|100000001\rangle + a_1b_0\|101110010\rangle + a_1b_1\|101000011\rangle +$ $a_2b_0\|110110100\rangle + a_2b_1\|110000101\rangle + a_3b_0\|111110110\rangle + a_3b_1\|111000111\rangle +$ $a_4b_0\|000111000\rangle + a_4b_1\|000001001\rangle + a_5b_0\|001111010\rangle + a_5b_1\|001001011\rangle +$ $a_6b_0\|010111100\rangle + a_6b_1\|010001101\rangle + a_7b_0\|011111110\rangle + a_7b_1\|011001111\rangle$
101	0	$a_0b_1\|101000000\rangle + a_0b_1\|101110001\rangle + a_1b_0\|100000010\rangle + a_1b_1\|100110011\rangle +$ $a_2b_0\|111000100\rangle + a_2b_1\|111110101\rangle + a_3b_0\|110000110\rangle + a_3b_1\|110110111\rangle +$ $a_4b_0\|001001000\rangle + a_4b_1\|001111001\rangle + a_5b_0\|000001010\rangle + a_5b_1\|000111011\rangle +$ $a_6b_0\|011001100\rangle + a_6b_1\|011111101\rangle + a_7b_0\|010001110\rangle + a_7b_1\|010111111\rangle$
101	1	$a_0b_1\|101110000\rangle + a_0b_1\|101000001\rangle + a_1b_0\|100110010\rangle + a_1b_1\|100000011\rangle +$ $a_2b_0\|111110100\rangle + a_2b_1\|111000101\rangle + a_3b_0\|110110110\rangle + a_3b_1\|110000111\rangle +$ $a_4b_0\|001111000\rangle + a_4b_1\|001001001\rangle + a_5b_0\|000111010\rangle + a_5b_1\|000001011\rangle +$ $a_6b_0\|011111100\rangle + a_6b_1\|011001101\rangle + a_7b_0\|010111110\rangle + a_7b_1\|010001111\rangle$

(continued)

Table A.6.1 (continued)

Alice's results	Bob's results	The collapsed results of qubits $b_0b_1b_2a_3cA_0A_1A_2B$
110	0	$a_0b_1\|110000000\rangle + a_0b_1\|110110001\rangle + a_1b_0\|111000010\rangle + a_1b_1\|111110011\rangle +$ $a_2b_0\|100000100\rangle + a_2b_1\|100110101\rangle + a_3b_0\|101000110\rangle + a_3b_1\|101110111\rangle +$ $a_4b_0\|010001000\rangle + a_4b_1\|010111001\rangle + a_5b_0\|011001010\rangle + a_5b_1\|011111011\rangle +$ $a_6b_0\|000001100\rangle + a_6b_1\|000111101\rangle + a_7b_0\|001001110\rangle + a_7b_1\|001111111\rangle$
110	1	$a_0b_1\|110110000\rangle + a_0b_1\|110000001\rangle + a_1b_0\|111110010\rangle + a_1b_1\|111000011\rangle +$ $a_2b_0\|100110100\rangle + a_2b_1\|100000101\rangle + a_3b_0\|101110110\rangle + a_3b_1\|101000111\rangle +$ $a_4b_0\|010111000\rangle + a_4b_1\|010001001\rangle + a_5b_0\|011111010\rangle + a_5b_1\|011001011\rangle +$ $a_6b_0\|000111100\rangle + a_6b_1\|000001101\rangle + a_7b_0\|001111110\rangle + a_7b_1\|001001111\rangle$
111	0	$a_0b_1\|111000000\rangle + a_0b_1\|111110001\rangle + a_1b_0\|110000010\rangle + a_1b_1\|110110011\rangle +$ $a_2b_0\|101000100\rangle + a_2b_1\|101110101\rangle + a_3b_0\|100000110\rangle + a_3b_1\|100110111\rangle +$ $a_4b_0\|011001000\rangle + a_4b_1\|011111001\rangle + a_5b_0\|010001010\rangle + a_5b_1\|010111011\rangle +$ $a_6b_0\|001001100\rangle + a_6b_1\|001111101\rangle + a_7b_0\|000001110\rangle + a_7b_1\|000111111\rangle$
111	1	$a_0b_1\|111100000\rangle + a_0b_1\|111010001\rangle + a_1b_0\|110100010\rangle + a_1b_1\|110010011\rangle +$ $a_2b_0\|101100100\rangle + a_2b_1\|101010101\rangle + a_3b_0\|100100110\rangle + a_3b_1\|100010111\rangle +$ $a_4b_0\|011101000\rangle + a_4b_1\|011011001\rangle + a_5b_0\|010101010\rangle + a_5b_1\|010011011\rangle +$ $a_6b_0\|001101100\rangle + a_6b_1\|001011101\rangle + a_7b_0\|000101110\rangle + a_7b_1\|000011111\rangle$

Table A.6.2 Applying X unitary operation

Alice's results	Bob's results	Unitary operation on qubits $b_0b_1b_2a_3$
000	0	$I \otimes I \otimes I \otimes I$
000	1	$I \otimes I \otimes I \otimes X$
001	0	$I \otimes I \otimes X \otimes I$
001	1	$I \otimes I \otimes X \otimes X$
010	0	$I \otimes X \otimes I \otimes I$
010	1	$I \otimes X \otimes I \otimes X$
011	0	$I \otimes X \otimes X \otimes I$
011	1	$I \otimes X \otimes X \otimes X$
100	0	$X \otimes I \otimes I \otimes I$
100	1	$X \otimes I \otimes I \otimes X$
101	0	$X \otimes I \otimes X \otimes I$
101	1	$X \otimes I \otimes X \otimes X$
110	0	$X \otimes X \otimes I \otimes I$
110	1	$X \otimes X \otimes I \otimes X$
111	0	$X \otimes X \otimes X \otimes I$
111	1	$X \otimes X \otimes X \otimes X$

Table A.6.3 The *X*-basis measurement results and the corresponding collapsed state

Alice's results	Bob's results	The collapsed results of qubits $b_0b_1b_2a_3c$
+ +	+ +	$a_0b_1\|00000\rangle + a_0b_1\|00011\rangle + a_1b_0\|00100\rangle + a_1b_1\|00111\rangle + a_2b_0\|01000\rangle + a_2b_1\|01011\rangle + a_3b_0\|01100\rangle + a_3b_1\|01111\rangle +$ $a_4b_0\|10000\rangle + a_4b_1\|10011\rangle + a_5b_0\|10100\rangle + a_5b_1\|10111\rangle + a_6b_0\|11000\rangle + a_6b_1\|11011\rangle + a_7b_0\|11100\rangle + a_7b_1\|11111\rangle$
+ +	+ −	$a_0b_1\|00000\rangle - a_0b_1\|00011\rangle + a_1b_0\|00100\rangle - a_1b_1\|00111\rangle + a_2b_0\|01000\rangle - a_2b_1\|01011\rangle + a_3b_0\|01100\rangle - a_3b_1\|01111\rangle +$ $a_4b_0\|10000\rangle - a_4b_1\|10011\rangle + a_5b_0\|10100\rangle - a_5b_1\|10111\rangle + a_6b_0\|11000\rangle - a_6b_1\|11011\rangle + a_7b_0\|11100\rangle - a_7b_1\|11111\rangle$
+ +	- +	$a_0b_1\|00000\rangle + a_0b_1\|00011\rangle - a_1b_0\|00100\rangle - a_1b_1\|00111\rangle + a_2b_0\|01000\rangle + a_2b_1\|01011\rangle - a_3b_0\|01100\rangle - a_3b_1\|01111\rangle +$ $a_4b_0\|10000\rangle + a_4b_1\|10011\rangle - a_5b_0\|10100\rangle - a_5b_1\|10111\rangle + a_6b_0\|11000\rangle + a_6b_1\|11011\rangle - a_7b_0\|11100\rangle - a_7b_1\|11111\rangle$
++	- -	$a_0b_1\|00000\rangle - a_0b_1\|00011\rangle - a_1b_0\|00100\rangle + a_1b_1\|00111\rangle + a_2b_0\|01000\rangle - a_2b_1\|01011\rangle - a_3b_0\|01100\rangle + a_3b_1\|01111\rangle +$ $a_4b_0\|10000\rangle - a_4b_1\|10011\rangle - a_5b_0\|10100\rangle + a_5b_1\|10111\rangle + a_6b_0\|11000\rangle - a_6b_1\|11011\rangle - a_7b_0\|11100\rangle + a_7b_1\|11111\rangle$
+ −	+ +	$a_0b_1\|00000\rangle + a_0b_1\|00011\rangle + a_1b_0\|00100\rangle + a_1b_1\|00111\rangle - a_2b_0\|01000\rangle - a_2b_1\|01011\rangle - a_3b_0\|01100\rangle - a_3b_1\|01111\rangle +$ $a_4b_0\|10000\rangle + a_4b_1\|10011\rangle + a_5b_0\|10100\rangle + a_5b_1\|10111\rangle - a_6b_0\|11000\rangle - a_6b_1\|11011\rangle - a_7b_0\|11100\rangle - a_7b_1\|11111\rangle$
+ −	+ −	$a_0b_1\|00000\rangle - a_0b_1\|00011\rangle + a_1b_0\|00100\rangle - a_1b_1\|00111\rangle - a_2b_0\|01000\rangle + a_2b_1\|01011\rangle - a_3b_0\|01100\rangle + a_3b_1\|01111\rangle +$ $a_4b_0\|10000\rangle - a_4b_1\|10011\rangle + a_5b_0\|10100\rangle - a_5b_1\|10111\rangle - a_6b_0\|11000\rangle + a_6b_1\|11011\rangle - a_7b_0\|11100\rangle + a_7b_1\|11111\rangle$
+ −	− +	$a_0b_1\|00000\rangle + a_0b_1\|00011\rangle - a_1b_0\|00100\rangle - a_1b_1\|00111\rangle - a_2b_0\|01000\rangle - a_2b_1\|01011\rangle + a_3b_0\|01100\rangle + a_3b_1\|01111\rangle +$ $a_4b_0\|10000\rangle + a_4b_1\|10011\rangle - a_5b_0\|10100\rangle - a_5b_1\|10111\rangle - a_6b_0\|11000\rangle - a_6b_1\|11011\rangle + a_7b_0\|11100\rangle + a_7b_1\|11111\rangle$
+ −	− −	$a_0b_1\|00000\rangle - a_0b_1\|00011\rangle - a_1b_0\|00100\rangle + a_1b_1\|00111\rangle - a_2b_0\|01000\rangle + a_2b_1\|01011\rangle + a_3b_0\|01100\rangle - a_3b_1\|01111\rangle +$ $a_4b_0\|10000\rangle - a_4b_1\|10011\rangle - a_5b_0\|10100\rangle + a_5b_1\|10111\rangle - a_6b_0\|11000\rangle + a_6b_1\|11011\rangle + a_7b_0\|11100\rangle - a_7b_1\|11111\rangle$
− +	+ +	$a_0b_1\|00000\rangle + a_0b_1\|00011\rangle + a_1b_0\|00100\rangle + a_1b_1\|00111\rangle + a_2b_0\|01000\rangle + a_2b_1\|01011\rangle + a_3b_0\|01100\rangle + a_3b_1\|01111\rangle -$ $a_4b_0\|10000\rangle - a_4b_1\|10011\rangle - a_5b_0\|10100\rangle - a_5b_1\|10111\rangle - a_6b_0\|11000\rangle - a_6b_1\|11011\rangle - a_7b_0\|11100\rangle - a_7b_1\|11111\rangle$
− +	+ −	$a_0b_1\|00000\rangle - a_0b_1\|00011\rangle + a_1b_0\|00100\rangle - a_1b_1\|00111\rangle + a_2b_0\|01000\rangle - a_2b_1\|01011\rangle + a_3b_0\|01100\rangle - a_3b_1\|01111\rangle -$ $a_4b_0\|10000\rangle + a_4b_1\|10011\rangle - a_5b_0\|10100\rangle + a_5b_1\|10111\rangle - a_6b_0\|11000\rangle + a_6b_1\|11011\rangle - a_7b_0\|11100\rangle + a_7b_1\|11111\rangle$
− +	− +	$a_0b_1\|00000\rangle + a_0b_1\|00011\rangle - a_1b_0\|00100\rangle - a_1b_1\|00111\rangle + a_2b_0\|01000\rangle + a_2b_1\|01011\rangle - a_3b_0\|01100\rangle - a_3b_1\|01111\rangle -$ $a_4b_0\|10000\rangle - a_4b_1\|10011\rangle + a_5b_0\|10100\rangle + a_5b_1\|10111\rangle - a_6b_0\|11000\rangle - a_6b_1\|11011\rangle + a_7b_0\|11100\rangle + a_7b_1\|11111\rangle$
− +	− −	$a_0b_1\|00000\rangle - a_0b_1\|00011\rangle - a_1b_0\|00100\rangle + a_1b_1\|00111\rangle + a_2b_0\|01000\rangle - a_2b_1\|01011\rangle - a_3b_0\|01100\rangle + a_3b_1\|01111\rangle -$ $a_4b_0\|10000\rangle + a_4b_1\|10011\rangle + a_5b_0\|10100\rangle - a_5b_1\|10111\rangle - a_6b_0\|11000\rangle + a_6b_1\|11011\rangle + a_7b_0\|11100\rangle - a_7b_1\|11111\rangle$
− −	+ +	$a_0b_1\|00000\rangle + a_0b_1\|00011\rangle + a_1b_0\|00100\rangle + a_1b_1\|00111\rangle - a_2b_0\|01000\rangle - a_2b_1\|01011\rangle - a_3b_0\|01100\rangle - a_3b_1\|01111\rangle -$ $a_4b_0\|10000\rangle - a_4b_1\|10011\rangle - a_5b_0\|10100\rangle - a_5b_1\|10111\rangle + a_6b_0\|11000\rangle + a_6b_1\|11011\rangle + a_7b_0\|11100\rangle + a_7b_1\|11111\rangle$
− −	+ −	$a_0b_1\|00000\rangle - a_0b_1\|00011\rangle + a_1b_0\|00100\rangle - a_1b_1\|00111\rangle - a_2b_0\|01000\rangle + a_2b_1\|01011\rangle - a_3b_0\|01100\rangle + a_3b_1\|01111\rangle -$ $a_4b_0\|10000\rangle + a_4b_1\|10011\rangle - a_5b_0\|10100\rangle + a_5b_1\|10111\rangle + a_6b_0\|11000\rangle - a_6b_1\|11011\rangle + a_7b_0\|11100\rangle - a_7b_1\|11111\rangle$

$--$	$-+$	$a_0b_1\|00000\rangle + a_0b_1\|00011\rangle - a_1b_0\|00100\rangle - a_1b_1\|00111\rangle - a_2b_0\|01000\rangle - a_2b_1\|01011\rangle + a_3b_0\|01100\rangle + a_3b_1\|01111\rangle - a_4b_0\|10000\rangle - a_4b_1\|10011\rangle + a_5b_0\|10100\rangle + a_5b_1\|10111\rangle + a_6b_0\|11000\rangle + a_6b_1\|11011\rangle - a_7b_0\|11100\rangle - a_7b_1\|11111\rangle$
$--$	$--$	$a_0b_1\|00000\rangle - a_0b_1\|00011\rangle - a_1b_0\|00100\rangle + a_1b_1\|00111\rangle - a_2b_0\|01000\rangle + a_2b_1\|01011\rangle + a_3b_0\|01100\rangle - a_3b_1\|01111\rangle - a_4b_0\|10000\rangle + a_4b_1\|10011\rangle + a_5b_0\|10100\rangle - a_5b_1\|10111\rangle + a_6b_0\|11000\rangle - a_6b_1\|11011\rangle - a_7b_0\|11100\rangle + a_7b_1\|11111\rangle$

Table A.6.4 Applying Z unitary operation

Alice's results	Bob's results	Unitary operation on qubits $b_0b_1b_2a_3$
+ +	+ +	$I \otimes I \otimes I \otimes I$
+ +	+ −	$I \otimes I \otimes I \otimes Z$
+ +	− +	$I \otimes I \otimes Z \otimes I$
++	− −	$I \otimes I \otimes Z \otimes Z$
+ −	+ +	$I \otimes Z \otimes I \otimes I$
+ −	+ −	$I \otimes Z \otimes I \otimes Z$
+ −	− +	$I \otimes Z \otimes Z \otimes I$
+ −	− −	$I \otimes Z \otimes Z \otimes Z$
− +	+ +	$Z \otimes I \otimes I \otimes I$
− +	+ −	$Z \otimes I \otimes I \otimes Z$
− +	− +	$Z \otimes I \otimes Z \otimes I$
− +	− −	$Z \otimes I \otimes Z \otimes Z$
− −	+ +	$Z \otimes Z \otimes I \otimes I$
− −	+ −	$Z \otimes Z \otimes I \otimes Z$
− −	− +	$Z \otimes Z \otimes Z \otimes I$
− −	− −	$Z \otimes Z \otimes Z \otimes Z$

Table A.6.5 Charlie's measurement results, the corresponding collapsed state, and appropriate unitary operation

Charlie's results	The collapsed results of qubits $b_0b_1b_2a_3$	Unitary operation on qubits $b_0b_1b_2a_3$
+	$a_0b_1\|0000\rangle + a_0b_1\|0001\rangle + a_1b_0\|0010\rangle + a_1b_1\|0011\rangle +$ $a_2b_0\|0100\rangle + a_2b_1\|0101\rangle + a_3b_0\|0110\rangle + a_3b_1\|0111\rangle +$ $a_4b_0\|1000\rangle + a_4b_1\|1001\rangle + a_5b_0\|1010\rangle + a_5b_1\|1011\rangle +$ $a_6b_0\|1100\rangle + a_6b_1\|1101\rangle + a_7b_0\|1110\rangle + a_7b_1\|1111\rangle$	$I \otimes I \otimes I \otimes I$
−	$a_0b_1\|0000\rangle - a_0b_1\|0001\rangle + a_1b_0\|0010\rangle - a_1b_1\|0011\rangle +$ $a_2b_0\|0100\rangle - a_2b_1\|0101\rangle + a_3b_0\|0110\rangle - a_3b_1\|0111\rangle +$ $a_4b_0\|1000\rangle - a_4b_1\|1001\rangle + a_5b_0\|1010\rangle - a_5b_1\|1011\rangle +$ $a_6b_0\|1100\rangle - a_6b_1\|1101\rangle + a_7b_0\|1110\rangle - a_7b_1\|1111\rangle$	$I \otimes I \otimes I \otimes Z$

6.4 Enhancing Robustness of Noisy Qutrit Teleportation with Markovian Memory

Quantum teleportation is the fundamental communication unit in quantum communication. Here, a three-level system is selected for storing and transmitting quantum information, due to its unique advantages, such as lower cost than a higher-level system and higher capacity and security than a two-level system. It is known that the key procedure for perfect teleportation is the distribution of entanglement through quantum channel. However, amounts of noise existing in the quantum channel may interfere the entangled state, causing the degradation of quantum entanglement. In the physical implementations of quantum communication schemes, noise acting on the carriers of successive transmissions often exhibits some correlations, which is the so-called quantum memory channel. In this section, a memory channel model during the entanglement distribution phase is constructed, and the uniform expression of the evolution of a two-qutrit entangled state under different kinds of correlated noise is derived. Finally, Pauli noise and amplitude-damping noise as the typical noise source are considered to analyze the influence of memory effects of noise on qutrit teleportation. It is expected to show that three-level teleportation under these two types of channels can generally enhance the robustness to noise if the Markovian correlations of quantum channel are taken into consideration.

6.4.1 The Model of Quantum Memory Channel During Entanglement Distribution

In this subsection, the quantum channel is modeled as the combination of two classes of channels, which is a public channel and two private channels. The diagram of this simple channel model is depicted in Fig. 6.14. The public channel is assumed to be a lossy photon channel where the two photons consisting of entangled photonic state are transmitted in sequence with the separation time scale of two photons much lower than the finite relaxation time of the channel environment. Two private channels are assumed to be ideally noiseless channels that act on the two photons independently and identically.

Although two types of channels are presented, only the lossy photon channel affects the quality of information transmission in the entanglement distribution. Much lower separation time of two photons transmitted through the lossy optical medium could suppress the finite relaxation time of local environment interacted with the external reservoir, leading to the output of the second input photon is affected by both input two photons. It is mentioned that such effects exhibit some classical correlations, not the quantum coherence. To account for such effects, a common memory system M is introduced. Therefore, the corresponding representation of the quantum memory channel can be described as the interaction of the carrier ρ_Q with a common memory system M and local environment e_i, that is,

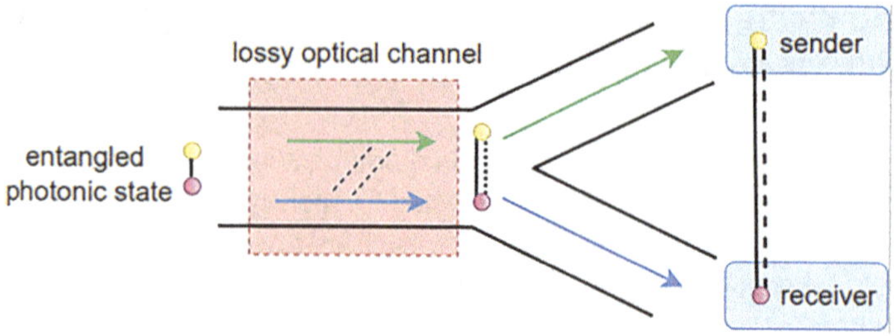

Fig. 6.14 The diagram of the information transmission in the entanglement distribution. The green arrow represents the transmission of the first photon and the blue arrow describes the transmission of the second photon after the separation time τ_s. The dotted line is the correlation of noise between two uses of the lossy channel, which originates from the much higher rate of consecutive transmission making the local environment retain a memory. The whole transmission process is the entangled photonic state transmitted through lossy memory channel produces a two-photon state exhibiting the coherence and the part of classical correlation and then transmitting the two-photon state through private noiseless channel to the sender and receiver

$$\Phi^{\otimes 2}(\rho_Q) = \mathrm{Tr}_{ME}\left[U_{q_2Me_2}U_{q_1Me_1}\left(\rho_Q \otimes w_M \otimes w_E^{\otimes 2}\right)U^{\dagger}_{q_1Me_1}U^{\dagger}_{q2}Me_2\right] \tag{6.63}$$

where Φ is a completely positive and trace-preserving map (CPTP) that transforms states of the sender's end of the channel into states on the receiver's end and ρ_Q, w_M, and w_E are entangled photonic state, memory state, and local environment state, respectively. $U_{q_iM_{e_i}}$ is the unitary operation that each input photon couples to the memory system and local environment. The graphical sketch of (6.63) is shown in Fig. 6.15. It is intuitive that each channel use is directly or indirectly affected by the previous one, which means the correlated source of noise is presented in the successive uses of channel.

According to existing research [32, 33], (6.63) can be expressed as the concatenation of two CPTP maps, that is,

$$\Phi^{\otimes 2}(\rho_Q) = \mathrm{Tr}_M\left[\Phi_{q_2M}\circ\Phi_{q_1M}\left(\rho_Q \otimes w_M\right)\right] \tag{6.64}$$

where single CPTP map Φ_{q_iM} describing the interaction of a particle of the entangled state with memory system through the unitary dilation is shown as

$$I\circ\Phi_{q_iM}\left(\rho_Q \otimes w_M\right) = I\circ\mathrm{Tr}_{e_i}\left[U_{q_iMe_i}\left(\rho_Q \otimes w_M \otimes w_{e_i}\right)U^{\dagger}_{q_iMe_i}\right]. \tag{6.65}$$

With the assumption that the action of past channel can only affect that of the current channel and is irrelevant to the future channel state, exhibiting the behavior of the Markov chain, (6.64) can be transformed into the following:

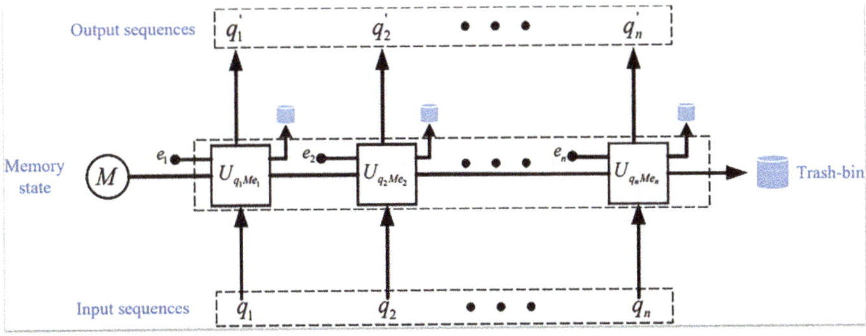

Fig. 6.15 Pictorial representation of the model of memory quantum channel with the n input sequences

$$\Phi^{\otimes 2}(\rho_Q) = \sum_{i_1, i_2} p_{i_2|i_1} p_{i_1} \Phi_{q_2}^{(i_2)} \otimes \Phi_{q_1}^{(i_1)}(\rho_Q) \tag{6.66}$$

where $\left\{\Phi_{q_j}^{(i)}\right\}$ is a set of CPTP maps operating on the jth carrier, p_{i_1} is the initial probability distribution, and $p_{i_2|i_1}$ is a conditional probability.

Furthermore, based on the assumption that the two light pulses will experience the same local environment if the transmission rate of light pulse is much higher than the inverse of the relaxation time of local environment coupled to the external reservoir, the conditional probability $p_{i_2|i_1}$ representing the effect of action of past channel on the current channel can be written as

$$p_{i_2|i_1} = (1-\mu)p_{i_2} + \mu\delta_{i_1, i_1} \tag{6.67}$$

where $\mu \in [0, 1]$ is the degree of memory. It is noted that $\mu = 0$ means the lossy channel is memoryless and $\mu = 1$ means a perfect memory channel where no input information will be revealed to local environment.

Combined with the condition probability described in (6.67), (6.68) can be rewritten as

$$\Phi^{\otimes 2}(\rho_Q) = (1-\mu)\sum_{i_1, i_2} p_{i_2} p_{i_1} \Phi_{q_2}^{(i_2)} \otimes \Phi_{q_1}^{(i_1)}(\rho_Q) + \mu \sum_{i_1} p_{i_1} \Phi_{q_2}^{(i_1)} \otimes \Phi_{q_1}^{(i_1)}(\rho_Q). \tag{6.68}$$

Therefore, the dynamical evolution of two-qutrit entangled state under Pauli noise and amplitude-damping noise with Markovian memory can be characterized by (6.68) when the operators of describing both noises are presented.

6.4.2 The Kraus Operators for Describing Correlated Amplitude-Damping Channel

The time evolution of a three-level atom in the V configuration is given by the master equation $\dot{\rho} = -i[H, \rho] + \Lambda\rho$ [3], in which ρ is the density matrix of the three-level atom; H describes the Hamiltonian of the total system; $-i[H, \rho]$ represents the Hamiltonian part, which is a unitary operation; and the damping term $\Lambda\rho$ describing the atom decay by the emission of a photon is defined as

$$\Lambda\rho = \frac{\gamma_{20}}{2}(2\sigma_{02}\rho\sigma_{20} - \sigma_{22}\rho - \rho\sigma_{22}) + \frac{\gamma_{10}}{2}(2\sigma_{01}\rho\sigma_{10} - \sigma_{11}\rho - \rho\sigma_{11}) \tag{6.69}$$

where γ_{10} and γ_{20} are the two different spontaneous decay constants that the atom transits from excited states $|1\rangle$ and $|2\rangle$ to ground state $|0\rangle$, respectively. σ_{ij} is an atomic transition operator taking an atom from level $|i\rangle$ to $|j\rangle$, which is defined as $\sigma_{ij} = |i\rangle\langle j|$. As amplitude-damping channel is used to describe the phenomenon of energy dissipation, it is obvious that the dynamical evolution of a qutrit state under amplitude-damping channel can be represented by the Lindblad equation $\dot{\rho} = \Lambda\rho$.

For characterizing the Markovian correlation of the interaction of a compound system ρ^c composed of two qutrit states with correlated amplitude-damping noise, it is necessary to calculate the Lindblad equation of the two-qutrit state, which is $\dot{\rho^c} = \Lambda^c\rho^c$. The damping term is written as

$$\Lambda^c\rho^c = \frac{\gamma_{20}}{2}(2A_{02}\rho^c A_{20} - A_{22}\rho^c - \rho^c A_{22}) + \frac{\gamma_{10}}{2}(2A_{01}\rho^c A_{10} - A_{11}\rho^c - \rho^c A_{11}) \tag{6.70}$$

where $A_{ij} = \sigma_{ij} \otimes \sigma_{ij}$. The method to solve this equation is making use of a special basis of left, $\{L_i\}$, and right, $\{R_i\}$, damping eigenoperators for a Lindblad superoperator $\tilde{\Lambda}$ [34]. It is assumed that the density matrix of the three-level compound system is defined as

$$\rho^c = \begin{bmatrix} \rho_{00} & \rho_{01} & \rho_{02} & \rho_{10} & \rho_{11} & \rho_{12} & \rho_{20} & \rho_{21} & \rho_{22} \\ \rho_{03} & \rho_{04} & \rho_{05} & \rho_{13} & \rho_{14} & \rho_{15} & \rho_{23} & \rho_{24} & \rho_{25} \\ \rho_{06} & \rho_{07} & \rho_{08} & \rho_{16} & \rho_{17} & \rho_{18} & \rho_{26} & \rho_{27} & \rho_{28} \\ \rho_{30} & \rho_{31} & \rho_{32} & \rho_{40} & \rho_{41} & \rho_{42} & \rho_{50} & \rho_{51} & \rho_{52} \\ \rho_{33} & \rho_{34} & \rho_{35} & \rho_{43} & \rho_{44} & \rho_{45} & \rho_{53} & \rho_{54} & \rho_{55} \\ \rho_{36} & \rho_{37} & \rho_{38} & \rho_{46} & \rho_{47} & \rho_{48} & \rho_{56} & \rho_{57} & \rho_{58} \\ \rho_{60} & \rho_{61} & \rho_{62} & \rho_{70} & \rho_{71} & \rho_{72} & \rho_{80} & \rho_{81} & \rho_{82} \\ \rho_{63} & \rho_{64} & \rho_{65} & \rho_{73} & \rho_{74} & \rho_{75} & \rho_{83} & \rho_{84} & \rho_{85} \\ \rho_{66} & \rho_{67} & \rho_{68} & \rho_{76} & \rho_{77} & \rho_{78} & \rho_{86} & \rho_{87} & \rho_{88} \end{bmatrix}. \tag{6.71}$$

After converting the density matrix ρ^c in Hilbert space into Fock-Liouville space [35], the Lindblad superoperator $\tilde{\Lambda}$ is derived as a matrix of *81* × *81*, which is related to γ_{20} and γ_{10}. Then according to the formula $\tilde{\Lambda}R_i = \lambda_i R_i$, the symbols λ_i for eigenvalues and R_i for right eigenstates are able to be calculated. Here, the values of λ_i are displayed:

$$\lambda = (0)_{49}, \left(-\frac{\gamma_{10}}{2}\right)_{14}, \left(-\frac{\gamma_{20}}{2}\right)_{14}, \left(-\frac{\gamma_{10}}{2} - \frac{\gamma_{20}}{2}\right)_{2}, -\gamma_{10}, -\gamma_{20}. \tag{6.72}$$

It is worth noting that the subscripts of eigenvalues indicate the frequency of the appearance of each eigenvalue and two eigenvalues without subscripts means they both appear once. Besides, the left eigenvectors L_i as the duals of right eigenvectors can be achieved on the condition that $Tr(L_iR_j) = \delta_{ij}$. Eventually, the dynamical evolution of the three-level compound systems ρ^c can be calculated in accordance with the formula $\Phi(\rho^c) = \sum_i tr(L_i\rho^c)\exp(\lambda_i t)R_i$, which is shown as follows:

$$\tilde{\rho} = \begin{bmatrix} \tilde{\rho_{00}} & \rho_{01} & \rho_{02} & \rho_{10} & e^{-\frac{\gamma_{10}}{2}t}\rho_{11} & \rho_{12} & \rho_{20} & \rho_{21} & e^{-\frac{\gamma_{20}}{2}t}\rho_{22} \\ \rho_{03} & \rho_{04} & \rho_{05} & \rho_{13} & e^{-\frac{\gamma_{10}}{2}t}\rho_{14} & \rho_{15} & \rho_{23} & \rho_{24} & e^{-\frac{\gamma_{20}}{2}t}\rho_{25} \\ \rho_{06} & \rho_{07} & \rho_{08} & \rho_{16} & e^{-\frac{\gamma_{10}}{2}t}\rho_{17} & \rho_{18} & \rho_{26} & \rho_{27} & e^{-\frac{\gamma_{20}}{2}t}\rho_{28} \\ \rho_{30} & \rho_{31} & \rho_{32} & \rho_{40} & e^{-\frac{\gamma_{10}}{2}t}\rho_{41} & \rho_{42} & \rho_{50} & \rho_{51} & e^{-\frac{\gamma_{20}}{2}t}\rho_{52} \\ e^{-\frac{\gamma_{10}}{2}t}\rho_{33} & e^{-\frac{\gamma_{10}}{2}t}\rho_{34} & e^{-\frac{\gamma_{10}}{2}t}\rho_{35} & e^{-\frac{\gamma_{10}}{2}t}\rho_{43} & e^{-\gamma_{10}t}\rho_{44} & e^{-\frac{\gamma_{10}}{2}t}\rho_{45} & e^{-\frac{\gamma_{10}}{2}t}\rho_{53} & e^{-\frac{\gamma_{10}}{2}t}\rho_{54} & e^{-\frac{\gamma_{10}+\gamma_{20}}{2}t}\rho_{55} \\ \rho_{36} & \rho_{37} & \rho_{38} & \rho_{46} & e^{-\frac{\gamma_{10}}{2}t}\rho_{47} & \rho_{48} & \rho_{56} & \rho_{57} & e^{-\frac{\gamma_{20}}{2}t}\rho_{58} \\ \rho_{60} & \rho_{61} & \rho_{62} & \rho_{70} & e^{-\frac{\gamma_{10}}{2}t}\rho_{71} & \rho_{72} & \rho_{80} & \rho_{81} & e^{-\frac{\gamma_{20}}{2}t}\rho_{82} \\ \rho_{63} & \rho_{64} & \rho_{65} & \rho_{73} & e^{-\frac{\gamma_{10}}{2}t}\rho_{74} & \rho_{75} & \rho_{83} & \rho_{84} & e^{-\frac{\gamma_{20}}{2}t}\rho_{85} \\ e^{-\frac{\gamma_{20}}{2}t}\rho_{66} & e^{-\frac{\gamma_{20}}{2}t}\rho_{67} & e^{-\frac{\gamma_{20}}{2}t}\rho_{68} & e^{-\frac{\gamma_{20}}{2}t}\rho_{76} & e^{-\frac{\gamma_{10}+\gamma_{20}}{2}t}\rho_{77} & e^{-\frac{\gamma_{20}}{2}t}\rho_{78} & e^{-\frac{\gamma_{20}}{2}t}\rho_{86} & e^{-\frac{\gamma_{20}}{2}t}\rho_{87} & e^{-\gamma_{20}t}\rho_{88} \end{bmatrix} \tag{6.73}$$

where $\tilde{\rho_{00}} = \rho_{00} + (1 - e^{-\gamma_{10}t})\rho_{44} + (1 - e^{-\gamma_{20}t})\rho_{88}$.

As mentioned above, $\sum_k E_{kk}\rho_Q E_{kk}$ represents the correlated part of the evolution of two-qutrit entangled state ρ_Q under amplitude-damping channel with Markovian memory; therefore, it is apparent that the Kraus operator E_{kk} can be achieved on the condition that $\sum_i tr(L_i\rho^c)\exp(\lambda_i t)R_i \sum_k E_{kk}\rho^c E_{kk}$, which has been introduced in ref. [36]. Combining (6.71) and (6.73), the operators E_{00}, E_{11}, and E_{22} can be finally obtained.

6.4.3 Noisy Quantum Teleportation Scheme in Three Dimensions

In what follows, the standard quantum teleportation in three dimensions is presented.

Step 1: The qutrit state to be transmitted from Alice to Bob is defined as

$$|\psi\rangle_{in} = \sum_{i=0}^{2} \alpha_i |i\rangle = cos\,\theta_0|0\rangle + \sin\theta_0\, cos\,\theta_1 e^{i\phi_1}|1\rangle + \sin\theta_0\, cos\,\theta_1 e^{i\phi_2}|2\rangle \quad (6.74)$$

where $0 < \theta_i \leq \frac{\pi}{2}$ and $0 < \phi_j \leq 2\pi$.

Step 2: Suppose a prior shared maximally entangled state among Alice and Bob is $|\Phi\rangle_{AB} = \frac{1}{\sqrt{3}}\sum_{j=0}^{2}|j\rangle_A \otimes |j\rangle_B$. After distributing the two particles A and B of the pure state $|\Phi\rangle_{AB}$ through quantum memory channel to communication parties Alice and Bob, respectively, the two-qutrit mixed entangled state M can be obtained with (6.74).

Step 3: Taking advantage of entanglement, Alice performs a joint measurement on her particles, and the corresponding measurement basis [37] is $|\phi_{nm}\rangle = \frac{1}{\sqrt{3}}\sum_{k=0}^{2} e^{2\pi ikn/3}|k\rangle \otimes |(k+m)\text{mod}3\rangle$ with $\{n, m\} \in \{0, 1, 2\}$. Once learning from Alice that she has received the measurement result (n, m), Bob performs a corresponding unitary transformation $U_{nm} = \sum_{l=0}^{2} e^{2\pi iln/3}|l\rangle\langle(l+m)\text{mod}3|$ on his previously entangled particle (particle B) for recovering the information carried by the initial state $|\psi\rangle_{in}$. Then, the final state of Bob's particle can be written in $\rho_{nm} = U_{nm} Tr_A\left[|\phi_{nm}\rangle\langle\phi_{nm}| \otimes I_B\left(|\psi\rangle_{in}\langle\psi| \otimes M\right)\right]U_{nm}^{\dagger}$.

Step 4: Quantum information is inevitably distorted by the noisy quantum channel, and the fidelity utilized to estimate the feasibility of the protocol is described as $F_{nm}(\theta, \phi) = \langle\psi|\rho_{nm}|\psi\rangle / Tr\rho_{nm}$. It is worth noting that $Tr(\rho_{nm})^{-1}$ referred as the normalization factor of fidelity is equal to the probability that the result of measurement is (n, m). Taking all possible teleported ρ_{nm} into consideration, the mean fidelity is estimated by $F(\theta, \phi) = \sum_{nm} F_{nm}(\theta, \phi)$.

In order to analyze the effect of noise on the fidelity of qutrit teleportation, it is necessary to construct relations between fidelity and noise intensity without regard to the probability amplitude of the initial state. Therefore, the average fidelity calculated by double integrals of the mean fidelity is eventually chosen to evaluate the transmission quality of quantum teleportation scheme [35], which is written in (6.82).

6.4.4 Dynamical Evolution of Two-Qutrit Entangled State Under Different Correlated Noise Environment

The channel model for entanglement distribution is constructed under the assumption that the light pulse will be transferred through a lossy photon channel and a noiseless channel to the communication party in sequence. The memory effects of lossy photon channel may emerge when two photons of the entangled photonic state are successively transmitted through the channel with the separation time scale of two consecutive transmissions much lower than the finite relaxation time of local environment interacted with the external reservoir. The diagram of the channel model is shown in Fig. 6.14.

Now, assume the preshared two-qutrit entangled photonic state is represented by the density matrix ρ_Q, and then the dynamical evolution of state ρ_Q transmitted through the lossy memory channel that has been derived in Sect. 6.4.1 can be written as

$$\Phi^{\otimes 2}(\rho_Q) = (1-\mu)\sum\nolimits_{i_1,i_2} p_{i_2} p_{i_1} \Phi_{q_2}^{(i_2)} \otimes \Phi_{q_1}^{(i_1)}(\rho_Q) + \mu \sum\nolimits_{i_1} p_{i_1} \Phi_{q_2}^{(i_1)} \otimes \Phi_{q_1}^{(i_1)}(\rho_Q) \tag{6.75}$$

where the parameter $\mu \in [0, 1]$ is the degree of memory describing the overlap ratio between two consecutive transmissions of photons. $\Phi_{q^j}^{(i)}$ is an arbitrary CPTP map on system state with the corresponding probability p_i. It is worth noting that the derivation of (6.75) is based on two assumptions: one is the Markovian behavior of lossy photon channel that is the action of past channel can only influence the current channel use and is irrelevant to the future channel state and the other is assume that two photons of the entangled photonic state will subject to the same local environment with much higher use rates of the lossy channel for determining the mathematical representation of the channel's Markovian behavior. Next, based on the model of quantum memory channel during the entanglement distribution, the evolution of two-qutrit entangled state under the following two typical noise environments is demonstrated for analyzing the influence of Markovian memory channel on qutrit teleportation.

6.4.4.1 Pauli Noise

If the lossy photon channel exhibits Pauli errors, the evolution of maximally two-qutrit entangled state transmitted through correlated Pauli channel in three dimensions could be directly characterized by replacing the map $\Phi_{q_1(q_2)}^{i_1(i_2)}$ with the Weyl operators $U_{nm} = \sum_{l=0}^{2} e^{2\pi i l n/3} |l\rangle\langle (l+m) \mathrm{mod} 3|$, $\{n, m\} \in \{0, 1, 2\}$, which is written as

$$\Phi^{\otimes 2}(\rho_Q) = (1-\mu)\sum_{\substack{e,f \\ j,k}} p_{ef}p_{jk}U_{ef}\otimes \mathrm{U}_{jk}\rho_Q \mathrm{U}_{ef}^{\dagger}\otimes \mathrm{U}_{jk}^{\dagger} + \mu\sum_{\mathrm{r,s}} \mathrm{p}_{rs}\mathrm{U}_{rs}\otimes \mathrm{U}_{rs}\rho_Q \mathrm{U}_{rs}^{\dagger}\otimes \mathrm{U}_{rs}^{\dagger}. \tag{6.76}$$

The corresponding probability pnm for performing the Weyl operator has been defined in ref. [35]. It is noticed that different value of $\{n, m\}$ corresponds to different types of Pauli noise, such as trit-flip region, $n = 0$ and $m \in \{0, 1, 2\}$; phase-flip region, $\{n, m\} \in \{0, 1, 2\}$ and $m = (0, 0)$; trit-phase-flip region, $\{n, m\} \in \{0, 1, 2\}$ and $(n, m) = (0, 0)$; and depolarizing noise region, $\{n, m\} \in \{0, 1, 2\}$.

6.4.4.2 Amplitude-Damping Noise

Amplitude-damping channel can be described by a process of energy dissipation of a quantum system to the environment. The transition of excitation occurs between the excited state and the ground state with a finite probability. Herein, it is assumed that the transitions between two excited states are not allowed. The Kraus operators describing amplitude-damping noise in three dimensions are concluded in refs. [38–41], which are represented in the standard computational basis

$$\begin{aligned} E_0 &= |0\rangle\langle 0| + \sqrt{1-P}(|1\rangle\langle 1| + |2\rangle\langle 2|) \\ E_1 &= \sqrt{P}|0\rangle\langle 1| \\ E_2 &= \sqrt{P}|0\rangle\langle 2| \end{aligned} \tag{6.77}$$

where P is the magnitude of the energy dissipation. The evolution of two-qutrit maximally entangled state interacted with the correlated amplitude damping can be calculated by (6.75), which is shown as

$$\Phi^{\otimes 2}(\rho_Q) = (1-\mu)\sum_{i,j} E_i\otimes \mathrm{E_j}\rho_Q \mathrm{E_i}^{\dagger}\otimes \mathrm{E_j}^{\dagger} + \mu\sum_{\mathrm{k}} \mathrm{p_k}\mathrm{E}_{kk}\rho_Q \mathrm{E}_{kk}^{\dagger}. \tag{6.78}$$

Combining (6.77) and (6.78), it is found that E_{kk} in the correlated term cannot be directly written like the Pauli noise due to the non-unitary property of amplitude damping. Therefore, the Lindblad equation describing quantum Markov channel is introduced. A three-level atomic system in V configuration is the typical case for demonstrating the spontaneous decay of atoms. The Kraus operators E_{kk} can be derived only if we solve the Lindblad equation of the compound system of two three-level atomic systems in V type. The details of derivation are demonstrated in Methods and the results are given as follows:

$$E_{00} = \begin{bmatrix} I_4 & & & \\ & \sqrt{1-P_1} & & \\ & & I_3 & \\ & & & \sqrt{1-P_2} \end{bmatrix}_{9\times 9} \tag{6.79}$$

$$E_{11} = \left[\begin{array}{c|cccc} & \sqrt{P_1} & 0 & \cdots & 0 \\ & 0 & 0 & \cdots & 0 \\ & \vdots & \vdots & \ddots & \vdots \\ & 0 & 0 & \cdots & 0 \\ \hline O_4 & & & & \end{array}\right]_{9\times 9} \tag{6.80}$$

$$E_{22} = \begin{bmatrix} 0 & \cdots & 0 & \sqrt{2} \\ 0 & \cdots & 0 & 0 \\ \vdots & \ddots & \vdots & \vdots \\ 0 & \cdots & 0 & 0 \end{bmatrix}_{9\times 9} \tag{6.81}$$

with $P_1 = 1 - e^{-\gamma_{10}t}, P_2 = 1 - e^{-\gamma_{20}t}$. I_m is the identity matrix of $m \times m$, and O_n is the null matrix of $n \times n$. According to the matrix form of E_{11} and E_{22}, we can directly achieve the Kraus operators E_1' and E_2' describing the amplitude-damping noise in the three-level system, which are described as $E_1' = \sqrt{P_1}|0\rangle\langle 1|$ and $E_2' = \sqrt{P_2}|0\rangle\langle 2|$; P_1 and P_2 are the attenuation coefficients. $E_0' = |0\rangle\langle 0| + \sqrt{1-P_1}|1\rangle \times \langle 1| + \sqrt{1-P_2}|2\rangle\langle 2|$ can be derived with the completeness relation $\sum_{i=0}^{2} E_i'^{\dagger} E_i' = I$.

Comparing our results with the Kraus operators of describing amplitude-damping noise in (6.77), it can be found that they are equivalent if the two attenuation coefficients are assumed in the same value. It is clear that (6.77) can be described as the special case for the Kraus operators of amplitude-damping noise in the three level system, which means the mathematical representation we derived about the correlated amplitude-damping noise is general applicably. Finally, putting all known Kraus operators into (6.78), the dynamical evolution of entangled photonic state under the lossy memory channel with the loss of photons can be calculated.

6.4.5 *Qutrit Teleportation Based on the Noisy Quantum Channel with Memory Effects*

Quantum information to be teleported cannot be completely recovered from the receiver for the presence of noise in quantum channel. The differences between input state and output state can be quantitatively measured by fidelity [12]. Here, the average fidelity $\langle F \rangle$ is introduced for eliminating the effects of the input state, which is the function of the noise intensity and the degree of memory and can be defined by

$$\langle F \rangle = \frac{2}{\pi^2} \int_0^{2\pi} \int_0^{2\pi} \prod_{\pi=0}^{1} d\theta_i d\phi_{i+1} \sin^{3-2i}\theta_i \cos\theta_i F(\theta, \phi). \tag{6.82}$$

The meanings of mathematical symbols are shown in Methods.

According to (6.82), the characteristics of qutrit teleportation under correlated Pauli channel and amplitude-damping channel are investigated. For the sake of convenience, we stipulate that the name of average fidelity is substituted into fidelity in the following elaboration:

6.4.5.1 Transmitting Information Under Correlated Pauli Channels

In the case of the maximally two-qutrit entangled quantum state interacted with uncorrelated Pauli-like noise, the fidelity of qutrit quantum teleportation under trit-flip, phase-flip, trit-phase-flip, and depolarizing noise is ordinally computed as

$$\begin{aligned} \langle F_T^u \rangle &= \langle F_P^u \rangle = 1 - \frac{3}{4}\left(2P - \frac{3}{2}P^2\right) \\ \langle F_{TP}^u \rangle &= 1 - \frac{3}{4}\left(2P - \frac{5}{4}P^2\right) \\ \langle F_D^u \rangle &= 1 - \frac{3}{4}\left(\frac{16}{9}P - \frac{8}{9}P^2\right) \end{aligned} \tag{6.83}$$

where P is the noise intensity satisfying $0 \leq P \leq 1$. The relations between fidelity and noise intensity are shown in Fig. 6.16, from which we can find that taking no account of the value of $P = 23$, the communication quality of qutrit quantum teleportation under trit-flip noise and phase-flip noise is always superior to classical communication of which the fidelity is $\frac{1}{2}$ [42]. In addition, the information transmitted through depolarizing channel causes greater distortion than phase-flip channel with the noise intensity increasing.

Considering the interaction of two-qutrit entangled quantum state with correlated Pauli noise, the fidelity of qutrit quantum teleportation is calculated as

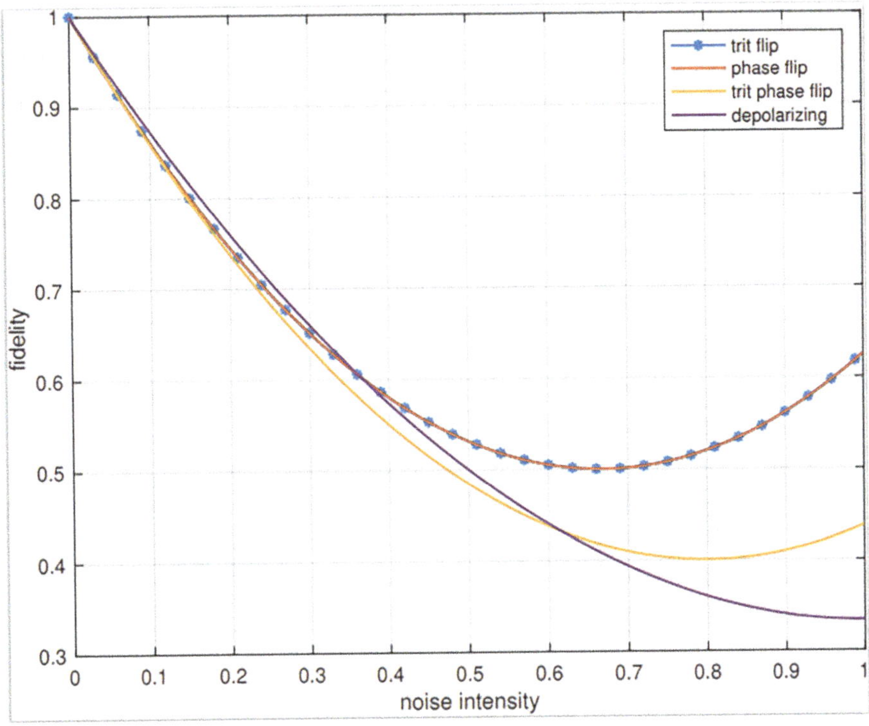

Fig. 6.16 A diagram drawing of the relations between fidelity and noise intensity of Pauli channel

$$\begin{aligned}
\langle F_T^c \rangle &= 1 - \frac{3}{4}(1-\mu)\left(2P - \frac{3}{2}P^2\right) \\
\langle F_P^c \rangle &= 1 - \frac{3}{4}\left(2P - \frac{3}{2}P^2 - \mu\left(P - \frac{3}{2}P^2\right)\right) \\
\langle F_{TP}^c \rangle &= 1 - \frac{3}{4}\left(2P - \frac{5}{4}P^2 - \mu\left(P - \frac{5}{4}P^2\right)\right) \\
\langle F_D \rangle &= 1 - \frac{3}{4}\left(\frac{16}{9}P - \frac{8}{9}P^2 - \mu\left(\frac{10}{9}P - \frac{8}{9}P^2\right)\right)
\end{aligned} \quad (6.84)$$

The pictures describing the relations between the fidelity, degree of memory, and noise intensity under the four types of Pauli noise are exhibited in Fig. 6.17. It is clear that when the preshared quantum state is coupled with correlated trit-flip noise, the fidelity $\langle F_T^c \rangle$ will improve to 1 with the increase of the degree of memory no matter what the noise intensity is.

Taking advantage of the memory effects of quantum channel, the fidelity of quantum information that can be reliably teleported through Pauli channel in two-dimensional Hilbert space is greatly enhanced [43]. However, the results of qutrit teleportation under correlated Pauli noise are intuitively not so satisfactory.

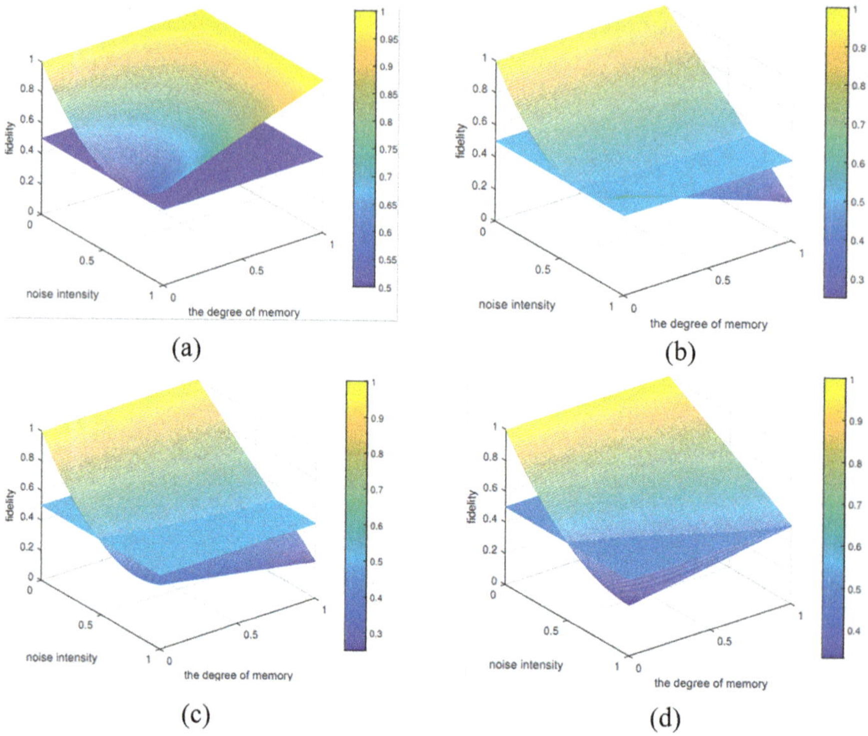

Fig. 6.17 The relations between the fidelity, noise intensity, and degree of memory under trit-flip, phase-flip, trit-phase-flip, and depolarizing noise are displayed in sequence

Here, we estimate whether the memory effects of quantum channel can improve the transmission quality by numerically analyzing the relationship between noise intensity and the degree of memory on the condition that the communication quality of qutrit quantum teleportation gains an advantage over classical communication.

For the special one, trit-flip channel, the inequation connecting noise intensity and the degree of memory is

$$\mu > 1 - \frac{1}{3}\left(\frac{1}{P} + \frac{3}{4-3P}\right), \forall P \in [0, 1]. \tag{6.85}$$

It is obvious that critical state $1 - \frac{1}{3}\left(\frac{1}{P} + \frac{3}{4-3P}\right)$ is always less than or equal to zero so that the degree of memory always belongs to (0, 1) regardless of the noise intensity, which signifies that the fidelity of qutrit quantum teleportation through correlated trit-flip channel is constantly higher than that of classical communication with the increase of the degree of memory whatever the noise intensity is.

In the case of correlated phase-flip channel, the relation can be calculated as the following inequations:

$$\forall \mu \in [0, 1], 0 \leq P < \frac{2}{3} \tag{6.86}$$

$$0 < \mu < 1 - \frac{2}{3P}, \frac{2}{3} \leq P \leq 1 \tag{6.87}$$

which means that the fidelity of qutrit quantum teleportation is irrelevant to the correlated coefficient μ in the condition of $P \in \left[0, \frac{2}{3}\right)$, but when P belongs to the interval $\left[\frac{2}{3}, 1\right]$, the qutrit quantum teleportation would be performed better than classical communication if μ is less than $1 - \frac{2}{3P}$. Nevertheless, with the degree of memory gradually increasing to 1, the quantum advantages of communication will disappear, which means the teleportation quality under correlated phase-flip channel performs worse than that under the memoryless channel.

For correlated trit-phase-flip channel and depolarizing channel, we can derive the critical conditions of the relations between noise intensity and the degree of memory, which can be, respectively, described as

$$\begin{aligned} \mu_{tr\_ph} &= 1 - \frac{2}{3}\left(\frac{1}{P} - \frac{1}{4 - 5P}\right) \\ \mu_{dep} &= 1 - \frac{3}{5}\left(\frac{1}{P} - \frac{1}{5 - 4P}\right) \end{aligned}. \tag{6.88}$$

The constraint graph is drawn in Fig. 6.18. In order to guarantee the fidelity of qutrit quantum teleportation is greater than 12, the noise intensity and the degree of memory existing in correlated trit-phase-flip channel and depolarizing channel separately satisfy the following criteria:

$$\begin{cases} \mu_{tr\_ph} \in [0, 1], 0 \leq P < 0.4734 \\ \mu_{tr\_ph} > 1 - \frac{2}{3}\left(\frac{1}{P} - \frac{1}{4 - 5P}\right), 0.4734 \leq P \leq 0.6667 \end{cases} \tag{6.89}$$

$$\begin{cases} \mu_{dep} \in [0, 1], 0 \leq P < 0.5 \\ \mu_{dep} > 1 - \frac{3}{5}\left(\frac{1}{P} - \frac{1}{5 - 4P}\right), 0.5 \leq P \leq 1 \end{cases} \tag{6.90}$$

Combining the constraint graph and the criterion, it can be discovered that the noise intensity can enlarge to a specific threshold with the improvement of the degree of memory in the condition of higher fidelity of qutrit quantum teleportation than classical communication. In other words, a three-level quantum teleportation may possess higher capacity to tolerate trit-phase-flip and depolarizing noise when considering the memory effects of quantum channel.

To conclude, distributing entanglement through Pauli channel with Markovian memory can enhance the anti-noise capacity of qutrit teleportation to some extent, except for the phase-flip channel.

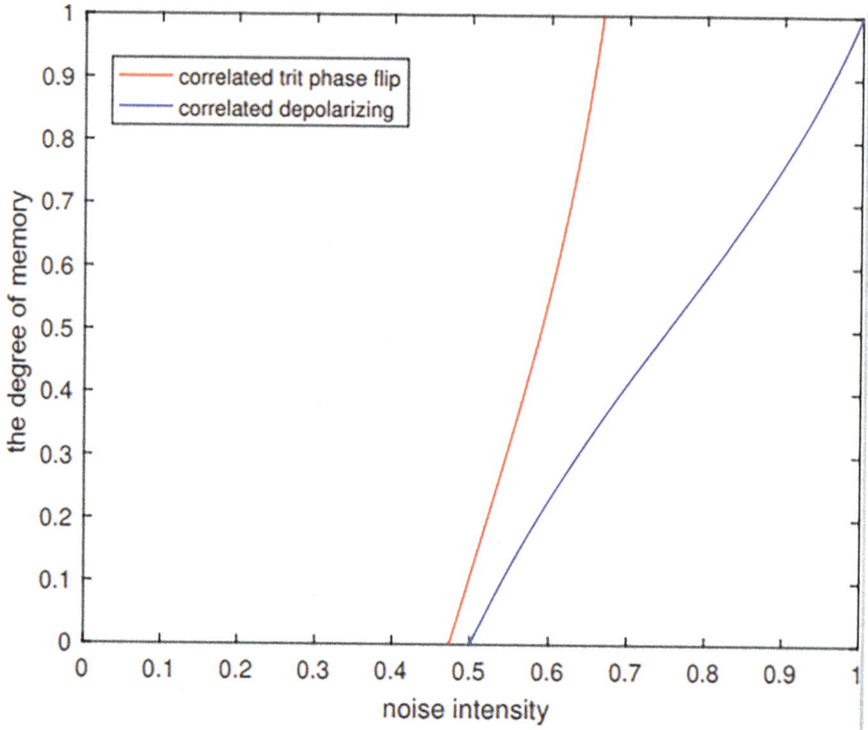

Fig. 6.18 The relations between noise intensity and the degree of memory in correlated trit-phase-flip channel and correlated depolarizing channel

6.4.5.2 Transmitting Information Under Correlated Amplitude-Damping Channel

The fidelity of qutrit teleportation under correlated amplitude-damping noise is derived as

$$\begin{aligned}\langle F_{am}^{c}\rangle &= \frac{2}{3}+\frac{\mu}{6}\left(\sqrt{(1-P_1)(1-P_2)}+\sqrt{1-P_1}+\sqrt{1-P_2}-1\right)\\ &\quad+\frac{1-\mu}{12}\left(2{P_1}^2+2{P_2}^2+2P_1P_2-6P_1-6P_2+4\right).\end{aligned} \tag{6.91}$$

Taking the case of $\mu=0$, the above formula will become

$$\langle F_{am}^{\mu}\rangle=\frac{2}{3}+\frac{1}{12}\left[(P_1-1)^2+(P_2-1)^2+{P_1}^2+{P_2}^2+(2P_1-2)(P_2-1)-2P_1-2P_2\right] \tag{6.92}$$

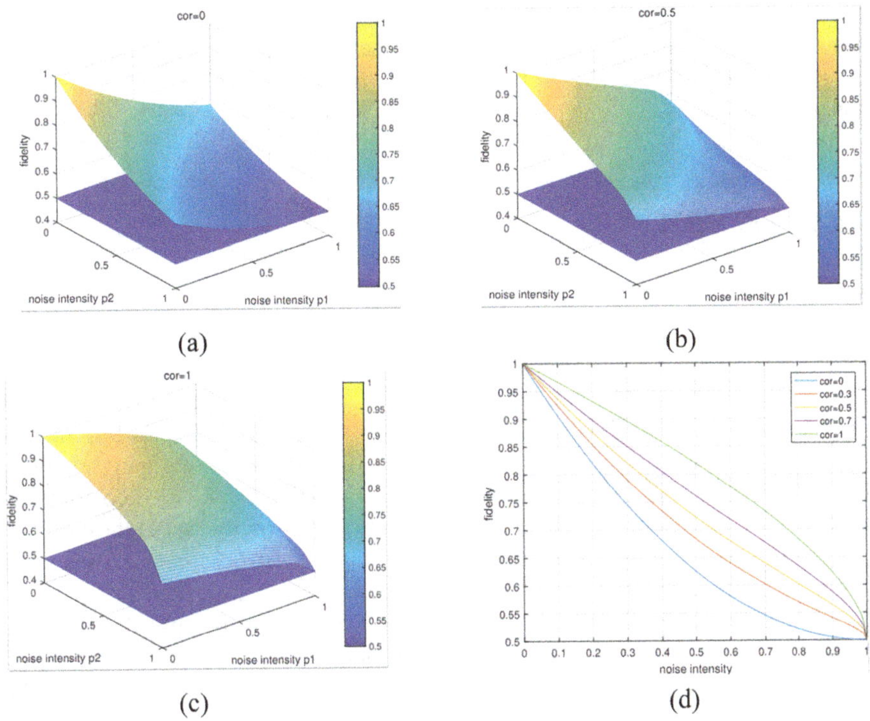

Fig. 6.19 The effect of the correlated amplitude-damping noise on fidelity of qutrit teleportation. The relations between fidelity and noise intensities P_1 and P_2 under the circumstances of $\mu = 0$, $\mu = 0.5$, and $\mu = 1$ are presented in (**a**), (**b**), and (**c**), respectively. In the condition of the equivalence of noise intensities P_1 and P_2, the connections between the fidelity of teleportation and noise intensity with different values of the degree of memory are established in (**d**)

which is utilized to describe the influence of uncorrelated amplitude-damping noise on teleportation. It is extremely complicated to numerically analyze the criteria that the fidelity of quantum teleportation is higher than that of classical communication. However, the relations between noise intensities P_1 and P_2 and fidelity $\langle F^c_{am} \rangle$ can be simulated with the distinct degree of memory μ, which are shown in Fig. 6.19.

The relations between fidelity and noise intensity in the condition of $\mu = 0$, $\mu = 0.5$, $\mu = 1$ presented in (a), (b), and (c) separately possess a general characteristic that the transmission quality will gradually diminish with the increase of noise intensities P_1 and P_2. Nevertheless, it is clear that the fidelity of teleportation under amplitude-damping channel with Markovian memory can be improved when the defined values of noise intensities P_1 and P_2 are defined.

Investigating the influence of the memory effects of amplitude-damping channel on the transmission quality of teleportation for convenience, we assume that the two attenuation coefficients P_1 and P_2 are identical to P. Then as shown in Fig. 6.19d, the fidelity can be enhanced with the increase of the degree of memory when the value of noise intensity is fixed except for the case of maximal attenuation. On the other hand, the teleportation can endure more noise with the stronger degree of memory if the fidelity is given. Therefore, taking Markovian memory of amplitude-damping channel into consideration is shown to be valuable for noisy teleportation to boost its transmission quality.

6.4.6 Summary

The realistic communication schemes where temporal rates at which signals are fed into the communication line are too rapid to allow for a resetting of the channel environment can lead to the presence of memory effects. Here, the Markovian model in three-level systems is introduced to construct correlated Pauli channel and amplitude-damping channel. The dynamical evolution of the preshared maximally two-qutrit entangled state under Pauli channel and amplitude-damping channel with Markovian correlations is derived by (6.75) where the operators representing Pauli errors and loss of photons are, respectively, based on two different kinds of basis, that is, a set of Weyl operators $\{U_{nm}\}$, $\{n, m\} \in \{0, 1, 2\}$, and the standard computational basis $\{0, 1, 2\}$. Particularly, the mathematical representation of describing correlated amplitude-damping noise is creatively solved by being connected with the Lindblad equation of the spontaneous emission of a three-level atom in the V configuration, where the transition between two excited states is dipole forbidden.

According to the mechanism of standard quantum teleportation, the relations between noise intensity and the degree of memory under correlated Pauli channel are numerically analyzed on the condition that teleportation gives the quantum advantages over classical communication. Compared with the teleportation under uncorrelated Pauli channel except for phase-flip channel, the remaining three types of noisy teleportation with Markovian memory are capable to tolerate stronger noise. Considering the phase-flip modeling the erosion of quantum information without loss of energy is immune to the Markovian behavior of quantum channel, it is a method to employ error correction to mitigate its noise on entanglement distribution in the next stage.

In the case of the preshared two-qutrit entangled state interacted with correlated amplitude-damping noise, the relations between fidelity and noise intensity can be constructed once the exact value of the degree of memory is given. It is expected that teleportation can withstand stronger noise with the increase of the degree of memory if the fidelity is given in a valid value.

Generally speaking, it is proved that qutrit teleportation can achieve stronger robustness to noise when taking account of Markovian memory into quantum channel. We are looking forward to apply this method to actual quantum information processing in the future.

References

1. Fortes R, Rigolin G (2015) Fighting noise with noise in realistic quantum teleportation. Phys Rev A 92(1):12338
2. Fortes R, Rigolin G (2016) Probabilistic quantum teleportation in the presence of noise. Phys Rev A 93(6):62330
3. Yang G, Lian BW, Nie M, Jin J (2017) Bidirectional multi-qubit quantum teleportation in noisy channel aided with weak measurement. Chin Phys B 26(4):40305
4. Ball J, Dragan A, Banaszek K (2004) Exploiting entanglement in communication channels with correlated noise. Phys Rev A 69(4):42324
5. Banaszek K, Dragan A, Wasilewski W, Radzewicz C (2004) Experimental demonstration of entanglement-enhanced classical communication over a quantum channel with correlated noise. Phys Rev Lett 92(25):257901
6. Bayat A, Burgarth D, Mancini S, Bose S (2008) Memory effects in spin-chain channels for information transmission. Phys Rev A 77:050306
7. Lupo C, Giovannetti V, Mancini S (2010) Capacities of lossy bosonic memory channels. Phys Rev Lett 104:030501
8. Lupo C, Giovannetti V, Mancini S (2010) Memory effects in attenuation and amplification quantum processes. Phys Rev A 82:032312
9. Zhou R-G, Qian C, Xu R (2020) A novel protocol for bidirectional controlled quantum teleportation of two-qubit states via seven-qubit entangled state in noisy environment. Int J Theor Phys 59(1):134–148
10. Ri-Gui Zhou; Ya-Nan Zhang; Ruiqing Xu; Chen Qian; Hou Ian. (2019) Asymmetric bidirectional controlled teleportation by using nine-qubit entangled state in noisy environment. IEE Access 7(1):75247–75264
11. Xu R, Rigui Z, Yaochong L, Shexiang J, Hou I (2022) Enhancing robustness of noisy qutrit teleportation with Markovian memory. EPJ Quantum Technol 9(1):17. ISSN: 2196-0763
12. Rehman J, Farooq A, Shin H (2020) Discrete Weyl channels with Markovian memory. IEEE J Sel Areas Commun 38(3):413–426
13. Liang X-T, Fan H-Y (2002) Entanglement-assisted classical capacities of some single qubit quantum noisy channels. Mod Phys Lett B 16:441–448. https://doi.org/10.1142/s0217984902003890
14. Sang M (2016) Huang: bidirectional quantum controlled teleportation by using a seven-qubit entangled state. Int J Theor Phys 55:380–383. https://doi.org/10.1007/s10773-015-2670-7
15. Duan YJ, Zha XW, Sun XM, Xia JF (2014) Bidirectional quantum controlled teleportation via a maximally seven-qubit entangled state. Int J Theor Phys 53:2697–2707. https://doi.org/10.1007/s10773-014-2065-1
16. Hong W (2016) Qin: asymmetric bidirectional controlled teleportation by using a seven-qubit entangled state. Int J Theor Phys 55:384–387. https://doi.org/10.1007/s10773-015-2671-6
17. Sadeghi Zadeh MS, Houshmand M, Aghababa H (2017) Bidirectional teleportation of a two-qubit state by using eight-qubit entangled state as a quantum channel. Int J Theor Phys 56:2101–2112. https://doi.org/10.1007/s10773-017-3353-3
18. Search H, Journals C, Contact A, Iopscience M, Address IP (2011) Enhanced multiparty controlled QSDC Using GHZ State. Commun Theor Phys 55(6):1007
19. Thapliyal K, Verma A, Pathak A (2015) A general method for selecting quantum channel for bidirectional controlled state teleportation and other schemes of controlled quantum communication. Quantum Inf Process 14(12):4601–4614
20. Zheng S-B (2004) Scheme for approximate conditional teleportation of an unknown atomic state without the bell-state measurement. Phys Rev A, Gen Phys 69(6):064302
21. Riebe M et al (2004) Deterministic quantum teleportation with atoms. Nature 429(6993): 734–737
22. Bouwmeester D et al (1997) Experimental quantum teleportation. Nature 390(6660):575–579

23. Liang X-T (2003) Classical information capacities of some single qubit quantum noisy channels. Commun Theor Phys 39(5):537
24. Zha YYX, Yu Y (2016) Asymmetric bidirectional controlled teleportation via seven-qubit cluster state. Int J Theor Phys 55(10):4197–4204
25. Zhou R, Xu R (2019) Bidirectional quantum teleportation by using six-qubit cluster state. IEEE Access 7:44269–44275
26. Jin YLX (2016) Bidirectional controlled teleportation by using nine-qubit entangled state in noisy environments. Quantum Inf Process 15(2):929–945
27. Choudhury BS, Samanta S (2017) Asymmetric bidirectional $3 \leftrightarrow 2$ qubit teleportation protocol between Alice and bob via 9-qubit cluster state. Int J Theor Phys 56(10):3285–3296
28. Zhi Yu L, Xin Sun S (2015) Teleportation of three-qubit state via six-qubit cluster state. Int J Theor Phys 54(5):1600–1603
29. Wang YLMSX, Nie Y (2016) Quantum teleportation of a four-qubit state by using six-qubit cluster state. Int J Theor Phys 55(8):3547–3550
30. Yang Y, Jiang M, Zhou L (2018) Improving the teleportation scheme of five-qubit state with a seven-qubit quantum channel. Int J Theor Phys 57(11):3485–3491
31. Zhou R, Qian C (2019) Cyclic and bidirectional quantum teleportation via pseudo multi-qubit states. IEEE Access 7(1):42445–42449
32. Wan KH, Dahlsten O, Kristjánsson H, Gardner R, Kim MS (2017) Quantum generalisation of feedforward neural networks. Npj Quantum Inf 3(1):36
33. Wang XL, Cai XD, Su ZE, Chen MC, Wu D, Li L et al (2015) Quantum teleportation of multiple degrees of freedom of a single photon. Nature 518(7540):516–519
34. Yeo Y, Skeen A (2003) Time-correlated quantum amplitude-damping channel. Phys Rev A 67(6):64301
35. Martin A, Guerreiro T, Tiranov A, Designolle S, Fröwis F, Brunner N et al (2017) Quantifying photonic high-dimensional entanglement. Phys Rev Lett 118(11):110501
36. Zhong-Xiao M, Zhan-Jun Z, Yong L (2004) Deterministic secure direct communication by using swapping quantum entanglement and local unitary operations. Chin Phys Lett 22(1): 18–21
37. Zhou RG, Xu R, Ian H (2019) Bidirectional quantum teleportation by using six-qubit cluster state. IEEE Access 7:44269–44275
38. Nielsen MA, Chuang IL. Quantum computation and quantum information. 2010
39. Ramzan M (2013) Three-player quantum Kolkata restaurant problem under decoherence. Quantum Inf Process 12(1):577–586
40. Schaeff C, Polster R, Huber M, Ramelow S, Zeilinger A (2015) Experimental access to higher-dimensional entangled quantum systems using integrated optics. Optica 2(6):523–529
41. Sun QC, Mao YL, Chen SJ, Zhang W, Jiang YF, Zhang YB et al (2016) Quantum teleportation with independent sources and prior entanglement distribution over a network. Nat Photonics 10(10):671–675
42. Valivarthi R, Davis SI, Peña C, Xie S, Lauk N, Narváez L et al (2020) Teleportation systems toward a quantum internet. PRX Quantum 1(2):20317
43. Grant E, Benedetti M, Cao S, Hallam A, Lockhart J, Stojevic V et al (2018) Hierarchical quantum classifiers npj. Quantum Inf 4:1

Index

R.-G. Zhou et al., *Design of Quantum Teleportation Schemes*, https://doi.org/10.1007/978-3-031-82725-9

www.ingramcontent.com/pod-product-compliance
Lightning Source LLC
Chambersburg PA
CBHW071548270425
25794CB00001BA/12
* 9 7 8 3 0 3 1 8 2 7 2 4 2 *